VEGETABLE PROTEIN PROCESSING

VEGETABLE PROTEIN PROCESSING

L.P. Hanson

NOYES DATA CORPORATION
Park Ridge, New Jersey   London, England
1974

Copyright © 1974 by Noyes Data Corporation
No part of this book may be reproduced in any form
without permission in writing from the Publisher.
Library of Congress Catalog Card Number: 74-82353
ISBN: 0-8155-0546-9
Printed in the United States

Published in the United States of America by
Noyes Data Corporation
Noyes Building, Park Ridge, New Jersey 07656

FOREWORD

The detailed, descriptive information in this book is based on U.S. patents relating to the processing of proteins of vegetable origin.

This book serves a double purpose in that it supplies detailed technical information and can be used as a guide to the U.S. patent literature in this field. By indicating all the information that is significant, and eliminating legal jargon and juristic phraseology, this book presents an advanced, technically oriented review of vegetable protein processing.

The U.S. patent literature is the largest and most comprehensive collection of technical information in the world. There is more practical, commercial, timely process information assembled here than is available from any other source. The technical information obtained from a patent is extremely reliable and comprehensive; sufficient information must be included to avoid rejection for "insufficient disclosure."

The patent literature covers a substantial amount of information not available in the journal literature. The patent literature is a prime source of basic commercially useful information. This information is overlooked by those who rely primarily on the periodical journal literature. It is realized that there is a lag between a patent application on a new process development and the granting of a patent, but it is felt that this may roughly parallel or even anticipate the lag in putting that development into commercial practice.

Many of these patents are being utilized commercially. Whether used or not, they offer opportunities for technological transfer. Also, a major purpose of this book is to describe the number of technical possibilities available, which may open up profitable areas of research and development. The information contained in this book will allow you to establish a sound background before launching into research in this field.

Advanced composition and production methods developed by Noyes Data are employed to bring our new durably bound books to you in a minimum of time. Special techniques are used to close the gap between "manuscript" and "completed book." Industrial technology is progressing so rapidly that time-honored, conventional typesetting, binding and shipping methods are no longer suitable. We have bypassed the delays in the conventional book publishing cycle and provide the user with an effective and convenient means of reviewing up-to-date information in depth.

The Table of Contents is organized in such a way as to serve as a subject index. Other indexes by company, inventor and patent number help in providing easy access to the information contained in this book.

15 Reasons Why the U.S. Patent Office Literature Is Important to You —

1. The U.S. patent literature is the largest and most comprehensive collection of technical information in the world. There is more practical commercial process information assembled here than is available from any other source.

2. The technical information obtained from the patent literature is extremely comprehensive; sufficient information must be included to avoid rejection for "insufficient disclosure."

3. The patent literature is a prime source of basic commercially utilizable information. This information is overlooked by those who rely primarily on the periodical journal literature.

4. An important feature of the patent literature is that it can serve to avoid duplication of research and development.

5. Patents, unlike periodical literature, are bound by definition to contain new information, data and ideas.

6. It can serve as a source of new ideas in a different but related field, and may be outside the patent protection offered the original invention.

7. Since claims are narrowly defined, much valuable information is included that may be outside the legal protection afforded by the claims.

8. Patents discuss the difficulties associated with previous research, development or production techniques, and offer a specific method of overcoming problems. This gives clues to current process information that has not been published in periodicals or books.

9. Can aid in process design by providing a selection of alternate techniques. A powerful research and engineering tool.

10. Obtain licenses — many U.S. chemical patents have not been developed commercially.

11. Patents provide an excellent starting point for the next investigator.

12. Frequently, innovations derived from research are first disclosed in the patent literature, prior to coverage in the periodical literature.

13. Patents offer a most valuable method of keeping abreast of latest technologies, serving an individual's own "current awareness" program.

14. Copies of U.S. patents are easily obtained from the U.S. Patent Office at 50¢ a copy.

15. It is a creative source of ideas for those with imagination.

CONTENTS AND SUBJECT INDEX

INTRODUCTION	1
GENERAL PROCESSES FOR VEGETABLE PROTEINS	2
Removal of Lipids	2
Countercurrent Extraction	2
Multizone Solvent Extraction	3
Protein Isolation Processes	6
Two-Step Extraction of Protein Isolates from Oilseeds	6
Protein Isolates by Aqueous Countercurrent Extraction	8
Molecular Sieve Removal of Lower MW Materials	10
Intracellular Protein Recovery Using Shock Waves	12
Hammermill Extraction of Proteins and Lipids	14
Apparatus for Separating Protein Enzymatically	17
Peroxide Treatment of Hydrolyzed Press Cake	19
Triple Enzyme Treatment of Vegetable Tissues	20
Recovery of Protein from Processing Wastes	21
Protein Treatment Processes	22
Encapsulated Protein Concentrates	22
Removal of Phytic Acid by Ultrafiltration	23
Bacterial Treatment for Improving Flavor and Odor	25
Detoxifying Proteins Containing Aflatoxin	26
Removal of Sulfur Compounds	27
Cation Exchange for Purifying Proteins	29
Protein Hydrolysates	30
Hydrolysis by *Hansenula montevideo,* n.sp. Yeast	30
Partial Hydrolysates by Controlled Acid Hydrolysis	31
Extraction of Undesirable Flavors from Hydrolysates	33
Blended Protein Hydrolysates	33
Dustproofing Hydrolysates with Polyoxyethylene Sorbitan Monoesters	35
Caramel Flavored Hydrolysates	37
PROCESSING WHOLE SOYBEANS	39
Wet Debittering Processes	39
Ammonium Salt Treatment	39
Steam Dehulling and Debittering Beans	40
Boiling Water-Saline Soaking Treatment	42
Heat Processing	44
Infrared Heating of Raw Soybeans	44

Two-Step Dry Heat-Hot Water Treatment	46
Dry Heating with Infrared	49
Canning Soybeans	53

PROCESSING FULL FAT SOY PRODUCTS 54

Debittering by Heat or Steam	54
Steam and Pressure Debittering Process	54
Alkaline Cooking at 80°C	56
Cooking Moisturized Full Fat Meal	57
Steam Jet Dehulling and Debittering Process	58
Detoxification with Alkylene Glycol	60
Applications of Full Fat Soy Products	62
Soybean Beverage Powders	62
Full Fat Soy-Cereal Products	63

DEFATTED SOY PRODUCTS 65

Oil Extraction Methods	65
Defatting Soybeans with Aqueous Alkali Metal Sulfites	65
Defatting Soybeans Using Ethanol	66
Solvent Removal Processes	67
Apparatus for Steam Treating Defatted Meal	67
Desolventizing and Toasting Apparatus	70
Enzyme and Yeast Debittering Processes	72
Flavor Improvement Using Yeast	72
Enzymatic Debittering Methods	74
Proteolytic Enzyme Treatment	75
Other Flavor-Enhancing Processes	77
Acid Digestion to Remove Flavor and Color	77
Acid and/or Peroxide Treatment of Soy Flour	79
Countercurrent Alcohol Extraction	81
Modification of Defatted Soy Flours	83
Heat and Humidity Conditioning for Soy Food Supplements	83
Increased Water Solubility of Soy Flour Using SO_2	84
Increasing Water Absorption of Soy Meal	86
Noncaking Soybean Meals	87

SOY PROTEIN CONCENTRATES 89

Extraction Processes for Producing Concentrates	89
Countercurrent Extraction Yielding Concentrated Wheys	89
Two-Phase Extraction of Full Fat Soy Products	91
Two Step Extraction of Defatted Soy Products	92
Aqueous Polysaccharide Extraction	94
Undenatured Soy and Peanut Concentrates	95
Shearing Forces to Liberate Protein	97
Urschel Mill Rupture of Protein Cells	97
Use of High Intensity Shearing Forces	98
Reconstituted Soy Concentrates	99
Soy Concentrates from Isolates and Fibrous Residue	99
Soy Protein Deposited on Cellular Material	101
Modification of Soy Concentrates	103
Reducing Microorganism Count	103
Partially Hydrolyzed Concentrates	105

SOY PROTEIN ISOLATES 107

Preparation of Isolates	107
Aqueous Extraction of Heat Treated Soy Protein at pH 6 to 8	107
Alcohol Treated Defatted Meals for Preparing Isolates	108
Reverse Osmosis for Isolating Proteins	110
Low Viscosity Soy Isolates	111

Isolates Having Improved Solubility or Dispersibility	113
Water-Soluble Soy Concentrates, Isolates and Whey	113
Catalase Improvement of Dispersibility and Flavor in Concentrates and Isolates	115
Soluble Denatured Soy Proteins	117
Dispersible Isolates by Jet Cooking	118
Pressurized Heating for Preparing Soluble Transparent Isolates	120
Gelable Soy Isolates	121
Thermoreversible Gels from Isolates	121
Heat Gelable Soy Protein	122
Heat Coagulable Isolates and Concentrates	124
Gelable Isolates from Soy Milk	125
Isolates for Specific Product Use	127
Soy Isolates for Baby Foods Using Enzyme Digestion	127
Soy Isolates for Gel-Like Meat Products	128
Alkaline Glycinin for Meat Processing	129
Copper Salts in Isolates for Baking	131
Partial Hydrolysis of Isolates for Cereals Using Mixed Enzymes	131

SOY HYDROLYSATES 133

Enzymatic Hydrolysis	133
Protease Hydrolysis of Oilseeds	133
Soluble Soy Protein Using Phytase	135
Dual Enzyme Solubilization and Hydrolyzation	136
Hydrolysis Under Acid Conditions	137
Controlled Incomplete Acid Hydrolysis	137
Acid Enzymatic Hydrolysis	139
Nutrient Amino Acid Compositions	140
Yeast-Hydrolyzed Protein Product	140
Palatable Balanced Amino Acid Compositions	141

COTTONSEED PROTEIN 144

Cottonseed Processing	144
Hexane-Acetone Extraction	144
Dynamic Gaseous Separation of Reduced Pigment Flour	145
Liquid Cyclone Process for Protein Concentrates	147
Gossypol Removal or Detoxification	149
Alkali-Protein Treated Meal	149
Soda Ash-Soda Soapstock Treatment	150
Alkali-Solvent Treatment of Meal	151
Amine Extraction	152
Amide Extraction	152
Fermentation by Rumen Microorganisms	153
Alkali-Peroxide Process	154
Removal of Solvents	155
Removal of Mixed Solvents from Marc	155
Removal of Water	157

GRAIN PROTEINS 159

High Protein Flours	159
Wheat Flour of Bran-High Protein Endosperm Blends	159
High Protein Wheat Endosperm Fraction Blended with Soy Flour	161
Air Fractionation at Critical Cuts of Milled Cereal Flour	161
Agglomerates of High Protein Flour	163
Protein Addition to Wheat Flour	165
Isolation of Grain Proteins	166
Aqueous Ammonia Extraction of Gluten	166
Glyceride Oils for Extracting Gluten	167
Protein from Ruptured Cereal Grain Germ	169
Protein from Cereal Endosperm Using Buffered Solutions	171

Two-Stage Extraction Process for Wheat Gluten	172
Gluten Modification	174
Carbon Dioxide for Drying Gluten	174
Rehydrating Gum Gluten	175
Lipid Coating of Undenatured Wheat Gluten	176
Dispersible Agglomerates of Gluten	177
Frothable Wheat Gluten as Egg White Substitute	179
Upgrading Odor and Taste of Gluten	180
Zein	181
Partially Deaminated Zein	182
Deaminated Zein by Alkaline Hydrolysis	183
Solvent Extraction of Zein	184

PROCESSING OTHER VEGETABLE PROTEINS — 186

Sunflower Meal	186
Purification by Acid Washing	186
Purification by Membrane Ultrafiltration	188
Safflower Seeds	190
Food Products from Oil-Free Safflower Seed Residue	190
Wet Processing of Safflower Seeds	191
Sesame Seeds	193
Seasame Seed-Flour Product	193
Sesame-Whey Combinations	193
Separation of Oil and Protein from Sesame Seeds	194
Castor Beans	196
Autodigestion to Remove Ricin from Castor Beans	196
Deallergenizing Castor Beans with Ammonium Hydroxide	198
Crystallization of Bean Protein from Dilute Salt Solutions	199
Peanuts or Groundnuts	201
Process for Making Peanut Flakes	201
Proteins from Seeds of Brassicaceae Family	202
Proteins from Mustard and Rape Seeds	202
Protein from Rape Seed	204
Crambé Proteins	205
Alfalfa Protein by Alkaline-Pancreatin Treatment	208
Coconut Protein	209
Mistletoe Protein	210
Extraction of Mistletoe Protein	210
Purifying Mistletoe Proteins by Ultracentrifugation	213
Electrophoresis of Mistletoe Protein	213

EXTRUDED FIBER PROCESSING — 215

Boyer Process	216
Processes for Extruded Fibers	217
Continuous Fiber Preparation from Protein Slurry	217
Gaseous Coagulation of Protein Fibers	218
Heat Binding Extruded Fibers	220
Additives for Fiber Processes	220
Acid Coagulating Bath Containing Sulfur Dioxide	220
Sulfites in Protein Coagulation Baths	222
Sulfur Compounds in Soy Fibers Extruded at High Temperatures and Pressures	222
Extruded Fibers from Protein Blends	224
Fibers from Blends of Casein and Soy Proteins	224
Fibers of Wheat Gluten and Soy Flour Blends	225
Fibers from Soy-Keratin Mixtures	226
Fibers from Protein-Carbohydrate Polymer Blends	227
Mixed Fibers Having a Polymeric Carbohydrate Gel Precursor	227
Aluminum Modified Protein-Alginate Fibers	228
Preservation of Spun Protein-Alginate Fibers	229

Contents and Subject Index

Protein-Thermogelable Polysaccharide Fibers	230
Miscellaneous Fiber-Forming Compositions	232
Protein Fibers from Safflower Seed Meal	232
Polyacrylic Acid Additive for Protein Fibers	233
Fibers from Protein-Lipid Extracts	234

OTHER PROTEIN FIBER PRODUCTION METHODS — 236
- Coagulation or Precipitation of Fibers — 236
  - Coagulation of Alginate-Protein Suspensions — 236
  - Protein Coagulation with Divalent Cations — 237
  - Soy Whey Protein-Polysaccharide Complex — 239
- Heat Expansion Fiber-Forming Processes — 240
  - Textured Fibers by Pressurized Heating — 240
  - Wheat Gluten Fibers by Heat Expansion — 242
  - Shredded Protein Texture — 244
- Protein Curd Frozen into Fibrous Sponge-Like Mass — 246

TEXTURED PROTEIN GELS AND EXPANDED PRODUCTS — 248
- Expanded Textured Proteins — 248
  - Expanded Hydratable Proteins — 248
  - Use of Flow Inducing Salts for Expanded Proteins — 250
  - α-Cellulose in Expanded Protein Products — 252
  - Expansion Using Microwave Energy — 253
  - Extruded Expanded Proteins Containing Mg or Ca Oxides — 254
  - Exit Gate for Texturizing Apparatus — 255
  - Textured Products from Defatted Cooked Soybean Flour — 256
  - Mechanically Tempering Prior to Heating — 257
- Textured Protein Gels — 259
  - Chewy Protein Gel Binders for Fibers — 259
  - Chewy Gel from Denatured Proteins — 259
  - Conversion of Fibers to Chewy Gel in Coagulation Bath — 261
- Extruded Granules of Gluten-Soy Flour Blends — 262

CONSUMER PRODUCTS — 264
- Simulated Milk Products — 264
  - High Yield Process from Soybeans — 264
  - Simulated Cow's Milk from Mixed Proteins — 265
  - Soy Milk from Sprouted Beans — 266
  - Deodorizing Soybean and Peanut Milks — 267
  - Wet Milling of Vegetable Protein — 268
  - Dispersible Soy Protein for Milks — 269
  - Soy-Milk Combinations — 271
  - Dual Inoculation for Flavor Improvement of Soy Milk — 272
  - Lipoxygenase Inactivated Full-Fat Soy Flour for Milk — 272
- Other Dairy-Type Products — 274
  - Soy Yogurt — 274
  - Heat Stable Cheese Curd from Soy Milk — 275
  - Blue Cheese from Soy Milk — 276
  - Cheese from Peanut (Groundnut) Protein — 277
- Paste Spreads — 278
  - Paste from Moisturized Soybeans — 278
  - Attrition Milled Soy Protein and Oil — 279
- Tofu, Miso and Tempa Products — 280
  - High Frequency Sterilization of Tofu — 280
  - Soybean Grits in Miso Preparation — 283
  - Steam Treatment for Preparing Soy and Miso Pastes — 284
  - Tempa Production in Perforated Bags — 286
  - Tempa from Cereal Grains — 288
  - Powdered Tempa — 290

Flavoring Materials　291
   Alcoholic Extraction of Soybeans in Making Soy Sauce　291
   Flavor Concentrates from Steffen Filtrate　292
   Granular Food Seasonings from Protein Hydrolysates　293
Miscellaneous Products　294
   Surface Active Protein Products　294
   Whipping Agents from Soy Protein　295
   Production of Dry Neutralized Proteinate　296
   Malt Treatment of Soy for Use in Cereals　298
   Soy Protein-Soy Lecithin Mixtures　299

COMPANY INDEX　302
INVENTOR INDEX　304
U.S. PATENT NUMBER INDEX　307

INTRODUCTION

Food shortages continue to increase, particularly in India and the recent drought zone in Africa. Proteins, which are essential to growth and good health are also needed in many more of the underdeveloped countries throughout the world. Meat protein, which is the choice of most individuals, is becoming more and more expensive to produce. Shortages, plus high prices have recently caused restriction of meat proteins in the diet of many families in the United States.

Vegetable proteins offer great potential as a direct food for human consumption. However, many of the proteins such as those obtained from soy, cottonseed, rapeseed and the like require processing to provide a food material having acceptable organoleptic properties for human consumption. Often toxic ingredients have to be removed from the protein before it can be used in animal feeds or in human food products.

In recent years many processes have been developed which provide texturized proteins, that is protein products which can be used as meat extenders or in preparing processed meat substitutes. These textured materials can be fibers, expanded porous granules or chewy gels. Textured meat extenders now appear on the supermarket shelves and protein-extended hamburgers appear on the menu of a large chain of family restaurants.

The U.S. Department of Agriculture has projected that by 1980, vegetable proteins will replace 20% of processed meats and 8% of total red meat sold in the U.S. Other forecasts also predict that by 1980, 9% of the world supply of soybeans could be consumed by humans and this would be higher if more acceptable products were developed.

This review covers 224 patents and one government report, mostly issued since 1960. The subject coverage has been aimed at the processing of agriculturally produced vegetable proteins primarily to make them usable as food products. Patents which merely use vegetable proteins in baked goods, or combine a textured protein with other ingredients to form a simulated sausage, as chicken loaf, were not included in this survey. Processes on production of protein by growth of bacterial or yeast cultures which could be classed by some as vegetable proteins are also not reviewed.

Soybean protein ranks first among the vegetable proteins studied. Processes for cottonseed protein and grain proteins (wheat, corn and the like) rank next in number of patents with other vegetable protein sources trailing in number of patented processes.

GENERAL PROCESSES FOR VEGETABLE PROTEINS

This chapter reviews those processes which can be used for several varieties of vegetable and/or animal proteins. These have been grouped under four headings covering the removal of fats or lipids, isolating the proteins, treatment of protein materials, and preparation and treatment of protein hydrolysates.

REMOVAL OF LIPIDS

Countercurrent Extraction

Fat-free and flavor-free protein products are obtained by a countercurrent extraction process developed by *J.C. Cavanagh; U.S. Patent 3,295,985; January 3, 1967.* The process removes water, fats, coloring matter, flavors, odors, oils, etc. without damage of proteins or oils from a wide range of flavored or odoriferous vegetable materials. A valuable solid product of good keeping quality is produced containing protein and/or carbohydrates and/or other solid matter in a substantially tasteless or odorless condition suitable for use as a food for humans and/or animals. The fats and other components may be recovered as principal or as by-products. This is done economically by the counterflow treatment of the material in a comminuted form with an organic solvent or mixture of solvents by adding water to the dry material before treatment or adjusting the water content at least the first stage of the treatment to insure that the solvent will have opportunity to contact and to extract the flavor components. The solvent alone without water will not dissolve the flavor components.

The selection of solvent may be made within wide limits. Acetone is a suitable solvent, as is a mixture of equal parts by volume of acetone, ethyl acetate, and ethyl alcohol. This latter mixture is particularly suitable for insuring that substantially all the flavor components are removed from the material. It is preferable to avoid the use of cyclic hydrocarbons, chlorinated hydrocarbons, or other solvents which may be harmful to the health of operating personnel, or which may leave undesirable residues in the end-products.

Using acetone, it is found that the flavor components may be fairly substantially removed in the counterflow treatment in the early stages during which the acetone is mixed with water in proportions of the order of 30% to 10% by volume. Components not removed in the early stages may then still remain in the solid product during and after the latter stages, in which the water content of the acetone may diminish from 10% down to about 2%. On the other hand, a solvent such as an equal mixture by volume of acetone, ethyl alcohol and ethyl acetate, is found to dissolve flavor components not only during the early stages when admixed with 30% to 10% by volume of water but, also, in the latter stages when admixed with 10% down to 2% or less by volume of water. A solvent of this kind is particularly valuable if it is desired to get very thorough extraction of flavor components. The degree of comminution of the material to be treated should be sufficient to enable the solvent to penetrate into the particles to achieve the desired degree of extraction of the fat or oil and/or other components.

The temperature of the materials is important to the extraction of oil by the solvent. For example, tests have indicated that the solubility of oil in solvent containing 10% water is on the order of 0.25% at ambient temperature while at 50°C the solubility is of the order of 2.5%. Solubility of the oil in aqueous-solvent mixtures indicates good solubility in stages 3, 4, 5, and 6 in a six-stage process. The temperature of the material should thus be maintained at around 50°C throughout all stages of the process for optimum results.

Example: There was used as raw material, shredded copra, in a six-stage countercurrent treatment, the solvent being acetone. Each stage consisted of agitating the mixture of copra with the solvent liquor in a vessel for approximately 5 minutes at a temperature of approximately 50°C, then filtering and squeezing the liquor from the solid matter, and moving the solid matter forward to the subsequent stage vessel, and passing the liquor backwards to the previous stage vessel. Each quantity of forward flow material entering the first stage was 30 grams of copra and to it was added 15 ml of water during the first stage of the treatment.

Each quantity of acetone entering the last stage as backflow was approximately 45 ml of acetone in relation to each 30 grams of copra forward flow. The countercurrent system was brought into equilibrium by rejecting the first several batches of outgoing solid product and of outgoing liquor. In the fourth stage, the mixture of solid matter and liquor was transferred for finer comminution to a disintegrator with high speed rotating cutter blades and then transferred back to the fourth stage vessel.

The outgoing solid product from eight batches, equivalent to 240 grams of ingoing copra, was combined and, after removal of entrained solvent by evaporation, weighed approximately 53 grams. This product was milled into a fine powder, almost pure white, and it had no detectable flavor of copra. The outgoing extraction liquor from eight batches was combined, and it consisted of two phases, the one being all or substantially all oil and the other being all or substantially all solvent-water with dissolved flavor components. The total volume of liquor was approximately 400 ml and to this was added a further 200 ml of water.

The solvent was removed from the two-phase liquor by fractional distillation, leaving a two-phase liquor consisting of oil and water, the latter with dissolved flavor components and a small amount of suspended solid matter. The oil was separated by centrifuging and was then filtered. It weighed approximately 164 grams and was almost water-white and had the appearance and character of high grade coconut oil. The aqueous phase containing the dissolved and suspended matter was concentrated by evaporation to yield a brown mixture of syrup and solid matter weighing approximately 21 grams and which had an intense flavor resembling coconut.

Other examples of the process cover the extraction of materials such as peanuts, cocoa nibs, safflower seeds, soybeans, cottonseed kernels, ground nut kernels and rice pollard.

Multizone Solvent Extraction

Protein concentrates are produced by a multizone solvent extraction in a process disclosed by *R.M. Henderson and W.F. Habermann; U.S. Patent 3,520,868; July 21, 1970; assigned to Beloit Corporation.* The separation process involved in each zone consists of mixing a quantity of solvent obtained from the next succeeding zone with protein-containing materials to extract the undesirable soluble materials, separating the solvent from the protein, withdrawing the solvent from that zone and recycling a portion of that solvent back into the zone while also passing the remaining portion of the withdrawn solvent back to the preceeding zone.

Any number of zones may be used depending upon the desired purity and the particular solvents and food for protein sources. The number of zones employed is defined as being the letter n where n is at least one. Most often, satisfactory separation of the protein can be effected in two or three zones, although any number of zones may be employed.

A wide variety of protein-containing materials may be processed such as: whole fish or frozen fish, meat and meat offal, brewers' grain, distillers' grain, malt grains, poultry and poultry offal, cotton seed, sunflower seeds, and other seed types, soybeans and other legumes, such as peas, beans, and the like.

A number of solvents may be used to remove the oils, fats, sugars, starches, and other undesirable portions from the protein to yield a protein concentrate. For example, water, isopropyl alcohol, ethyl alcohol, methyl alcohol, acetone, amyl acetate, ethyl acetate, hexane, t-butyl alcohol, trichloroethylene, methyl ethyl ketone, and many of the common commercial solvents. Mixtures of two or more solvents are sometimes quite effective in removal of soluble materials from protein-containing foods.

The particular steps of this process consist of introducing the protein material into each succeeding zone along with a quantity of solvent from the next succeeding zone. The solvent and the dissolved material are then separated from the protein in this zone. The protein is passed on to the next succeeding zone for treatment. As the protein is withdrawn from the zone, the solvent containing the dissolved fats, sugars, starches, and the like are also withdrawn from the zone, and a portion of this withdrawn solvent is recycled back into the zone at the point of addition of the protein-containing material. The balance of the solvent is then passed back to the next preceding zone for use therein. Of course, the solvent introduced into the nth zone is normally fresh solvent, and the solvent withdrawn from the first zone is usually removed from the system for recovery, purification, or disposal.

FIGURE 1.1: PROTEIN CONCENTRATION BY EXTRACTION WITH A SOLVENT

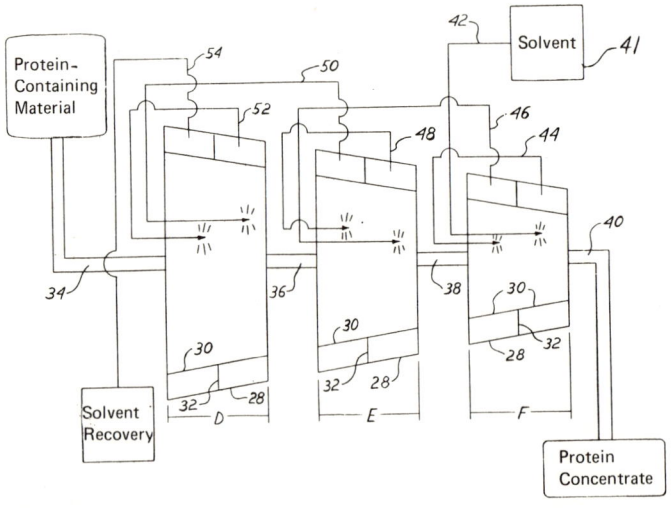

Source: R.M. Henderson and W.F. Habermann; U.S. Patent 3,520,868; July 21, 1970

The above drawing is a schematic flow diagram showing a preferred embodiment of the process in which the n number of zones is 3. The substance is passed through 3 centrifuges 28 which are represented by the zones D, E, and F. The protein material is introduced into zone D through line 34 and withdrawn from zone D and introduced into zone E through pipe 36. Likewise, the protein-containing material is withdrawn from zone E and introduced into zone F through line 38. Finally, the protein concentrate is withdrawn from zone F at point 40.

Fresh solvent is introduced from solvent supply 41 through pipe 42 into the interior of zone F at a point near the exit 40. The solvent passes through the screen 30 after having mixed with and extracted soluble materials from the protein-containing substance. The solvent is collected in the area between the screen 30 and the outside of the centrifuge 28, and is maintained near the exit half of the centrifuge by the retaining wall 32. This solvent, containing the last amount of material extracted from the substance is recycled back into the zone through pipe 44 to a point near the entrance of zone F.

In a similar manner, this solvent passes through the protein material extracting additional soluble fats and is collected in the area defined by the screen 30 and the outside of the centrifuge 28. Again the retaining wall 32 insures that the solvent will remain near the front portion of the zone F. The solvent is then transferred to the next preceding zone through pipe 46 where it is introduced into the centrifuge at a point near the exit of zone E in pipe 38. This solvent extracts additional solubles and is collected between the screen 30 and the wall of the centrifuge 28 in retaining wall 32. This solvent after being collected is recycled through pipe 48 back into the front portion of zone E where it again passes through the protein-containing material collecting even more of the soluble matter. Pipe 50 then passes the solvent back to the next preceding zone, zone D.

The solvent coming through pipe 50 is introduced into the centrifuge of zone D at a point near its exit 36 where it passes through the substance and is collected and recycled through pipe 52. This solvent is introduced into the interior of the zone D near the front part of the centrifuge where the final extraction is performed. At this point, the most easily extractable portions of the protein-containing material are subjected to action by the solvent when it already contains a large portion of soluble material. This solvent is withdrawn from the system through pipe 54.

Thus it can be seen that solvent extraction takes place twice in each zone with the use of a retaining wall 32 between the two portions of the means for collecting the solvent already separated from the protein-containing material. The protein-containing material is washed or treated with solvent with successfully purer solutions in direct proportion to the purity of the protein-containing material. The pure solvent in pipe 42 contacts the protein concentrate at its purest state while the solvent containing the highest concentration of dissolved material, coming through pipe 52, passes through the raw protein-containing material which contains the highest amount of soluble matter.

R.M. Henderson and W.F. Habermann; U.S. Patent 3,538,069; November 3, 1970; assigned to Beloit Corporation have modified the process of U.S. Patent 3,520,868, wherein the protein is first contacted with a heated gas to remove volatile materials. The particular steps of this modified process consist of introducing the protein material into a stream of heated gas for sufficient time to reduce the volatile material, such as water, light oils, and the like. Such a procedure is known generally as flash drying and is normally carried out by inserting the material being treated into a stream of hot air which has been generated from the heat source. The hot air carries the particle being treated through a passage. Adjustment of the length of time of passage, the temperature of the air and other variables will adequately regulate the amount of volatile material removed.

In most instances, it is desirable to remove at least 50% of the volatile material contained in the raw food product. In a more preferred embodiment, 80% or more of the volatile material should be removed. By the term volatile material, it is meant those materials contained in the protein source which are capable of rapid evaporation or volatilization at temperatures below about 300°F. In most instances, a source of hot gas at temperatures ranging above 300°F will contact the protein for a short period of time sufficient to raise the temperature of the material to a point where water and other highly volatile materials will flash off of the particle. The contact time is sufficient to volatilize the liquids, but it is not sufficient to raise the temperature of the solid material to a point where oxidation or other adverse occurrences will become a factor. Many devices are provided for carrying out the initial step of this process. All that is needed essentially is a source of hot gas, preferably air, and a passageway through which the air and the particles being heated may pass. Once a substantial amount of the volatile material, the protein material is treated as described in U.S. Patent 3,520,868.

PROTEIN ISOLATION PROCESSES

Two-Step Extraction of Protein Isolates from Oilseeds

W.H. Martinez and L.C. Berardi; U.S. Patent 3,579,496; May 18, 1971; assigned to the U.S. Secretary of Agriculture have developed a process for extracting selectively protein fractions from oilseed flour. This process comprises a preliminary extraction with water or a dilute (0.008 M) polyvalent cationic salt solution and subsequently with aqueous base (0.015 N sodium hydroxide).

The protein isolates thus extracted differ in composition, average molecular weight and solubility characteristics. With certain oilseeds, this process provides for the recovery of more of the total nitrogen from the flour being processed and for an improved flavor and color in the major isolate than does a conventional single-step alkaline extraction process.

The separate extractions provide for the separation of primarily the low molecular weight proteins precipitable at pH 3.8 to 4.2 (Isolate-I) and the high molecular weight, alkali-soluble proteins precipitable over a pH range which varies with the particular oilseed meal being processed (Isolate-II). With certain oilseeds, Isolate-I contains as much as 17% of the total N of the flour. Isolate-II can contain, depending upon the oilseed, from 45% to 72% of the total nitrogen. The isolates differ in composition, average molecular weight, and in solubility characteristics. With certain oilseeds, the two isolates account for slightly more of the total N of the flour than does an isolate obtained from a single-step, alkaline extraction. In addition, the color and flavor of the major isolate (Isolate-II) is improved.

The quality and yields of the isolates obtained with the double extraction process are relatively insensitive to time of extraction beyond 30 minutes, to extraction temperature within the range of 25° to 75°C, type of acid used for precipitation, or extraction ratio (flour to solvent) within the range of 1:20 and 1:10. However, in the specific case of cottonseed an increasing extraction temperature increases gland rupture and the dispersion of pigments in the water extract. The increased temperature and dispersion subsequently results in an increased content of total and bound (total gossypol minus free gossypol) gossypol in the water extracted fraction and a decreased content of free and total gossypol in the alkali-extracted fraction.

The oilseed kernel is composed primarily of cotyledonous tissue and contains varying amounts of protein. The major portion of this protein is located in the protein bodies or aleurone grains which fill the cells of the cotyledons. The protein bodies range in size from 1 to 20 microns and each is surrounded by a unit membrane. In some oilseeds, embedded within the matrix of the protein body is another particle called the globoid. Globoids are the storage sites of the phytin of the seed and contain certain metal ions in addition to the inositol phosphate. Protein bodies remain intact during defatting, grinding, and appropriate extraction. With certain oilseeds such as cottonseed and sunflower, the protein bodies do not rupture on suspension in water.

Extended suspension, beyond several hours will cause the protein bodies to swell and perhaps rupture but will not bring about solubilization of the major storage or protein body proteins. With other oilseeds, such as soybean and peanut, the protein bodies do rupture on suspension in water permitting dispersion of the proteins. The use of dilute polyvalent cationic salt solutions with this type of oilseed stabilizes the protein bodies and permits selective extraction. Such dilute divalent salt solutions can be considered isotonic salt solution for the protein bodies of the seed. The following diagram outlines the process as used in the example of cottonseed flour.

FIGURE 1.2: SELECTIVE EXTRACTION USING WATER OR DIVALENT METAL SALTS

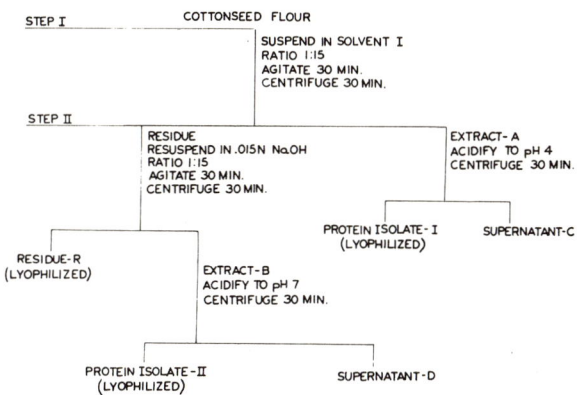

Source: W.H. Martinez and L.C. Berardi; U.S. Patent 3,579,496; May 18, 1971

In this process, distilled water, 0.008 M calcium chloride, 0.008 M calcium acetate, 0.008 M magnesium chloride, 0.008 M magnesium sulfate or other salts of polyvalent cations can be used as Solvent I. See table below for condensed data of the use of various solutions as Solvent I.

Oilseed flour	Step-1 Solvent	Percent of Total N	Step-2 Solvent	Percent of Total N	Product	Percent of Total N	Composition Nitrogen	Phosphorus	Calcium	Magnesium
Cottonseed	H₂O	25.9	.015 N NaOH	65.7	Isolate I	16.0	13.68	3.22	0.01	
					Isolate II	55.1	15.93	0.48	0.02	
	.008 M CaCl₂	20.0	.015 N NaOH	56.6	Isolate I	5.6	14.73			
					Isolate II	51.1	16.27	0.30	0.04	
	.008 M MgSO₄	21.9	.015 N NaOH	66.6	Isolate I	10.0	13.65	2.14		0.38
					Isolate II	54.2	15.54	0.49		0.31
Soybean	.008 M CaCl₂	14.0	.015 N NaOH	54.0	Isolate I	6.0	12.98	1.28	0.08	
					Isolate II	50.0	15.32	0.96	0.28	
	.008 M MgCl₂	15.0	.015 N NaOH	55.2	Isolate I	6.0	12.01	1.31		0.28
					Isolate II	49.1	14.82	0.61		0.10

Extractions were normally conducted at a ratio of 1 gram of flour per 15 ml of solvent. Permutations of 1:15, 1:10, and 1:10, 1:10 in Steps 1 and 2, respectively, were also investigated with cottonseed, but as noted above were without significant effect. The ratio in Step 2 is based upon the original weight of flour. Step 1 was conducted at temperatures of 25° and 75°C and within this range temperature was found to exert little, if any, effect. Extracts were centrifuged at 7,800 g, decanted and filtered through No. 4 filter paper to remove any small particles dislodged during decantation. Extracts were stirred constantly during acidification with 1 N HCl. Isolates were centrifuged at 7,800 g.

The pH of maximum precipitation was determined by titrating aliquots of the extracts related to specific pH values, centrifuging under the appropriate conditions of isolation and determining the nitrogen content of the supernatant.

As noted above, either water or solutions containing an isotonic concentration of polyvalent cationic salt such as 0.008 M calcium chloride, calcium acetate, magnesium chloride or magnesium sulfate can be used to extract the relative low molecular weight, nonprotein-body proteins. The high molecular weight protein-body proteins are solubilized by reextraction of the water extracted flour with alkali (0.015 N NaOH). If the criteria of protein acceptance is nitrogen content, then the polyvalent salt, dilute alkali procedure is the preferred process. If both recovery of total nitrogen and total weight of recovered nitrogen are important, then the water, dilute alkali procedure is the preferred process with oilseeds such as cottonseed and sunflower.

Protein Isolates by Aqueous Countercurrent Extraction

H. Bock; U.S. Patent 3,402,165; September 17, 1968; assigned to Protein-Compagnie GmbH, Switzerland has provided an improved method for treating oil mill residues for producing high yields of a soluble protein fraction suitable for use as an animal feed.

It was found that oilseed meals can be broken down into a protein fraction, a soluble component fraction and into fibers by treating an oilseed meal-water mixture, mixed in a ratio of oilseed meal to water of 1:1.5 to 1:20 and preferably of 1:3 to 1:8, by the following steps carried out in a closed circulation:

(1) Mashing the starting meal-water mixture by contacting it with a solution-suspension LA1, which contains in addition to dissolved substances, insoluble protein.

(2) Breaking down the mash in a fiber separation T1 into a fiber fraction FT1 and a protein emulsion fraction ET1,

(3) Breaking down the protein emulsion fraction in a protein separation T2 into a protein concentrate ET2 and a solution LT2;

(4) Possibly discharging the solution LT2 after evaporation thereof from the circulation;

(5) Washing the protein concentrate ET2 in a protein washing A2 by passing the protein concentrate in countercurrent to fresh water W, introduced into the circulation from without, to thereby form the solution LA2 and the protein fraction E;

(6) Separating the protein fraction E and discharging it from the system;

(7) Washing the fiber material FT1 in a fiber washing out A1, bypassing the same in countercurrent to the solution LA2, whereby the solution-suspension LA1 and the fiber fraction F are formed;

(8) Separating the fiber fraction F and discharging the same from the system;

(9) Returning the solution-suspension LA1 for use in the mashing of further starting material.

The following symbols having the meanings indicated below have been employed for describing the process.

```
M = mash
T1 = fiber separation step
FT1 = fiber fraction recovered from the fiber separation step
ET1 = protein emulsion fraction recovered from the fiber separation step
T2 = protein separation step
ET2 = protein concentrate recovered from the protein separation step
LT2 = solution of soluble components recovered from the protein separation step
A1 = fiber washing step
F = fibers recovered from the fiber washing step
LA1 = solution-suspension recovered from the fiber washing step
A2 = protein washing step
E = protein
LA2 = solution recovered from the protein washing step
L = soluble fraction
```

The process is carried out within a completely closed and continuous system. The starting material and water mixed in a specific ratio are introduced into the system. The amount of solution which is drawn off from the system depends on the amount of fresh water, which is added for washing the protein, and on the water content of the protein and fiber materials which are taken out of the system (water lost to the system). Thus, in accordance with this process, a proper and critical relationship exists and must be maintained between the fresh water introduced into the system and the starting material; maintaining the relationship being dependent on insuring that the amount of water present in the system is a constant.

In place of the solution-suspension LA1, which contains large quantities of undissolved protein, there may be conveniently used for the mashing the solution LT2 or a mixture of both solutions LA1 and LT2. In this case, the solution-suspension LA1 or the residue (unused portion) of the same is prior to or after the fiber separation step T1 added to the protein emulsion fraction ET1.

The starting materials which may be used include the oilseed meals, cakes and expellers, as, for example, soybean meal, peanut meal, palm kernel meal, peanut expeller, soybean expeller and cottonseed expeller. The expellers and cakes are obtained by the continuous and/or discontinuous pressing of oilseeds. The so-called meals are obtained by extraction of oilseeds with benzene or hexane, by the customary procedures. If fractions having high swelling capacity and solubility are desired, then the debenzening of the meals must take place under mild conditions, so that the protein does not coagulate.

There may also be used solvent-moist rough grinds as starting material, in which case the removal of the solvent takes place by heat treatment carried out within the closed system. By using vacuum and equipment, such as flow evaporators, rotation evaporators, or thin-layer evaporators, it is possible to carry out the evaporation process quickly and carefully, so that no damage of the protein takes place and a highly soluble product results. The following example serves to illustrate the process.

Example: In accordance with the process there are introduced 400 kg of soybean meal and 1,800 l of water into the closed system. The pH of the water is adjusted with sulfurous acid and dilute hydrochloric acid to a value of 4.5. The temperature is maintained at 40°C. The water which is introduced is calcium and magnesium cation-free. The water which is used constitutes water which was introduced at the end of the protein washing step A2, namely, into the terminal part of the washing arrangement for use in washing the protein fraction.

The water is passed in countercurrent flow through the washing arrangement A2 becoming enriched with soluble constituents. This enriched water is thereafter used for washing of the fiber constituents A1, which washing is also carried out by passing the fibers in countercurrent flow contact with the water. The water is thereby further enriched with soluble components and simultaneously takes up any protein still adhering to the fibers following the fiber separation T1. The solution-suspension LA1 thereby formed is used for the mashing of the starting material which is continuously introduced into the installation.

There is added to the mixer 4,000 l/hr of solution LT2 in order to produce a convenient concentration for the further processing. The mash consisting of fibers, protein, soluble components, water and intact starting material is passed through a curved sieve in order to sift out the coarse material. The portion remaining on the sieve is ground further using a pin-mill and added to the portion which is passed through the sieve, and the combined mash is conducted through a second curved sieve having a slit width of 50μ FT1. The protein emulsion which has passed through the sieve and which contains 38 g protein per l is introduced into a separator arrangement T2 and, as a result, a protein fraction is recovered having a content of 180 g protein per l. The solution simultaneously recovered is free of any solid constituents and contains 11% of dissolved components. The solution is withdrawn from the system in a quantity of 1,360 l per hr. 4,000 l of this withdrawn solution are utilized for mashing the raw material (above). The fiber fraction which is recovered FT1 is subjected to a washing treatment utilizing the water recovered from the protein washing treatment A1 and the fibers following the washing are freed from any water in a sieve centrifuge and discharged from the system.

The protein emulsion which is recovered from the separator arrangement T2 and which contains 180 g per l of protein is washed, using the arrangement as above described. The protein fraction recovered from the last of the 3 separators is treated in a suction filter and the filter cake washed with 1,800 l of water having a temperature of 40°C. The water used for this washing has a pH value of 4.5 which pH has been obtained by addition of dilute hydrochloric acid to the water.

This water, which is introduced in an amount of 1,800 l at the end of the process for a final washing of the protein, is used for preliminary washing of the protein and thereafter as washing agent for the fibers. Finally, this water is introduced in its entirety, 1,800 l for the mashing of the 400 kg soybean meal. The analysis of the starting soybean meal is as follows:

	Percent
Protein	44.8
Ash	5.4
Raw fiber	7.0
Oil	0.2
Carbohydrates	29.6
Water	12.5

The following are the yields obtained when 400 kg of the soybean product having the above analysis is subjected to the process.

	Percent
Protein (159.2 kg)	39.8
Fibers (98.4 kg)	24.6
Soluble components (142.4 kg)	35.6

The protein which is recovered has purity of 90.7% calculated with respect to the dry product. When the above example was repeated using soybean meal derived in the processing of dehulled soybeans having a protein content of 51.3%, there was obtained a yield of 44.7% of protein.

Molecular Sieve Removal of Lower MW Materials

Molecular sieve techniques have been used in the separation of edible protein from the salts of skim milk and cheese whey. In the brochure entitled *Gel Filtration, a New Way of Recovering Protein from Milk and Whey* (S1092E/3) of Alfa-Laval Co., a cheese whey is initially concentrated to about 23% solids and is fed onto the bed of molecular sieve grains. After the feed is absorbed into the bed, water is used to elute the protein molecules from the entrapped salts. Although effective separations can be realized this process has several disadvantages, including a tendency for the feed to cause a plugging of the bed, inefficient use of the molecular sieve material, etc.

C.S. Dienst and J.M. Attebery; U.S. Patent 3,547,900; December 15, 1970 have found that at least some of these disadvantages in protein separation from salts can be overcome by the use of a thin bed of molecular sieve particles capable of separating protein from salt molecules, using a bed thickness (or height) of from 10 to 100 mm (preferably 10 to 60 mm), and a bed loading, (i.e., volume of proteinaceous liquid feed/total bed volume, including both particles and bed voids, expressed as percent) from above 25 to 65% (preferably 30 to 50% and most preferably above 30%). Although the solids content of the liquid feed solution is limited only by the need for a viscosity sufficient to permit the feed to pass through the bed of molecular sieve material (which will depend on molecular sieve particle size, bed packing, pressure drop considerations, etc.), the feed solutions generally contain a solids content up to about 45% total weight of dry solids (preferably with a minimum of at least 20%, most preferably at least 25%) when cheese wheys are used. The thin bed process improves the operating efficiency of the protein separation process while retaining the capability of producing a high protein yield or protein purity.

By varying the total volume of eluate collected as it emerges from the bed, it is possible to control the percent protein yield or the percent protein purity, (i.e., amount of recovered protein in the total solids in the eluate). This control can be conveniently accomplished by measuring or monitoring the conductivity of the eluate, which is an indication of the salt concentration, or by ultraviolet absorbance, which is an indication of the protein concentration, as will be discussed later.

In the operation of this process the aqueous proteinaceous feed solution is introduced into a bed of molecular sieve material having pores of a size permitting the penetration only of molecules smaller than the protein, entrapping those smaller molecules in the molecular sieve material, forcing the dissolved protein through the bed in the liquid outside the molecular sieve material, eluting the protein selectively from the bed, and recovering the protein-containing eluate. With the thin beds it is important to carefully introduce the feed onto the bed to avoid channeling or bed disruption.

Use of a perforated distribution plate is one means for maintaining bed integrity and uniformly bringing the feed solution into contact with the input surface of the bed, although different techniques may be used with various equipment, e.g., centrifuge, column, vacuum or pressure filter, etc. Further detailed description of equipment useful for molecular sieve separation appears in the literature. As in any molecular sieve separation process, means for accurately sampling and collecting the eluate must be provided. With the relatively brief cycles of the thin beds it is even more essential that the collection of eluate can be essentially instantaneously terminated to achieve optimum fractionation.

Although this process is applicable to any aqueous protein solution having dissolved material of lower molecular weight than the protein (such as salts), skim milk and cheese wheys, e.g., cottage cheese whey and cheddar cheese whey, are readily available at low cost and are therefore advantageously employed in the thin bed process. Brewers' yeast, soy whey, animal blood, fermentation media containing protein from microbiological action, and torula yeast may also be treated by this process. Tables 1 and 2 present comparative data on the effect of bed thickness in whey treatment.

TABLE 1

Run	Diameter (cm.)	Height (cm.)	Bed volume (liters)	Charge/cycle as percent of bed volume	Surface area (cm.²)	Charge volume (cm.³×10³)	Surface area per charge (cm.³/l.)	Thickness of charge liquid (cm.)	Surface area/bed depth	Height/diameter
1	40	60	75	25.4	1,255	19	66.0	15.1	21.0	1.50
2	40	100	125	24.8		31	40.5	24.7	12.6	2.50
3	40	150	185	25.4		47	26.7	37.5	8.4	3.80
4	60	60	170	25.3	2,810	43	64.0	15.3	46.8	1.00
5	60	100	280	25.0		70	40.2	24.9	28.1	1.67
6	60	150	420	25.0		105	26.8	37.4	18.8	2.50
7	80	60	300	25.0	5,020	75	67.0	14.9	83.7	0.75
8	80	100	500	25.0		125	40.2	24.9	50.2	1.25
9	80	150	750	25.1		188	26.7	37.5	33.5	1.88
10	180	100	2,500	25.0	25,450	625	40.7	24.6	254.5	0.56
11	*70.87	3	15.1	30.0	5,023	4.53	1,109	0.902	1,674	0.042
12	*70.87	4	20.1	30.0	5,023	6.03	833	1.20	1,256	0.056
13	*70.87	3	15.1	30.0	5,023	4.53	1,109	0.902	1,674	0.042
14	*70.87	3	15.1	25.0	5,023	3.77	1,332	0.75	1,674	0.042
15	*70.87	3	15.1	36.0	5,023	5.44	923	1.08	1,674	0.042

*70.87 cm. square cross section.

Conventional bed thicknesses were used in runs 1 through 10, and the thin beds were used in runs 11 through 15. In runs 11 through 15 a square bed cross section (70.87 x 70.87 cm) was used. Runs 1 through 12 used cheese whey feed, at a solids content of about 23% for runs 1 through 10 and at a solids content of about 40% for runs 11 and 12. The feed for runs 11 and 12 were penetrated to remove lipids and lactose. Raw cottage cheese whey at 23% solids was used as feed in run 13. The feed in runs 14 and 15 were cheddar cheese whey at 20% solids and raw soy whey at 20% solids, respectively. The data for runs 1 through 10 has been obtained from the Alfa-Laval sales information bulletin discussed earlier.

TABLE 2

Run	1	2	3	4	5	6	7	8	9	10	11	12	13	14	15
Diameter (cm.)	40	40	40	60	60	60	80	80	80	180	*70.87	*70.87	*70.87	*70.87	*70.87
Height (cm.)	60	100	150	60	100	150	60	100	150	100	3	4	3	3	3
Kg. dry resin per bed	15	25	37	34	56	84	60	100	150	500	1.51	2.01	1.51	1.51	1.51
Kg. charge per cycle	20.4	33.3	50.5	46.2	75.3	112.9	80.6	134.4	202.1	671.9	5.35	7.12	4.80	4.03	5.82
Kg. protein per cycle	0.51	0.83	1.26	1.16	1.88	2.82	2.02	3.36	5.05	16.8	0.58	0.77	0.12	0.202	0.27
Kg. solids per cycle	4.65	7.59	11.5	10.5	17.2	25.7	18.4	30.6	46.1	153.2	2.30	3.06	1.09	0.806	1.16
Kg. solids/kg. resin per cycle	0.31	0.30	0.31	0.31	0.31	0.31	0.31	0.31	0.31	0.31	1.52	1.52	0.722	0.534	0.768
Kg. protein/kg. resin per cycle	0.034	0.033	0.034	0.034	0.034	0.034	0.034	0.034	0.034	0.034	0.384	0.383	0.079	0.134	0.179
Cycles per hour	5.90	2.81	1.57	4.53	2.50	1.29	4.13	2.36	1.22	2.38					
Kg. solids/kg. resin per hour	1.83	0.84	0.49	1.40	0.78	0.40	1.28	0.73	0.38	0.74	1.85(a) 3.04(b) 4.56(c) 6.08(d) 0.47(a)	1.85(a) 3.04(b) 4.56(c) 6.08(d) 0.47(a)	0.881(a) 1.44(b) 1.70(c) 2.17(d) 2.89(e)	0.651(a) 1.07(b) 1.26(c) 1.60(d) 2.14(e)	0.937(a) 1.54(b) 1.81(c) 2.30(d) 3.07(e)
Kg. protein/kg. resin per hour	0.200	0.096	0.053	0.154	0.027	0.014	0.044	0.025	0.013	0.025	0.77(b) 0.91(c) 1.15(d) 1.54(e)	0.77(b) 0.91(c) 1.15(d) 1.54(e)	0.158(b) 0.186(c) 0.237(d) 0.316(e)	0.096(a) 0.268(b) 0.316(c) 0.402(d) 0.536(e)	0.218(a) 0.358(b) 0.422(c) 0.537(d) 0.716(e)
Kg. protein/hour per 500 liter bed volume								7.92			23.6(a) 38.6(b) 45.6(c) 57.9(d) 77.2(e)	23.4(a) 38.3(b) 45.3(c) 57.5(d) 76.6(e)	4.83(a) 7.94(b) 9.37(c) 11.92(d) 15.9(e)	8.14(a) 13.4(b) 15.8(c) 20.1(d) 26.7(e)	10.9(a) 17.9(b) 21.1(c) 26.8(d) 35.7(e)
Kg. solids hour per 500 liter bed volume								72.2			93.6(a) 153.2(b) 179.8(c) 229.8(d) 306.4(e)	92.8(a) 152.3(b) 179.6(c) 228.4(d) 304.5(e)	44.0(a) 72.2(b) 85.0(c) 108.2(d) 144.3(e)	32.5(a) 53.3(b) 62.9(c) 80.1(d) 106.6(e)	49.0(a) 76.8(b) 90.7(c) 115.2(d) 153.6(e)

*70.87 cm. square cross section.
(a) = Results at 1.22 cycles/hr.
(b) = Results at 2 cycles/hr.
(c) = Results at 2.36 cycles/hr.
(d) = Results at 3 cycles/hr.
(e) = Results at 4 cycles/hr.

A comparison of runs 7 through 9 with 11 through 13 illustrates the difference in the operating results with approximately the same surface area of molecular sieve material (Sephadex G-50 medium) in the bed. For example, with beds having a surface area of 5,023 cm², i.e., runs 8 and 11 through 13) and at 2.36 cycles per hour those runs using thin beds, (i.e., runs 11 through 13) show from 18% to about 500% increase in kilograms of protein per hour per 500 liters of bed volume over that of the comparable conventional bed, (i.e., run 8), depending on the feed composition, and the increase in kilograms protein per kilogram resin per hour ranges from 644 to 3,500%. By appropriate selection of the eluate fraction the protein purity of product obtained from the thin bed can be as high as that of the thicker bed. Since the number of cycles per hour can be more readily increased with the thinner beds, these advantages can be enhanced accordingly.

It is apparent that the thin bed process constitutes a significant improvement over the conventional process and offers the added flexibility and efficiency so important to commercial success.

Intracellular Protein Recovery Using Shock Waves

Intracellular proteins in plant and animal cells are recovered for nutritional use by a process of *M. Allen; U.S. Patent 3,389,997; June 25, 1968; assigned to General Electric Company* by subjecting the source material to electrohydraulic shock waves which break the material open and release the intracellular proteins into the surrounding liquid medium.

Broadly speaking, the two basic types of protein are the intracellular proteins and the extracellular proteins. The intracellular proteins are found in both plant and animal cells. Those proteins present in plants include albumins, globulins, glutelins, glutamines, nucleoproteins, protamines and cytoplasmic proteins. The intracellular animal protein, by way of comparison, comprises albumins, globulins, histones, protamines and nucleoproteins. Generally, the process for recovering proteins from materials having intracellular proteins comprises placing a mixture of the material with a suitable liquid into a reaction chamber where it can be subjected to a shock wave created by an electrohydraulic arc discharge located in operative relationship to the protein bearing material liquid mixture. Operation of the electrode creates a shock wave or pressure gradient that travels through the material-liquid mixture causing proteins to be separated from the parent material.

The process can best be described by referring to Figure 1.3. Specifically, the apparatus comprises a container **10** which defines a reaction chamber **11** divided into upper and lower parts (**11a, 11b**) by a separating diaphragm **12**. The container **10** is conveniently constructed of two sections **13** and **14** so that the diaphragm can be fitted between them. The sections are held together by appropriate means such as the nut and bolt combinations **15** that extend through the flanges on the two container sections. The upper end of **13** is closed by an end cap **20** and the lower end of section **14** by end cap **21**.

An electrode **25** extends into the upper portion **11a** through an opening located in end cap **20**. This electrode comprises an inner electrode **26**, an outer electrode **27**, and an insulating layer **28** which separates the inner and outer electrodes. Upon connecting the electrode to a source of electricity, such as a capacitor bank, arcing occurs between electrodes **26** and **27** to create a shock wave within the medium in container **10**. Also shown within **10** is receptacle **30** which has perforate bottom **31**. The protein-bearing material may be forced through the openings of **31** and physically disintegrated. This physical action further aids in the recovery of protein.

FIGURE 1.3: INTRACELLULAR PROTEIN RECOVERY BY ELECTROHYDRAULIC SHOCK WAVES

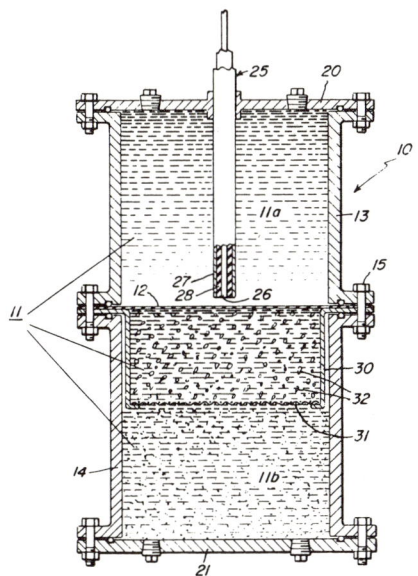

Source: M. Allen; U.S. Patent 3,389,997; June 25, 1968

As an example of the effectiveness of this process, an apparatus similar to that shown in Figure 1.3 was charged with a mixture of decorticated whole peanuts in a 10% saline solution. The ratio of solid to liquid was 1:5, e.g., 200 grams of peanuts in 1,000 ml of 10% saline solution. Control samples were prepared identically but received no electrohydraulic discharge. The following operating conditions were used to effect the process:

(1) 10 kv
(2) 24 microfarads
(3) 1,200 joules
(4) 1/8" spark gap
(5) the electrode was located 1.7 cm from the diaphragm.

The following table shows the number of discharges to which the mixture was subjected and the percentage of protein material (in terms of initial peanut weight) removed:

No. of Discharges	% Protein Extracted
120	11
180	28
240	58
300	65

Since a peanut contains about 65% protein, almost total recovery is obtained at about 300 discharges. Controls, which were not subjected to shock, have only 4 to 9% protein extracted in 1.5 to 2.5 hours of mixing time.

The presence of the perforate member **31** can further improve the recovery of protein from the parent material. To illustrate this, decorticated, defatted peanut meal was suspended in a 10% sodium chloride solution (pH 8.0) in the proportion of one part of solid to five parts of liquid. Electrohydraulic operation was at 15 kv, 5/16" spark gap with a gap resistance of 7,400 ohms. The sample chamber was separated from the arcing chamber by a suitable diaphragm. Within the sample chamber was a 1/24" grid plate through which the peanut solids were forced by the action of the shock wave. The total energy input at all capacitances was 67,500 joules. The results were as follows:

	% of Available Protein Extracted at Indicated Capacitance		
	12 μf	18 μf	24 μf
Peanut meal passed through grid	62	70	71
Peanut meal not passed through grid	52	62	63

Hammermill Extraction of Proteins and Lipids

I. Chayen; U.S. Patent 2,928,821; March 15, 1960; assigned to C.C.D. Processes N.Y. Ltd. has used a hammermill to isolate a protein-lipid complex from vegetable proteins. These complexes will not be created simply by mixing proteins and lipids even in alkaline aqueous media, nor will they be created when raw vegetable materials are processed by customary mechanical disintegration steps. For example, if such materials are ground, ballmilled, pebblemilled, crushed, flaked, pressed, mortared, rolled dry with a liquid, or subjected to ultrasonic vibrations by electric piezo crystals or other vibrating elements, this complex will not be formed.

The conditions under which the hammermill operates are critical, one of the criteria being that there must be at least a certain liquid-to-solid weight ratio. More particularly, the weight of liquid is at least 3 times the weight of the raw vegetable material, although, under special circumstances, e.g., with raw vegetable material having very small cells, it is possible to secure usable results where the weight of liquid is only twice the weight of the raw vegetable material. If less than the critical amount of liquid is present, there is a substantial loss in the efficiency of operation of the process, and, moreover, the liquid then only acts as a lubricant and binds the solid material in the form of a paste or emulsion. Furthermore, the dwell time of the raw vegetable material in the hammermill is of the utmost importance. A dwell time of as little as $\frac{2}{3}$ of a second is preferred and a dwell time of as long as 3 seconds gives desirable results. About one second customarily yields the best results. It is not usually desirable to extend the dwell period further than 3 seconds since the longer the retention time of the raw vegetable material in the hammermill, the finer will be the resultant size of the vegetable debris. The fine debris tends to encourage emulsification, to promote colloidization, and to be difficult to remove.

The preferred manner of setting the retention time in the hammermill is by proper selection of the screen or grating covering the hammermill outlet and through which the material is discharged. A hammermill used for this process is a continuous flow mill rather than a batch mill. Changing the fineness of the screen or grating openings on the available outlet area will change the retention time.

Still further, the amount of energy imparted to the raw vegetable material is another critical feature since if insufficient energy is utilized, the desired specific interreactivity of the proteins and lipids will not be secured. The proper energy is imparted if the tip speed of the hammers is at least 5,000 ft/min, although at slightly lower tip speeds, e.g., in the order of 4,000 ft/min, some useful results are obtained with weak wall cells. It is preferable not to use such low tip speeds inasmuch as the extraction is too incomplete and the process is run inefficiently. With present day materials and mechanical strength limitations 20,000 ft/min is a practical upper limit.

The high speed multiple rapidly repeated hydrodynamic shock waves are characterized by the fact that they do not attrite the cell walls as by abrasion or mechanical rubbing to leave cell debris of smaller than cell size. Rather, they burst the walls of masses of cells while leaving the masses as single structures, i.e., as large groups of ruptured coherent cells, so that the solid material left in the liquid carrier will, under magnification, have a honeycomb-like porous structure whereby the presence of fine solid particulate material which is highly detrimental to further treatment is avoided.

Considerable control may be exercised over the protein-to-lipid ratio in the ultimate complex. This can be accomplished, for example, by varying the pH at which the hammermill treatment is effected from the acid side through neutral to the alkaline side, or by extracting part of the lipids at some step of the process. In this fashion he has been able to change the lipid contents of the complex from as little as 2 to as much as 60% by weight when using certain oilseeds as starting materials and to change the lipid contents from as little as 2 to as much as 16% by weight of the complex when starting with leaf and grass as the raw vegetable materials.

The complexes are unusually stable. They resist high speed commercial centrifugation up to as much as 10,000 grams, although emulsions and simple mixtures of proteins and lipids are broken far below this point. The protein-lipid complex after being prepared will readily redissolve time after time at a pH of about 8.0 to 8.5 and reprecipitate at a pH of about 4.5 or lower. The reprecipitation point in all instances is specific to the protein of the plant material of origin and the lipid simply reprecipitates as a part of the protein. No matter how many times this process is repeated, the lipids and proteins will not separate nor will their ratio change once the complex has been formed. This indicates an extremely tight binding of the two materials and is a clear sign that a complex has been formed.

Although a variety of raw vegetable materials may be used in the process, emphasis is placed on peanuts in the examples, but the patent also has examples using rye grass, alfalfa and sugar beet tops. The following example shows the extraction of the complex from peanuts.

Example: 1,000 grams of decorticated (skinned) peanuts testing 4.58% nitrogen and 43.6% oil before treatment were passed through a 9" hammermill, i.e., a hammermill having a rotor 9" in diameter, running at a speed of 8,500 rpm and having fitted over its discharge outlet a plate perforated with 1/32" openings. The peanuts were fed into the mill along with five times their weight of an 0.1% caustic soda water solution. The caustic soda water solution was at a temperature of 45°F. The retention time of the material in the mill was about 1 second.

The mixture discharged from the hammermill was passed to a basket centrifuge to remove the solid debris in a well-known manner. The centrifuge operated at about 400 grams. The solids so removed then were washed with an equal weight of tap water to remove the entrained mother liquor which was added to the separate liquid secured from the first centrifugation in the basket centrifuge to form an alkaline water phase. Next the alkaline water phase was acidulated to bring its pH to about 4.8, hydrochloric acid being used for this purpose and being added at ambient temperature. This pH is below the isoelectric point of peanut protein so that such acidulation precipitated all the dissolved protein that was present.

However, as explained earlier, the protein and lipid extracted from the raw vegetable material in the manner specified and dissolved in water combined automatically to form a loosely bound synthetic protein-lipid complex, so that upon acidulation the complex, rather than the protein, precipitated. Acidulation was performed in a suitable receptacle, the hydrochloric acid being stirred into the liquid in the cold.

The precipitate was recovered by centrifuging in the same basket centrifuge, for example, the same one as that employed for removing the solid debris. The recovered complex was washed with acidulated water and recentrifuged. At this point the complex still was not highly stable. It had been dewatered by the centrifuging to a paste containing about 80% by weight of water. Alternately, the dewatering could have been accomplished by filtration or by gravity separation followed by decantation. It is not efficient, i.e., economically feasible, to perform this dewatering step by evaporation.

The recovered complex is then air dried, this operation taking place at a temperature not over 160°F, and desirably in an inert atmosphere, for instance, a nitrogen or carbon dioxide atmosphere. If desired, the drying could have taken place at subatmospheric pressures in order to encourage removal of water at lower temperatures. When dried to a moisture content slightly lower than 8%, the complex constituted 500 grams of a fine dry powder analyzing 50% peanut protein and 50% oil. The balance of the oil was present in the solid residue extracted in the first centrifugation. This oil was recovered by conventional means.

I.H. Chayen; U.S. Patent 3,090,779; May 21, 1963; assigned to C.C.D. Processes N.Y. Ltd. has modified this process to obtain material which is not a stable complex but if desired can be separated into separate protein and lipid fractions. The product obtained from the hammermilling is dried by azeotropic distillation in place of air dryings as in U.S. Patent 2,928,821. This drying step is shown below in the example. The material is obtained as in the example of U.S. Patent 2,928,821 up to the drying step.

The proteins and lipids are recovered separately by drying the mechanically dewatered paste complex by azeotropic drying. Toluene is added to the dewatered paste. More than enough toluene is added to form an azeotrope with all the water present. Preferably, about one-half is added at the beginning; subsequently, the other half is added during the boiling step. Sufficient additional toluene is added to maintain the complex as a paste after the azeotrope is boiled off. Typically, in toto, 8 lb of toluene is added initially for each pound of the 80% water paste. Enough toluene is used for the ultimate amount of toluene present after the azeotrope has been boiled off completely to be about the same as the amount of water originally present in the paste.

The boiling is continued with its temperature being carefully watched and with toluene being added to prevent the boil-off of any water per se until the temperature sharply rises to 108°C. At this time the only liquid remaining in a substantial quantity is toluene. The moisture (water) content at such time is between 4 to 6% by weight of the solids present. Next, the paste which consists of toluene and protein-lipid complex is washed with fresh toluene. This removes the nonpolar solubles, specifically, the toluene-soluble lipid fraction and the toluene-soluble chlorophyll fraction. This lipid fraction and chlorophyll fraction are recovered by distillation in the usual manner. Then the complex is washed with a polar solvent, specifically acetone which dissolves and removes the polar-soluble fraction of the lipids and of the chlorophyll. It also dissolves and removes the carotene. Finally, the remainder of polar solvent with which the protein is wetted is evaporated. Substantially all of the materials other than the protein were soluble in one or the other of the polar and nonpolar solvents, i.e., the toluene and the acetone. The lipid content is very greatly reduced by the aforesaid treatment, there being left only a tiny fraction in the order of 1% by weight of the solids.

Apparatus for Separating Protein Enzymatically

A disclosure by *D.A. Meadors; U.S. Patent 3,442,656; May 6, 1969* provides an apparatus and method for chemically processing vegetables and raw materials into a meat-like product. This apparatus is shown in Figure 1.4, which is a sectional view of the chemical separator. Great detail of the construction of the apparatus is given in the patent, but emphasis here is placed on the operation of the apparatus.

FIGURE 1.4: APPARATUS FOR SEPARATING PROTEIN ENZYMATICALLY

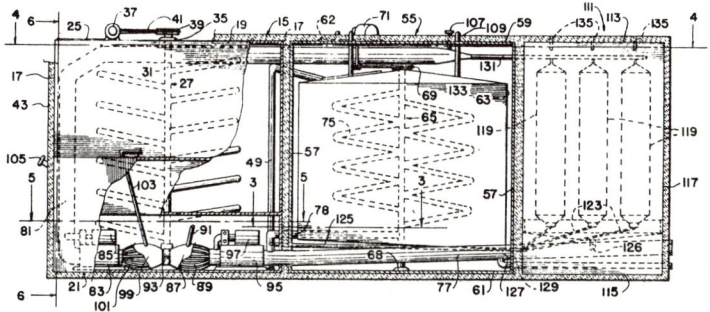

Source: D.A. Meadors; U.S. Patent 3,442,656; May 6, 1969

A mixture of finely ground vegetable raw materials is placed in the fermentation tank **15** through lid **25**. These raw materials can be of any suitable protein-containing vegetable matter, such as soybean, alfalfa, grains, and the like. The mixture of ingredients can be varied to provide a variety of end food products. Water is added to the mixture in tank **15** to create a mash which is sufficiently liquid to be agitated and pumped. The electric heater **45** and thermostat **47** are set to keep the temperature of the mixture at approximately 98°F. The motor **37** is started to rotate the agitator **27** during the fermentation process. This process should be continued for approximately eight hours.

During the fermentation process, natural bacteria are formed which create acids such as hydrochloric which further aid in chemically separating component parts of the raw materials. In this process the cellulose and pentosans in the mixture are broken down into organic acids, chiefly acetic and butyric. Glucose and other sugars are also formed through the bacterial action. Gases, chiefly carbon dioxide and methane are exhausted from tank **15** through vents provided.

After eight hours of agitation, the resultant mixture is transferred from tank **15** through transfer conduits **49** into the inner portion **63** of tank **55**. The mixture is transferred by pump **95**. The agitator **65** in processing tank **55** is started by starting the motor **71**. The temperature in tank **63** is maintained at 98°F. Enzymes are now added to the mixture in tank **63**. These enzymes are principally trypsin, amylase, and lipase. The trypsin attacks the proteins and creates proteoses, peptones, and some amino acids. The amylase changes the starches into malt, sugar, and the like. The lipase splits the fats in the mixture into fatty acids and glycerine. Suitable salts such as those found in bile are added to aid in breaking down the fats in the mixture. Erepsin is added which attacks the proteoses and peptones resulting in the production of amino acids therefrom. Invertase is added which hydrolyzes sugars in the mixture.

A liquor is placed in the storage tank **79**. The beginning solution of this liquor is primarily water and can be all water. However, faster results can be obtained by including 5% glucose, 1% salt and 1% soluble iron in the liquor. The liquor is pumped from the storage tank **79** by pump **83** through the liquor discharge line **81** in sufficient quantity to fill the space between the outside portion of tank **55** and the inner tank **63**. The valve **107** is closed. The liquor absorbs amino acids through the porous walls of inner tank **63** as well as some of the fatty acids. A portion of the liquor circulates through the hollow agitator **65** and absorbs fats, lipoids and the like through the porous materials of agitator **65**. The liquor drains into the return duct **77**, through pump **95** and into purging tank **87**, passing through the tubes **89** therein where carbon dioxide escapes from the liquor and is vented through vent **91**.

The liquor then passes into an oxidizing tank **99** and through the tubes **101**. Oxygen under pressure is supplied to the tank **99** through line **103** thereby passing oxygen through the porous tubes **101** and into the liquor. The liquor is then passed through pump **83** for recirculation through the discharge line **81**. During this circulating process the pump **83** pumps the liquor in surges during the circulation.

After circulating the liquor through the processing tank for six hours, valve **107** is opened allowing the lighter ends of the liquor containing the greater part of the nutrients extracted from tank **55**, to pass into the branch lines **131** and into the porous tubes **121** of the food containers **119** in tank **111**. Approximately one-third of the liquor enters the food containers **119**, the remaining portion thereof recirculating through tank **55**. As the liquor passes through the tubes **121** in a zigzag fashion, the nutrients which include amino acids, fats, oxygen, carbohydrates, minerals and vitamins pass through the porous walls of the tubes **121** and are deposited in the food storage container **119**. The tubes **121** are preferably held in place in the food containers **119** by a plurality of grids made of protein fibers. These fibers tend to become absorbed in the product and provide additional chewy texture. Fibers of this sort are well-known to the art and any suitable kind can be used.

Bacteria-killing catalysts are also passed through the tubes **121** and into the food compartment **119**. These catalysts are allowed to drain from the food container through drain lines **123**, the waste products therefrom being removed by filtration. The liquor passed through the tubes is returned to the return ducts **77** by duct **126** for recirculation.

After the food materials have been deposited from the food containers **119**, another chemical reaction occurs which is a metabolic and anabolic process that forms a cellulose-like product containing nucleic acid, a compound of heterocyclic nitrogenous bases, sugar, and phosphoric acid. During this process the carbonhydrates and fats are oxidized to carbon dioxide and water which are returned to the return ducts **77** in the circulating system. During the circulating process last described, the pump **95** is preferably operated at a speed of approximately 60 strokes per minute and pump **83** is preferably timed to operate at approximately 180 strokes per minute. This circulating process should be continued for ten to twelve hours or until the food containers **119** are filled.

Example: A preferred mixture of ingredients to produce a synthetic meat-like food having a beef flavor is as follows: The beginning in the fermentation tank contains 600 lb of alfalfa, 1,200 lb of corn or grain or sorghum, 180 lb of cottonseed oil, soybean meal or the like, $6^{2}/_{3}$ lb of defloriated phosphate, $6^{2}/_{3}$ lb of ground limestone, $6^{2}/_{3}$ lb of salt granules to which is added 1,200 gal of water. After processing this mixture, the resultant end product will have approximately 43.5% water, 15.7% proteins, 37.6% fats, and 3.2% minerals. The end product closely resembles color and taste of beef and can be similarly handled and treated.

Peroxide Treatment of Hydrolyzed Press Cake

Hydrogen peroxide has been used by *R.A. Johnson and P.T. Anderson; U.S. Patent 3,127,388; March 31, 1964; assigned to Food Techniques, Inc.* to increase the yield of protein obtained by the alkaline or alkaline-enzyme hydrolysis of vegetable press meals or cakes. Seeds such as those of cotton, sesame, soy rape, wheat, bran, coconut and safflower are frequently treated to remove the vegetable oil carried by the seeds. The by-product of this operation is a meal or press cake which is extremely high in protein content. The protein however, is associated with cellular and nonnutritive constituents, many of which are either unpleasant to taste or allergenic to humans, thus rendering the press cake unfit for human consumption.

Methods have been developed in the past to remove the protein from the parts of the press cake which were not fit for human consumption. Generally, these methods employ an alkaline hydrolysis or alkaline-enzyme procedure so that the protein in the press cake is hydrolyzed and thereby solubilized. The insoluble portion of the press cake could then be filtered. The protein in solution could then be precipitated and recovered to be used for any desired purpose. The most serious handicap attendant such methods is the low yield obtained—lowest in the straight alkaline hydrolysis.

In practicing this process the vegetable meal is treated with hydrogen peroxide. In all other respects the process generally follows the prior alkaline hydrolysis or alkaline hydrolysis with enzyme methods. Preferably, the hydrogen peroxide is added to the vegetable meal after the meal has been slurried. Thus, it will be present when the initial heating phase is conducted in alkaline solution. The peroxide may, however, be added later in the procedure if desired with satisfactory results. In the preferred procedure the initial heating and hydrolysis is continued in the presence of hydrogen peroxide until there is indication that the peroxide action has been completed. This indication may generally be found in the fact that foaming will cease when the peroxide action has ceased.

The amount of hydrogen peroxide to be added may be varied over a considerable range depending on the particular vegetable meal under consideration and whether or not the peroxide is to be used in conjunction with an enzyme treatment. It has been found that excellent results will be obtained if hydrogen peroxide in the quantity of 0.01 mol to 0.4 mol for each 100 grams of vegetable seed meal to be processed is used. It should be understood, however, that these quantities may be varied by making appropriate changes in the process. For example, if the peroxide is added later in the process after the first concentration of protein has been made by removing the cellular material and the protein concentration thereby increased, lesser quantities of peroxide may be used. Greater quantities of peroxide may be used without substantial deleterious effect but no advantage will in general be gained.

The foregoing figures are also based on a seed meal having the usual or most prevalent concentrations of protein found in seed meal of about 42 to 52%. Seed meals which do not fall in the foregoing category will require more or less peroxide depending on whether they are higher or lower in protein concentration respectively. As used here, however, the unit 100 grams of seed meal should be taken to mean seed meal which contains between 42 and 52% protein.

Example: A 20 gram sample of milled, sieved, straight screw pressed, cottonseed meal was slurried in 180 ml of distilled water. The meal had passed a 42 mesh screen and had a protein content of 48.8%. 1.6 ml of 19% sodium hydroxide was added. The slurry was heated to 190°F and 3 ml of 130 volume hydrogen peroxide was added with agitation. The slurry was held at 190°F for 30 minutes until the peroxide action was completed as evidenced by subsidence of the foam. The slurry was then cooled to 85°F and distilled water added to readjust to the original volume. The pH of the slurry was 7.32. A 0.5%

bromelin aqueous solution was added at the rate of 0.2% bromelin in relation to the protein substrate solids. Altogether, 3.9 ml of bromelin (0.0195 gram) was added. The slurry was then incubated at 85°F for 2 hours. After the incubation 1.6 ml of 19% sodium peroxide was added to the slurry to give a final pH of 10. The slurry was then heated to 160°F and held 30 minutes at this temperature to bring the available protein substrate into solution.

The slurry was then centrifuged to separate the insoluble material from the liquid. The remaining insoluble material was then washed with 120 ml of hot alkaline water (1 ml of 19% sodium hydroxide per liter of distilled water) and held at 160°F for 15 minutes with occasional stirring. The washed insoluble material was then recentrifuged as before.

The resultant combined liquids had a pH of 9.3. The total volumes of the solutions was 277 ml. Analysis of the solution gave 2.28% protein and 4.34% total solids. Calculation shows that 64% of the protein was extracted from the cottonseed meal by this method. This represents a 9.3% increase over a similar method which does not employ the peroxide. 264 ml of the resultant alkaline solution was heated to 75°F and the protein precipitated from the solution by the addition of 3.1% hydrochloric acid until a pH of 3.8 was reached. The mixture was then centrifuged and the precipitated protein washed twice, each time with 50 ml portions of acidified water (pH 4.5) and recentrifuged.

The total liquid including wash water was 381 ml. Analysis showed a protein content of 0.467%. The yield of precipitated protein was then calculated by difference. The overall protein recovery by this method was 45.4%. This represents an 11.6% increase over the enzyme method without benefit of peroxide. In addition, the protein sample had a very desirable light cream color.

Triple Enzyme Treatment of Vegetable Tissues

E.M.J. Blanchon; U.S. Patent 3,258,407; June 28, 1966 has developed a process for the extraction of proteins and other useful constituents contained in vegetable tissues, in particular, seeds and oleaginous vegetable tissues. The prior processes for extracting the proteins and other useful constituents contained in vegetable tissues have a serious disadvantage, mainly as regards the coefficient of digestive utilization of vegetable tissues, particularly oleaginous tissues. These various techniques aim only, in the case of oilcake, for example, at extracting the proteins, or in the case of whole oleaginous seeds, at extracting both the oil and the proteins, but were not capable of also extracting the other constituents useful for human food or animal feedstuffs, which include assimilable glucides, mineral salts, vitamins, oligo-elements and accompanying enzymes.

In this process for the extraction of the proteins and other useful constituents of vegetable tissues, vegetable tissues are subjected to the combined enzyme action of pectinolytic, amylolytic and proteolytic enzymes, the effect of which is to liberate the oils, sugars and other water-soluble substances contained in the cellulose cortical cells of the vegetable tissues. This enzyme action is followed by a separation process to remove the residual cellulose waste and by a process for recovering the oils, sugars and other water-soluble substances, as well as the proteins, liberated by the enzyme action.

The vegetable tissues may be subjected successively to the action of pectinolytic enzymes, then to that of amylolytic enzymes and then to that of proteolytic enzymes. Preferably, the vegetable tissues are subjected simultaneously to the action of pectinolytic, amylolytic and proteolytic enzymes.

Where the various kinds of enzymes referred to above, namely pectinolytic, amylolytic and proteolytic, are caused to act one after the other, the medium is adjusted at each stage to a pH value and a temperature which are most favorable to the action of the particular type of enzyme. Where the vegetable tissues are subjected simultaneously to the action of all the above three types of enzymes, the pH value and temperature are adjusted so as to permit this simultaneous action to take place in the most favorable possible conditions.

Example 1: The seeds are ground or the cake is pulverized in a hammermill or by some other means, dry or in water, and the material is then made into a paste by the addition of sufficient quantities of water, 5 volumes, for example. This paste is subjected to the action of pectinolytic enzymes. By way of example, preparations rich in pectin-esterases and in pectin-polygalacturonases of bacterial or fungic origin (of the type prepared industrially for clarifying fruit juices) may be used. The pH value is adjusted to 6 and the temperature to 20°C. The amount of dried enzymes (on a neutral carrier) used is 2 grams per liter of water used in making the paste treated. The duration of the enzyme action is approximately 1 hour.

At the end of that time, the paste so treated is subjected to the action of an amylolytic enzyme, to render the starch soluble. Again by way of example, one may use an α-amylase (α-glucosidase) of the type used for desizing textiles. The pH value is maintained at 6 to 6.5 and the temperature is raised to 50°C. The duration of this enzyme treatment is similarly one hour. The amount of enzyme used should preferably be of the order of 1 gram of dry enzyme per liter of water used for making the paste.

Next, the paste is subjected to the action of proteases. Again, by way of example, one may use an industrial protease of bacterial origin, such as is customarily used for preserving beer against cold weather clouding. The pH value is adjusted to 8.5 and the temperature is brought up to 55°C. The duration of this enzyme action is again one hour. The amount of enzyme used should preferably be of the order of 2 grams of dry enzyme per liter of water used for making the paste. The paste is then filtered, the residue is washed and refiltered and the two filtration liquids are added together.

If whole oleaginous products or oil-rich cake are being treated, this liquid is centrifuged so as to separate the oil before the treatment is continued. If oil-separated cake is being treated, this stage may be omitted, since there is no oil to separate. In either case, after this optional stage, the proteins are precipitated. For this, the pH value is adjusted to 4.8 by the addition of HCl, for example, or any other suitable means may be adopted. Centrifuging follows, to separate the proteins, and the residue is dried by atomizing, for example.

Example 2: The various enzymes used in three successive stages in Example 1 may be caused to act simultaneously on the vegetable tissues, the conditions as to pH value, temperature and the reaction time being modified.

The actions of the first and second enzyme stages, that is to say, the pectinolytic and amylolytic actions, can be carried out simultaneously at a pH value of 6.5 and a temperature of 37° to 40°C. The combined reaction time should be extended to about 1½ hours.

The third enzyme action, that of the proteases, may also be carried out at the same time as the above two, for example, at pH value of 7 and a temperature of 40°C, but the duration of the combined enzyme treatment should then be extended to 2¼ hours.

Recovery of Protein from Processing Wastes

The process of *I. Jantzen; U.S. Patent 3,390,999; July 2, 1968;* assigned to *Arthur C. Trask & Sons* produces a protein enriched animal feed by adding an aqueous solution of lignosulfonic acids to the aqueous protein-containing liquid to effect precipitation of combined protein-lignosulfonic acids and separating the precipitate.

Industrial waste materials, usually in the form of aqueous solutions and dispersions that contain considerable quantities of proteins that are ordinarily lost, are exemplified by wastewater from potato starch plants and wastewater from green animal food silos, in addition to animal protein wastes. In many of these instances, the presence of the protein impedes or prevents the recovery of other valuable products from these wastes. Thus, in potato starch water the proteins must be removed before the carbohydrates can be recovered from the waste. Furthermore, the presence of the proteins in the sewage is undesirable as they can decompose into amino acids which make the waste quite offensive. In this process these proteins are recovered in the form of combined protein-lignosulfonic acids and this combined form can be used as an animal feed component to enrich the feed with the protein. Thus, the protein-containing wastewaters can be treated with waste sulfite liquor as it comes from the digesters of the paper mill. Although the waste sulfite liquor can be used without previous treatment as it is tapped from the pulp mill digesters, this liquor can be further treated to separate out the α-lignosulfonic acids as the other lignosulfonic acids do not take part in the precipitation. A method for separating out these alpha acids is disclosed in U.S. Patent 2,838,483.

The waste sulfite liquor from the digesters contains approximately 40% alpha-lignosulfonic acids which form the precipitate with the protein. When the waste liquor or the separated alpha acids are mixed with the aqueous protein waste, the α-lignosulfonic acids and the protein combine in the proportions of 40 weight parts of acid to 100 weight parts of protein to give 140 parts of lignoprotein (protein and α-lignosulfonic acids in combined form).

Because of the presence of sulfur dioxide many countries limit the amount of this ingredient in animal feeds to a maximum of about 3%. Because this process forms the protein and lignosulfonic acids in combined form, which then precipitates out, free sulfur dioxide and loose combined sulfur dioxide are removed. This permits the use of large quantities of lignoproteins in animal feed without exceeding the maximum 3% of sulfur dioxide. Thus, the lignoprotein of this process may be used in animal feeds up to about 12%, and this will introduce only the maximum 3% sulfur dioxide into the feed. The lignoprotein can also serve as a binder to make animal feed pellets by usual pelletizing processes.

The protein-lignosulfonic acids after being precipitated as described above may be easily separated from the liquid medium by filtration. This protein-rich material may be rendered soluble by mixing with aqueous alkali solutions, such as aqueous ammonia of a pH of about 5. The resulting solution may be dried in the usual manner as by spray driers and the resulting dry powder is easily dissolved in water. Furthermore, it is nonhygroscopic when so dried.

Example: Wastewater from potato starch production containing 1% protein and 4% starch was treated with 0.4% α-lignosulfonic acids. The protein was precipitated in combined form with the alpha acids and was mixed with a standard animal feed material in an amount of about 12%.

PROTEIN TREATMENT PROCESSES

Encapsulated Protein Concentrates

Protein sources such as casein, soy, fish and others usually exhibit off-taste to foods containing them. They may also linger in the mouth as a chalky tasting precipitate. For this reason, many nutritious sources of protein have been left unexploited for human consumption.

D.T. Rusch; U.S. Patent 3,793,464; February 19, 1974; assigned to ICI America Inc. has made such protein concentrates more palatable by encapsulating the material with lipids derived from tallow, lard, soybean, cottonseed, and corn oils, etc., having iodine values of 1 to 90, and melting points above 70°F.

Practically any source of protein such as sodium or calcium caseinate, casein, soy protein, fish protein, gluten, and other recently discovered protein sources can be made palatable by encapsulation with edible lipids. These can be saturated or unsaturated vegetable oils or animal fats. Among the most suitable are cottonseed oil, safflower oil, corn oil, soybean oil, butter fat, coconut oil, peanut oil, lard, chicken fats, hydrogenated cottonseed oils, corn oils, soybean oils, peanut oils, olive oils, and coconut oils. These can be used in the form of solids and plastics. Those fats having an iodine value of 1 to 90 and having melting points above 70°F are particularly useful and preferred. The fat encapsulated protein can be dispersed in an aqueous medium using an emulsifying agent but in some instances the agent is not required. Examples of suitable emulsifying agents to be used alone or in combination are lecithin, glycerol esters, glycol esters, polyglycerol esters of fatty acid, as well as fatty acid esters of sorbitol, sorbitan, and mannitol and ethoxylated derivatives thereof.

In the practice liquid fat can be blended with solid proteinaceous material using an emulsifying agent if needed. The type and level of emulsifier is dependent upon the function desired in the finished product. In systems where water is a large portion of the product composition (such as in a beverage) care must be taken such that an excess of emulsifier does not contribute a detergent action which would strip the fat coating from the protein. A surfactant level of less than 15% (based on fat) will normally be appropriate for most preparations. In cases where the fat has a melting point above 70°F, the fat may be placed in a mixing bowl, heated to above the melting point wherein the protein can be admixed. The weight ratio of fat to protein can be from 4:5 up to 100:1, but usually a ratio of 1:1 to 4:1 is preferred.

The emulsified fat-protein blend may then be mixed with the aqueous base. Usually the emulsion of the fat-aqueous base intermixture is conducted at a temperature above the melting point of the fat mix. The conditions of the emulsifier and procedure are adjusted such that the materials form the desired oil-in-water or water-in-oil dispersion. The emulsion is easily stabilized by cooling as rapidly as possible to well below the congealing point and preferably to about 45°F. The faster the fat is congealed, the more stable the emulsion and the less chance the protein has of being leached into the aqueous phase and thereby losing its protective coating.

Example: 31.5 grams of coconut oil (MP 98° to 101°F) is liquified by heating to a temperature of 130°F and mixed with 3.5 grams of a molten emulsifier blend comprising 85% sorbitan monostearate and 15% by weight polyoxyethylene (20) sorbitan monostearate emulsifier. To this is added 35 grams of casein with vigorous agitation. This hot fat-emulsifier-casein blend is then added to 1,500 grams of an aqueous base heated to 140°F, and which contains 15% by weight sugar plus phosphoric acid and lactic acid such that the pH is adjusted to 2.8. The mix is then run through a homogenizer adjusted in the first stage at 2,000 psi and in the second stage at 500 psi to form an oil-in-water emulsion. It is thereafter cooled to 45°F and stored. Excessive foaming does not occur in the homogenizer since the reactivity of the protein is diminished.

Removal of Phytic Acid by Ultrafiltration

Phytic acid, the hexaphosphate ester of myoinositol, is a natural component of plant seeds where it occurs in the form of calcium and magnesium salts. During the isolation of proteins from defatted oilseed meals, a substantial amount of phytic acid coprecipitates with the proteins at the isoelectric point in the form of protein-phytic acid complexes in which the phytic acid is strongly bound to the basic groups of lysine, histidine and arginine of the proteins. For example, a soybean protein isolate usually contains 0.8 gram of phosphorus phytate per 100 grams of protein, an amount representing as much as 60% of the total phosphorus present in the seed.

The preparation of protein isolates free of phytic acid is desirable, particularly if the proteins are intended for human nutrition. Phytic acid is a strong chelating agent that can immobilize metals of biological importance such as calcium, magnesium, iron and zinc. When present in substantial amounts in the diet, phytic acid can interfere with the normal intestinal absorption of such metals, leading to a series of deficiency disorders. For example, the rachitogenic effect of cereal-based diets has been linked to the presence of phytic acid in cereals like wheat, corn and oats, thus requiring a supplement of both calcium and vitamin D to overcome that effect.

A process has been provided by *G.A. Iacobucci, D.V.B. Myers and K. Okubo; U.S. Patent*

3,736,147; May 29, 1973; assigned to The Coca-Cola Company and Kikkoman Shoyu Co., Ltd., Japan for preparing plant proteins having low phytic acid content by subjecting protein isolates to ultrafiltration in the presence of a suitable chemical reagent. The purification of seed proteins by this process can be carried out with any defatted seed meal, in particular soybean, peanut, cottonseed, sesame and sunflower, in a broad range of pH between 2.0 and 11.0, at any temperature between 4° and 70°C, and using either aqueous extracts or suspensions of the meals in water as the protein source.

The selection of a suitable chemical reagent depends upon the pH at which the protein source is subjected to ultrafiltration. For example, when the pH is approximately in the range 4.5 to 7.0 the enzyme phytase may be employed. At least a part of the required amount of phytase may be present as a component of the protein source. For the removal of phytic acid by ultrafiltration when the pH is in the range approximately 2.0 to 4.5, the ultrafiltration should be conducted in the presence of a large excess of divalent cations such as calcium and magnesium ions. For pH in the range of approximately 7.0 to 11.0, the ultrafiltration should be conducted in the presence of a strong chelating agent such as ethylenediaminetetraacetic acid (EDTA), nitrilotriacetic acid (NTA), 2,2'-ethylenedioxybis(ethyliminodiacetic) acid (EGTA), iminodiacetic acid (IDA) and the like, in concentrations in the range 0.05 M to 0.50 M.

Of course, the chemical reagent used in any particular process will depend upon the precise process conditions, and the selection of a suitable chemical reagent can be made readily by one skilled in the art. Also, it will be recognized that to obtain optimum removal of phytate it may be necessary to adjust the pH of the protein source.

The term ultrafiltration describes the process of dialysis under a pressure gradient across a semipermeable membrane, and is also referred to as diafiltration in the specialized literature. The fractionation by this procedure of a mixture of solutes dispersed in an appropriate solvent like water will depend on the molecular dimensions of the solutes in relation to the porosity of the membrane used for such purpose. For the purification of crude extracts of seed proteins, good protein retentions can be achieved by using membranes of molecular weight cut-off limits of 300,000 daltons to as low as 10,000 daltons. The extensive washing with water at the appropriate conditions of pH, temperature and pressure can remove from the system the permeable components like sugars, salts, nonprotein nitrogenous compounds and phytic acid, leaving over the membrane a retentate comprised of proteins of high chemical purity.

Example: Defatted soybean flakes (8.94% N, 0.80% P, 1.43 grams P/100 grams protein), 440 grams, were extracted with 15 parts of water, the suspension being made alkaline to pH 8.6 using 5 N NaOH and the insolubles removed by filtration of the suspension through a 100 mesh screen. The insolubles were reextracted with 5 parts of water with respect to the original meal, the second extract obtained in the same fashion as the first, and the two extracts combined. The pH of the pooled extract was lowered to 5.0 with 5 N HCl, and the resulting suspension kept at 4°C overnight.

The following day, over a period of 9.3 hours at 65°C, pH 5 and 20 psig, the suspension was concentrated by ultrafiltration from 8.57 liters to 4.1 liters over 0.25 ft^2 of cast cellulose acetate anisotropic membrane having a MW cut-off between 10,000 to 15,000. The membrane was embodied within the Amicon TC-1 apparatus. After concentration of the suspension to 4.1 liters, demineralized water (20.35 liters) was then passed through the suspension at pH 5, 65°C and 20 psig over a period of 39 hours.

The solids retained over the membrane were recovered by freeze-drying to afford 209 grams of isolate being 14.85% N, 0.35% P and 0.38 gram P/100 grams protein. Recovery in the isolate based on the meal was 47.5% by weight, 78.8% by nitrogen and 20.6% by phosphorus. A comparison of the essential amino acid composition of the isolate to that of a commercial isolate [Meyer, E.W., *Soy Protein Concentrates and Isolates*, p 147 in Proceedings of International Conference on Soybean Protein Foods, held at Peoria, Illinois, October 17–19, 1966, ARS 71-35 (1967)] is shown in the table below. Tryptophan, although an essential amino acid, has not been included in the table.

Essential Amino Acid Content of Soybean Protein Isolates

Amino Acid, g/16 g N	Soybean Protein Isolate	
	Commercial	Example
Lysine	6.0	6.9
Cystine (as cysteic acid)	0.9	1.4
Valine	5.0	5.2
Methionine	1.0	1.4
Isoleucine	4.9	5.3
Leucine	8.1	8.1
Phenylalanine	5.6	5.5

Other examples were included in the patent covering protein isolates and concentrates from cottonseed, sesame seeds, and peanuts.

Bacterial Treatment for Improving Flavor and Odor

Many vegetable proteins, such as soybeans, protein hydrolysates, and wheat gluten meal, possess characteristic beany or cereal flavors and odors which limit their sale and use because the flavor and/or odor is unappealing to large numbers of potential consumers. Efforts to remove these flavors and odors from such foods have in the past been chemical procedures primarily utilizing enzymes. While such procedures have been successful with some foods, they are not uniformly efficient in deflavoring other foods. In addition, such procedures are comparatively expensive.

T.M. Hoersch and J.L. Shank; U.S. Patent 3,364,034; January 16, 1968; assigned to Swift & Company have developed a process for deflavoring and odor removal from vegetable and animal food materials using bacteria. Generally, this process comprises contacting meat protein and vegetable protein products such as soybean meal, soybean flour, wheat gluten meal, corn meal, cottonseed meal, and hydrolyzed vegetable proteins, which have a characteristic flavor with nonpathogenic, saprophytic bacteria, proliferating the bacteria in the food and deactivating the bacteria after the bacteria have been incubated for a time sufficient to remove the flavor and/or odor from the food. The method is very inexpensive, yet highly effective in providing bland foods free of flavor and odor normally associated with the food undergoing the treatment.

Bacteria useful in this process include nonpathogenic organisms of the families Lactobacteriaceae, Micrococcaceae, Pseudomonodaceae and Enterobacteriaceae. Gram positive organisms of the genera Lactobacillus, Pediococcus, Leuconostoc and Streptococcus, are particularly useful in the process. Specific organisms of these genera which may be employed include *Lactobacillus lactis, bulgaricus, acidophilus, Leuconostoc citrovorum, Pediococcus cerevisiae* and *Streptococcus lactis.* Typical organisms of the family Micrococcaceae suitable for use include *Micrococcus citreus.* Gram negative organisms which can be employed as deflavoring agents include the nonpathogenic species of the family Enterobacteriaceae, genera Pseudomonads, Proteus and Aerobacter. Specific bacteria of these genera include *Pseudomonae fragi* and *Aerobacter aerogenes.*

The food undergoing treatment is inoculated with a culture of the above bacteria or mixtures thereof and the inoculated product is held under conditions of bacterial growth for 16 to 144 hours. The inoculated product is usually maintained at a temperature of 32° to 100°F and preferably at 75° to 90°F. Only a small amount of the culture is required to carry out the deflavoring. In most cases 0.5 to 2.0% of the culture on a volume to weight of substrates should be added to the product. In the preferred form of the process about 0.75 to 1.25% volume per weight is used. The culture is prepared by growing the bacteria in a culture medium.

Example: 500 grams of defatted, finely ground soybean flour was mixed with 1,500 ml of distilled water. 20 grams of sodium chloride and 200 mg of sodium nitrite were added to the mixture to inhibit bacterial action. The mixture was then autoclaved for 20 minutes at 250°F and 15 lb of pressure. After cooling the bacteria-free soybean slurry, 10 ml of an 18 hour Micrococcus culture was added to 1,010 grams of the slurry while the remaining 1,010 grams served as a noninoculated control. The Micrococcus culture was prepared by growing stock cultures on brain heart infusion agar (Difco) and transferred weekly. Prior to use transfers were made to brain heart infusion broth which was incubated for 18 to 24 hours at 86°F. A 1% inoculum on a volume to weight of substrate basis was employed.

Both the inoculated and noninoculated samples were allowed to remain at 78°F for 18 hours and at the end of this period the inoculated sample was devoid of soybean flavor, while the noninoculated sample had the characteristic soybean taste.

Other examples covered the treatment of wheat gluten meal, cottonseed meal, ground wheat plus animal protein products.

Detoxifying Proteins Containing Aflatoxin

It has been known that agricultural products may become infected with strains of the mold *Aspergillus flavus* which produce a group of highly toxic substances known collectively as aflatoxin. The extreme toxicity of aflatoxin is demonstrated by the fact that the LD_{50} of the B_1 component (believed to be the main toxic component of aflatoxin) is less than 30 micrograms for day-old ducklings. Moreover, aflatoxin has been shown to be toxic to many kinds of animals, including swine, calves, rabbits, fish, guinea pigs, pheasants and chickens, as well as turkeys and ducks.

The detoxification process described by *M.S. Masri, H.L.E. Vix and L.A. Goldblatt; U.S. Patent 3,429,709; February 25, 1969; assigned to the U.S. Secretary of Agriculture* is accomplished by contacting the contaminated material with ammonia. The treatment may be done at ambient temperatures; however, destruction of the toxic principles occurs more rapidly with increasing temperature. Consequently, if it is desired to expedite the detoxification, the treatment is accompanied by the application of heat, for example, the material plus added ammonia is held at an elevated temperature, for instance, at 200°F.

The ammonia may be applied as gaseous NH_3, as a gaseous mixture of NH_3 and steam or in aqueous solution, i.e., as ammonium hydroxide solutions. Where the treatment is conducted at or near room temperature a simple closed vessel will be appropriate while at higher temperatures a pressure-resistant vessel such as an autoclave is required to retain the ammonia vapors. The ammonia is employed in large excess as compared with the amount of aflatoxin in the material being treated. Usually, one uses a minimum of 0.3 gram of NH_3 per kilogram of material. To ensure adequate aflatoxin destruction it is preferred to use a large proportion of ammonia, for example, 10 to 30 grams of NH_3 per kg of material being treated.

The time of treatment will depend on various factors including the amount of aflatoxin in the starting material, concentration of ammonia in the system, the physical and chemical characteristics of the material (for example, the particle size of the material, its porosity or density, its content of fats, etc.) and particularly on the temperature employed. For example, in typical runs with peanut meal it has been found that the treatment may require as much as 7 to 14 days at room temperature whereas at 200°F, the same end is achieved in about 30 to 60 minutes.

The process can be used to detoxify materials of all kinds which are contaminated with aflatoxin. The process is particularly useful in the treatment of agricultural products, since such materials are especially likely to become contaminated with the *A. flavus* mold when exposed to conditions conducive to mold growth. Typical examples of such materials in the category of vegetative cellular materials are seeds and the residues remaining

General Processes for Vegetable Proteins

after extraction of oil therefrom, e.g., peanuts (shelled or in shell), soybeans, peanut meal, cottonseed meal, soybean meal, flaxseed meal, grains such as wheat, barley, rice, rye, oats, corn, and meals or flours prepared from any of these grains; forages such as alfalfa, clover, grasses, sorghum, bran, cowpeas, ensilage, mixed feeds, etc. Other examples of agricultural products include such materials as fish meal, tankage, dried blood, distillery and brewery residues, dried whey, dried milk, casein, dehydrated fruits, vegetables, meats, eggs, etc. The process is demonstrated by the following example.

Example: The starting material was a commercial hexane-extracted peanut meal known to be contaminated with aflatoxin. Chemical assay of this material indicated 1,300 ppb of aflatoxin B_1. Analysis of the meal was as follows:

	Percent
Moisture	6.7
Nitrogen	7.1
Crude fiber	13.6
Lipids	2.28
Ash	5.55

Concentrated ammonium hydroxide (28 to 30% NH_3, SG 0.9) was mixed with the peanut meal in a food mixer for 1 to 2 minutes, using 100 ml ammonium hydroxide per kg of peanut meal. The treated meal was placed in covered pans which were allowed to stand for two days at room temperature. After this holding period, the pans were uncovered for about two hours to permit escape of ammonia vapor and the products then dried in a forced air oven (about 158°F) to their original weight.

Samples of the untreated meal and the treated meal were then bioassayed, using two-day old ducklings as test animals. (Ducklings are especially sensitive to aflatoxin, hence are particularly useful for bioassay.) The diet supplied to the ducklings was as follows:

	Percent
Peanut meal (treated or untreated)	70
Casein	14
Corn oil	12
Salt mixture, USP XIV	2
Vitamin mixture	2

The results obtained are tabulated below:

Treatment of Peanut Meal with NH_4OH (2 Days at Room Temperature)

Sample	Mean body weight, grams, at end of—		Deaths at end of 2nd week
	1st week	2nd week	
Treated	108	191	*0/10
Untreated	84	95	6/10

*Number after oblique line designated number of ducklings in the feeding test.

Removal of Sulfur Compounds

T.J. Staron; U.S. Patent 3,803,328; April 9, 1974; assigned to Etablessement Public: Institut National de la Recherche Agronomique, France has developed a process for improving vegetable proteins whereby the crude protein cakes are macerated in an aqueous medium with strains of microorganisms, notably the yeast *Geotrichum candidum*. The cakes are thus freed from the sulfur-containing impurities and aflatoxins which contaminate them and limit their use at present. The cakes obtained have an improved nutritive value.

New, pure proteins can be isolated from the maceration liquids by precipitation to the isoelectric pH or in the presence of saline solution. The cakes to which the process can be applied are made from seeds of any vegetable origin. Colza, rape, groundnut, sunflower, soy, sesame, castor oil, cotton vinia sinensis, broad bean (*Faba vulgaris*) and other vegetable seeds can notably be mentioned.

The words maceration conditions relates here to a treatment consisting in placing cake in intimate contact with strains of the selected microorganism, the treatment being carried out in an aqueous medium. The conditions imply a temperature slightly higher than ordinary temperature, 30° to 45°C for example. With *Geotrichum candidum* the best results are obtained with maceration temperatures of 37° to 40°C. The pH of the maceration medium varies during the operation, but it generally remains between 4 and 6.5. The pH is at its highest value at the start and it decreases during the treatment.

The length of treatment varies with the temperature, but also depends on the results desired and other factors, such as the volume of the soaking tanks. With *Geotrichum candidum*, for example, sulfur-containing products are completely eliminated in between 30 and 40 hours at 30°C, and this time falls below 30 hours if the operation is carried out at 37° to 40°C. To obtain total extraction of cake proteins and to isolate them as pure proteins, the treatment should be carried out under the same temperature for a period of 60 to 90 hours. Details of the process are given in the following example on colza cake.

Example: Maceration of the colza cake was carried out in a 40 l pilot fermenter according to the following formulation:

Colza cake	6 kg
Geotrichum candidum culture	5 l
Tap water	19 l

Starting pH was 6.4, pH at extraction was 4. This lowering of the pH is essentially due to the freeing of proteins in the culture liquids. Culturing is carried out with slow stirring, or by a stationary technique, for 30 to 60 hours at 37°C without preliminary sterilization. Indeed, preliminary trials have shown that under the conditions described no contamination has ever occurred. Furthermore, stirring the medium and ventilation do not appear to be necessary when small amounts are macerated.

Test portions permit the freeing of 5-vinylthiooxazolidone, abbreviated to VTO in the maceration liquids and its progressive breakdown to be checked. An increase in temperature is seen to have a surprising effect. At 27°C, hydrolysis of thioglycosides occurs first but it is only after 35 hours culturing that isothiocyanates breakdown starts, and they are only completely destroyed after 85 hours maceration. At 37°C, on the other hand, hydrolysis of thioglycosides and breakdown of the isothiocyanates formed occur simultaneously.

During maceration, the *Geotrichum candidum* progressively solubilizes the colza cake proteins. Thus, when maceration is carried out for 60 to 80 hours at 37°C, 90% of the proteins are dissolved. During the first 30 hours there are freed a heteroprotein toxic to the mouse (protein α) and a protein containing large amounts of glutamic acid, proline, lysine and sulfurized amino acids (protein β). After 35 hours protein α is observed to disappear and soluble fractions are freed with an amino acid composition very similar to that of total colza cake (see table below). Fractions so obtained can easily be leached out by conventional chemical methods and contain between 65 and 80% proteins. The macerated cake was treated in various ways:

1. Recovery of the entire macerated mass and evaporation of water by atomization.
2. Isolation of the insoluble fraction by centrifugation and drying.
3. Separation of the soluble fraction and drying.
4. Fractionation of the soluble fraction by $(NH_4)_2SO_4$.

Operating under these conditions it is possible to distinguish the following fractions.

(1) Total macerated colza cake
(2) Macerated colza cake, insoluble fraction
(3) Macerated colza cake total soluble fraction
(4) Protein α which precipitates by the addition of 20% $(NH_4)_2SO_4$ by weight per volume
(5) Protein β which precipitates with 40% $(NH_4)_2SO_4$ by weight per volume

The amino acid composition of the proteins in the fractions obtained are shown in the following table, which also shows that *Geotrichum candidum* does not damage the composition of proteins in the colza cake during maceration. On the other hand, results indicate that it progressively frees protein fractions with a variable amino acid composition.

Protein β, rich in glutamic acid, proline, lysine, histidine and sulfurized amino acids, represents 25% of the total proteins. The cellulose content increases in the insoluble fraction during maceration because this element is not attacked, but it can, however, easily be removed. During maceration, the cake is found to lose about 10% of its weight. This occurs essentially to the detriment of carbohydrates.

The table below shows that maceration with *Geotrichum candidum* causes colza cake to become richer in most of the indispensable amino acids, which explains its good protein efficiency.

	Percent by dry weight				
	Colza cake control	Total atomized macerated colza cake	Macerated colza cake insoluble fraction	Macerated colza cake, total soluble fraction	Protein β
Protein content	35.3	45.2	Variable	64	10
Aspartic acid	7.0	6.9	7.0	7.0	2.4
Threonine	4.3	4.5	4.5	4.0	3.0
Serine	4.3	4.2	3.8	4.4	3.2
Glutamic acid	18.3	18.7	17.0	19.2	26.0
Proline	7.4	7.1	7.3	7.2	9.2
Glycine	5.2	5.2	5.3	5.2	4.1
Alanine	4.2	4.9	5.0	5.2	3.8
Valine	4.7	4.9	5.1	4.8	3.9
Cystine	2.6	2.7	2.3	3.2	4.8
Methionine	1.9	2.0	1.6	2.1	2.4
Isoleucine	4.6	4.5	4.6	4.3	3.2
Leucine	7.0	6.9	7.1	7.3	6.1
Tyrosine	2.5	2.6	3.1	2.9	1.4
Phenylalanine	4.2	4.0	4.1	3.9	2.8
Ammonia/liquor	1.8	1.7	1.6	1.6	2.3
Lysine	5.3	5.6	5.1	6.1	6.8
Histidine	2.7	2.9	3.0	3.2	3.5
Arginine	6.5	6.6	6.8	6.6	5.6
Tryptophan	1.2				0.9
Cellulose	14	17	Variable	2	0

Cation Exchange for Purifying Proteins

A cation exchange resin in water is used by Y. Kawamura, T. Tamuki, A. Nagata, S. Terashima, T. Matsumoto, S. Onishi, S. Konishi, K. Yoshimura, H. Yamamoto and S. Ikawa; U.S. Patent 3,099,649; July 30, 1963; assigned to Ajinomoto KK, Japan to remove impurities in oil-free vegetable seeds. Proteins are recovered while carbohydrates and ash are removed by the resin. The improvement of this process over prior processes are shown in the following.

An extraction vessel was charged with one part of oil-free soybeans and 5 parts of water and then 0.7 part of Diaion SK #1 (H type of the product prepared by Mitsubishi Kasei-Kogyo KK) in a bag of synthetic fiber was introduced into the extraction water. The mixture was agitated at room temperature to effect extraction. After 30 minutes, the bag containing Diaion SK #1 was removed from the mixture. The extraction liquid was separated from the oil-free soybeans by centrifuge. The result of this treatment and of conventional processes are set forth for comparison in the following table.

Treatment	Loss of dissolved protein, percent	Extraction rate of carbohydrates, percent	Extraction rate of ash, percent
This process	0.8-1.2	45-50	80-90
0.1 NHCl extraction [1]	4-5	45-50	50-60
60% methanol extraction [2]	2-3	35-40	20-30

[1] Oil-free soybeans 1 part, 0.1 NHCl 5 parts, room temperature, 30 minutes.
[2] Oil-free soybeans 1 part, 60% methanol aqueous solution 5 parts, room temperature, 30 minutes.

PROTEIN HYDROLYSATES

Hydrolysis by *Hansenula montevideo*, n.sp. Yeast

A process that hydrolyzes fish, animal and vegetable proteins, or mixtures of these has been disclosed by *V.H. Bertullo and C.R. Pereira; U.S. Patent 3,516,349; June 23, 1970*. Hydrolysis of animal or vegetable proteins, that is to say, passage to liquid phase of the products obtained by cleavage, such as polypeptides and amino acids may be obtained by using Bio-Proteo-Catenolysis (BPC). The term means biological breakdown of the protein chain on the basis of the action of a proteolytic yeast and will be referred to simply as proteolysis. The particular proteolytic yeast used in this process was classified as follows: family Endomycetaceae, subfamily Saccharomycetae, tribe Hansenula, subgenus Hansenula. The yeast will be named here as *Hansenula montevideo,* n.sp. (Bertullo). A deposit of the culture of this yeast has been made at the Veterinary School of Montevideo, University of the Republic of Uruguay, on April 26, 1966, and has been assigned the Number L 1.042.

This present process using *Hansenula montevideo* n.sp. does not cause crystallization of the hydrolyzed products as they are formed during hydrolysis, as occurs with other methods. In general, the process requires that for each metric ton of fish or any other protein it is intended to hydrolyze, apart from the addition of the fermentable source provided by a carbohydrate, for example, commercial sucrose, at least one liter of the cultures of *Hansenula montevideo* n.sp. with a yield of not less than two million viable cells per milliliter should be added, a greater amount being admissible with no inconvenience.

The cultures may be added either with the nutrient medium in which they were produced, or else prepared in the usual commercial manner, previously diluted so as to facilitate dispersion. Cultures can be maintained viable for a considerable period of time, the only requirement being that they should be kept at a suitable temperature, for example, not above 4°C.

The process serves to hydrolyze vegetable proteins brought to suitable humidity conditions, among them the soybean (*Glycine max*) or their flours, and other Leguminosae, such as peanut (*Arachis hypogaea*), chick pea (*Cicer arietinum*), lentil (*Lens esculenta*), kidney bean (*Phaseolus vulgaris*), broad bean *(Vicia faba)*, etc., or fresh or dried alfalfa (*Medicago sativa*) in combination with a fermentable carbohydrate, any of them either pure or mixed with fresh ground fish, in different proportions.

The overall process, here given for fish products, is outlined below.

   (1) Grinding of the product, namely fish or residues;
   (2) Adding of a fermentable carbohydrate in about 7 to 10% by weight of the total amount;
   (3) Mixing of the ingredients and placing them in a fermenter;
   (4) Adjusting the temperature to between 28° and 35°C;
   (5) Adding of the *Hansenula montevideo* n.sp. yeast culture, after reaching the right temperature;

(continued)

(6) Stirring of the mass at 40 to 55 rpm;
(7) Carrying on hydrolysis for 18 to 24 hours or more;
(8) Filtering and separating of solids;
(9) Degreasing by mechanical means;
(10) Concentrating the solids to 50%;
(11) Drying; and
(12) Packing in sacks and storing.

This process permits one to catenolize mixtures of animal and vegetable proteins. For example, equal quantities of fish and alfalfa are mixed, the fermentable carbohydrate and the yeast are added, and the resulting hydrolysis yields a final product with a total of 30 to 32% of crude proteins, up to 75% of which are hydrosoluble. The mixture dries readily and can be used as a protein concentrate for ruminants.

The process also catenolizes vegetable proteins. In the case of alfalfa, hydrolysis occurs between 18 and 24 hours, simply by adding to the ground fresh product the yeast and the fermentable carbohydrate. There is no loss of organic nitrogen and the product dries readily. For this reason, it is an excellent substitute for silo fermentation, where loss of nitrogen is considerable, spoiling of the fodder may occur, and large storage facilities and an extended fermentation time under high temperature are needed. In working according to the process, temperature never rises beyond 35°C, thereby insuring integrity of the protein of the cleavage products.

In the case of soy or soybean, it is necessary to hydrate the product, grind it and bring it to the fermentation conditions. The final product is a homogeneous mass with a bland ripe apple odor. If allowed to settle, the mass will separate into two phases: an amber colored liquid phase, and a creamy white solid phase. During hydrolysis, antitrypsic and rat growth-inhibiting factors disappear.

Partial Hydrolysates by Controlled Acid Hydrolysis

A partial hydrolysate of edible protein, used as an additive for flour, dough and baked products, has been prepared by *F.D. Vidal; U.S. Patent 3,655,403; April 11, 1972; assigned to Pennwalt Corporation*. In contrast with proteins themselves and in contrast with many amino acids that can be made by hydrolysis and fractionation of proteins, a product can be prepared by controlled, partial hydrolysis of naturally-occurring proteins such as soy protein, wheat gluten, fish protein, meat protein and the like, which when used with flour or in dough improves the bread-making properties of the flour or dough.

This potential of a partial hydrolysate of protein does not appear to have been known, and has been found to be peculiar to the partially hydrolyzed material, in comparison with the unhydrolyzed protein or with products of complete hydrolysis that exhibit no significant flour- or dough-improving function. Thus for example by subjecting soy protein to acid hydrolysis which is limited in that the hydrolyzing action is appreciably less than complete, the resulting product is found to constitute an effective improving agent for wheat flour.

In carrying out the flour treating procedure, the partial protein hydrolysate, whether in original and preferably neutralized solution or advantageously in dry, finely divided form, is introduced into the dough mixture, and then becomes effective for the described results. The improving action is readily attained in conventional bread procedures, including the sponge dough and straight dough types, and continuous dough-mixing processes. While the action appears to be of slower nature, somewhat similar to bromate as contrasted with fast-acting improvers such as azodicarbonamide or iodate, the time of action afforded in continuous methods, e.g., in the proofing and baking steps, is ample.

In defining the extent of partial hydrolysis of a protein, and thus of defining the hydrolysate agent of this process, the water-soluble hydrolysate is characterized by a content of α-amino nitrogen which is more than 35% and less than 95% (or preferably in a range of

40 to 93%) of the content of such nitrogen in the water-soluble hydrolysate material that would be obtained from the same starting protein material by complete hydrolysis. Unhydrolyzed protein, and likewise the result of hydrolysis appreciably below the lower limits mentioned above, shows no significant improving activity in flour or dough.

Example 1: The protein used was a commercial soy product, having a protein content of 52% by weight. In appropriate apparatus, 1.25 lb of such material and 3.96 lb of 2N, (i.e., 2-normal) hydrochloric acid were refluxed for 5 hours, the temperature thus being slightly over 100°C. At the end of this time the reaction liquor was filtered to remove insolubles (being a few percent, including the humin fraction) and was evaporated to dryness in a rotating vacuum evaporator. The product was obtained in 90% yield as a somewhat hygroscopic yellow-brown solid, which could be used as such, or which can be formulated with suitable diluents to give a free-flowing powder. The product, upon analysis, showed an α-amino nitrogen content equal to about 40% of the total nitrogen, it being conveniently significant to express the α-amino nitrogen analysis by such ratio (in percent). In comparison complete hydrolysis of the soy protein showed α-amino nitrogen as approximately 67% of the total nitrogen, whereas in the untreated protein the ratio is less than 1%.

Although the partial hydrolysate is usually employed in the acid form directly resulting from operation as above, the filtered solution can be neutralized to yield a salt form, e.g., an ammonium, sodium, other alkali metal or indeed any other appropriate salt, such salt formation being with respect to the carboxyl groups of the various fractions of the hydrolysate.

Example 2: Gluten, e.g., as obtained from wheat flour, was hydrolyzed for three hours by reflux in 2N hydrochloric acid, using a weight ratio of three parts of HCl to one part of protein. The resulting partial hydrolysate product was analyzed as having an α-amino nitrogen to total nitrogen ratio of 34%. When tested by the straight dough method with clear flour, the maturing activity was found to be optimum at about 100 mg per 100 g of flour and the maturing or improving effect in the baked loaves was rated as good.

Example 3: Tests were made of partial hydrolysates from various proteins, e.g., by following procedures essentially like that of Example 1, wherein the protein preparation was subjected to refluxing with 2N hydrochloric acid for three hours, and the resulting reaction liquor filtered and evaporated to yield a partial hydrolysate having a substantial content of α-amino nitrogen but significantly less than would be representative of complete hydrolysis. Batches of bread dough were made and processed by the straight dough method using a clear flour, and 100 gram loaves were baked. The results are tabulated as follows, including the dosage of the several partial hydrolysates in milligrams per 100 grams of flour. For brevity, the results of volume and texture examinations are given in summary as maturing effect.

Material	Dosage	Maturing Effect
Gluten hydrolysate	100 mg/100 g flour	Good
Soy hydrolysate	200 mg/100 g flour	Good
Fish meal hydrolysate	200 mg/100 g flour	Good
Corn protein hydrolysate	300 mg/100 g flour	Fair-Good
Meat protein colloid hydrolysate	400 mg/100 g flour	Fair-Good

Some other proteins were also tested, i.e., by conversion to partial hydrolysate, and in baking operations of the same sort. Thus the partial hydrolysate of lactalbumin used in amount of 150 mg per 100 grams of flour afforded fair to good improving action. Lesser or perhaps minimal improving effect, although of some utility, was achieved with partial hydrolysates of casein and compressed yeast respectively employed in amounts of 250 mg and 300 mg per 100 grams of flour. Combinations of the partial hydrolysates were also tested, in baking bread, with the attainment of good maturing activity, as by the use of

50 mg each of the fish meal and gluten hydrolysates, or a combination of 150 mg soy hydrolysate and 40 mg gluten hydrolysate per 100 grams of flour.

Extraction of Undesirable Flavors from Hydrolysates

Protein hydrolysates obtained through acid hydrolysis contain undesirable flavor substances which distinguish them unfavorably from the taste of broth made of fresh meat or meat extract.

A.W. Hack; U.S. Patent 3,493,385; February 3, 1970; assigned to Corn Products Company has found that the major part of these undesirable flavor substances in an acid hydrolysate obtained from vegetable or animal proteins can be removed by subjecting the hydrolysates to extraction with natural animal and/or vegetable fats and oils or synthetic esters of alcohols, especially low aliphatic and preferably polyvalent alcohols with 2 to 6 carbon atoms and straight-chained fatty acids.

The best way to do this is to partly neutralize the acid hydrolysates, which are freed, if necessary, of the humin substances through filtration prior to the extraction and extract in a pH range of 2.0 to 4.5, preferably 2.8 to 3.0. Treatment takes place at room temperatures or a slightly higher temperature which may exceed the melting point of the fat.

The weight ratio of aqueous phase and fat or oil during the extraction is preferably between 2:1 and 20:1. After the extraction, the hydrolysates are neutralized to 5.5 to 6.0 pH and subsequently filtered. Far-reaching removal of undesirable flavor substances can be attained through counterflow extraction. The fats used for the extraction can be subjected to refining, especially deacidification and deodorization in the customary manner for reuse.

Example 1: Seven parts by weight of protein hydrolysate obtained by hydrolyzing equal parts of wheat gluten and corn gluten with 20 to 25% hydrochloric acid at 120° to 130°C for 6 hours are neutralized to a pH of 2.5 to 3.5 by the addition of sodium carbonate. To the solution freed from the precipitated humin substances through filtration, 1 part by weight of soybean oil is added and finely dispersed through vigorous agitation. After separating the oil through sedimentation or centrifuging of the hydrolysate, the liquid freed from undesirable flavor substances is adjusted to the desired pH of 5.6 to 5.8 and filtered, if necessary.

Example 2: Five parts by weight of a hydrolysate prepared according to Example 1 through partial neutralization and filtration are thoroughly mixed with 1 part by weight of solid animal fat at a temperature exceeding the melting point of the fat. The fat used for the removal of undesirable flavor substances can be separated according to Example 1 through centrifuging or through removal of the solidified fat after cooling.

Blended Protein Hydrolysates

Protein from more than one vegetable protein source is colloidally solubilized in the process of *R.A. Johnson and P.T. Anderson; U.S. Patent 3,397,991; August 20, 1968* to provide a homogeneous blend of protein having a preselected assay of essential amino acids. The process is designed to permit a blend of several vegetable proteins to balance the lack of a particular essential amino acid or acids in the vegetable proteins involved and thereby produce an end product that is high in all of the essential amino acids or in certain ones of these acids of special interest.

A dry mixing of previously isolated proteins does not usually achieve a homogeneous blending of the materials because of the physical properties of dried protein. The low bulk density and electrostatic charges on the protein which may be imparted by the drying process prevents uniformity in the combined materials. The process prepares a homogeneous blend of protein having a preselected assay of essential amino acids by colloidally solubilizing protein contained in at least two different sources and combining the solubilized protein in the same aqueous solution. Each of the sources of protein is selected for its amino acid content and is combined in an amount so as to provide the desired assay of amino acids after solubilization. The aqueous solution of the solubilized protein is separated from the residue of the protein sources and the aqueous solvent is removed from the protein to yield the desired end product of uniform amino acid assay.

The isolation or extraction step of the process generally follows that for colloidally solubilizing protein and removing it from vegetable meal. This generally consists of forming an aqueous slurry with the vegetable meal or press cake and water. The slurry may be prepared so as to contain preferably 10% to 20% solids and in any event is made sufficiently liquid so as to render the slurry flowable for liquid transport. Sufficient liquid is preferably utilized so that contact is made by the liquid with all surfaces of the meal during agitation and treatment. Colloidal solubilization of the protein in the meal is accomplished in this aqueous slurry.

While any suitable process could be used, the preferred method contemplates a partial alkaline hydrolysis where the alkalinity is imparted by an alkali such as potassium or sodium hydroxide. Preferably sufficient alkali is added to adjust the pH of the slurry to 10.5 although a pH range of 9 to 12 may be used. This partial hydrolysis and colloidal solubilization of the protein is enhanced by heating the slurry between 150° and 200°F for a time necessary to achieve the requisite degree of hydrolysis.

Example: 37.5 lb of solvent extracted soy meal (protein content, 44.6%) and 37.5 lb of solvent extracted sesame meal (protein content, 46.8%) were slurried with 81 gallons of water with good agitation. The resultant slurry had a volume of 95 gallons. 4,750 ml of a sodium hydroxide solution (approximately 16%) were added to give an initial pH of 10.5. The slurry was heated to 166°F and held for 30 minutes with good agitation. The pH had lowered to 9.7. 1,300 ml of the same sodium hydroxide solution were added, raising the pH to 10.3 and insuring complete reaction of the hydrogen peroxide. 2,000 ml of 35% hydrogen peroxide were added and the temperature raised to 193°F by steam injection (approximately 3 minutes come-up time) and held at the temperature for 25 minutes until the hydrogen peroxide reaction was complete.

The slurry was cooled to 156°F and first passed through a pulper to remove the coarse solids and then through a desludging centrifuge to remove the rest of the solids. The coarse and fine solids from the pulper and centrifuge were composited. A volume of water equal to the volume of the liquid solution was added to the composited solids, the resultant slurry heated to 140°F and again run through the pulper and centrifuge to separate the solids from the liquid solution. This recycle is a washing procedure.

The two liquid portions were combined and the temperature adjusted to 80°F. The protein was precipitated by adding a hydrochloric acid solution (approximately 10% HCl) rapidly with good agitation until a pH of 4.0 was reached. The precipitated protein was allowed to settle and 65 gallons of the whey were decanted. 86 gallons of acidulated water (adjusted to pH 4.5 with the HCl solution) were added to the precipitated curd, the mixture well mixed and the curd allowed to settle. 75 gallons of liquid were then removed by decantation. 77 gallons of acidulated water were added to the precipitated curd, again remixed and the curd allowed to settle. 89 gallons of liquid were then removed by decantation. 90 gallons of acidulated water were added to the precipitated curd, well mixed and the curd again allowed to settle. 98 gallons of liquid were removed by decantation.

The precipitated protein was then passed over a vacuum drum filter to remove excess free water and the curd washed again on the drum with fresh water. The curd was then made into a dispersoid by raising the pH to 7 by addition of a sodium hydroxide solution (10% sodium hydroxide), the product heated to 130°F and spray dried with influent air at about 400° to 450°F. (The effluent air temperature in the spray dryer was approximately 180° to 190°F.) The spray dried protein isolate had a protein content of 85.5%. It represented 34.2% of the solids in the original material and 52.8% of the protein in the original material. Amino acid assay results of this material, labelled Product A are to be found in the table below. The dried sludge solids from the alkaline extraction step had a protein content of 27.0%. It represented 34.7% of the solids in the original material and 25.1% of the protein in the original material. Approximately 31.1% of the solids and 22.1% of the protein in the original material were present in the acidified whey and washes in the form of soluble material.

Amino Acid	Product A mg/g as is
Tryptophan	10.1
Phenylalanine	49.5
Isoleucine	42.8
Lysine	30.4
Valine	58.2
Threonine	41.2
Methionine	15.4
Leucine	64.8

Dustproofing Hydrolysates with Polyoxyethylene Sorbitan Monoesters

Powdered hydrolyzed protein product prepared from vegetable or animal raw material including, e.g., soy flour, wheat gluten, corn gluten, cotton seed meal, peanut meal, unextracted yeast, meat products, etc. is very dusty, making handling difficult in operations such as screening, blending, packaging, etc. The loss of hydrolyzed protein dust in these operations creates yield loss and housekeeping problems for both the manufacturer and user.

A former method to minimize the dustiness in hydrolyzed protein was to coat the powdered hydrolyzed protein with an edible oil, (e.g., cottonseed oil). However, this technique leaves much to be desired. These oils are not water-soluble or completely water-dispersible. It is difficult to maintain a uniform composition in the feed to a spray drier. Water solutions of the powdered hydrolyzed protein treated with oil are not clear and contain an oil film on the surface unless the solution is continuously and violently agitated. The oil also creates an undesirable taste and mouth feel in many food seasoning applications where the product is used either in the solution or powdered form.

It was found by *W.D. Roberson; U.S. Patent 3,656,963; April 18, 1972; assigned to Hercules Incorporated* that dustiness is eliminated or at least minimized by drying an aqueous mixture of hydrolyzed protein and an antidusting agent comprising a polyoxyethylene sorbitan monoester of a saturated or unsaturated fatty acid having 12 to 18 carbon atoms. Preferably the antidusting agent will be either polyoxyethylene (20) sorbitan monolaurate, polyoxyethylene (20) sorbitan monopalmitate, polyoxyethylene (20) sorbitan monostearate, polyoxyethylene (20) sorbitan monooleate, or mixtures thereof. In the examples and elsewhere percent is by weight unless otherwise indicated.

Examples 1 through 8: Variable: Amount of Antidusting Agent—In examples 1 through 8, 0.5 gallon of a soy protein acid hydrolysate at pH 5.4 and containing 3.5 lb dry solids per gallon was used in each experiment. On a dry basis, the hydrolyzed protein contained 30.2% protein (N x 6.25), 53.5% NaCl, and about 7% monosodium glutamate (MSG). The solution was agitated at room temperature while the stated quantity of polyoxyethylene (20) sorbitan monooleate was added as the antidusting agent. Stirring was continued for 15 minutes after completing the addition of antidusting agent. The solution was then dried in a Swenson spray drier (3 ft diam x 7 ft high) under the following conditions:

Drier inlet air temperature	320° - 350°F
Drier outlet air temperature	190° - 225°F
Two-fluid type nozzle (air and liquid)	0.093 in diam

(continued)

Liquid feed temperature 75°F
Feed pressure 30 - 40 psig
Atomizing (air) pressure 30 - 40 psig

The dried products were analyzed for moisture content on a Cenco moisture meter at a 90-volt setting. The dusting characteristics of all the samples were measured using a funnel-drop test. This apparatus consisted of a polyethylene funnel, 4⅛ inches in diameter at the top and ⅝ inch in diameter at the bottom. A length of $^{11}/_{16}$ inch i.d. rubber tubing was connected to the outlet of the funnel. The vertical height of the funnel was 4 inches. 25 grams of sample was allowed to flow from the funnel into a tared aluminum dish. The dish was centered directly below the funnel at a distance of 12 inches between the bottom of the rubber tubing and the bottom of the dish. The alignment of the funnel and aluminum dish was maintained by using a funnel plate and dish guide. The dish was 3¼ inches in diameter at the top, 2 inches in diameter at the bottom and 1 inch high. The percent dust lost due to drifting or splattering from the dish was determined as follows:

$$\frac{25 \text{ g} - \text{weight collected in dish (g)}}{25 \text{ g}} \times 100 = \text{Percent dust loss}$$

Further details appear in the table below.

Example No.	Antidusting Agent*, %**	Dust Loss, %	Product Moisture, %
1	None (control)	41.41	2.7
2	0.1	34.43	3.0
3	0.7	25.00	1.5
4	1.0	14.42	1.3
5	3.0	5.26	1.2
6	10.0	4.18	2.0
7	15.0	2.68	2.6
8	20.0	2.36	3.6

*Polyoxyethylene (20) sorbitan monooleate
**By weight of dry solids in hydrolysate (hydrolyzed protein solution) used

Examples 9 through 13: Variable: Type Antidusting Agent—The process used in Examples 9 through 13 was substantially the same as for Examples 1 through 8. In all cases (except the control where none was added) the amount of antidusting agent used was 0.7% by weight of the dry solids in the hydrolysate (hydrolyzed protein solution) used. Further details appear in the table below.

Example No.	Type Antidusting Agent	Dust Loss, %	Product Moisture, %
9	None (control)	41.41	2.7
10	Polyoxyethylene (20) sorbitan monooleate	12.78	2.3
11	Polyoxyethylene (20) sorbitan monolaurate	10.59	2.2
12	Polyoxyethylene (20) sorbitan monopalmitate	12.34	2.4
13	Polyoxyethylene (20) sorbitan monostearate	6.10	3.2

The examples show that amounts of antidusting agent within the range of 0.1 to 20% by weight of dry solids in the aqueous solution of hydrolyzed protein gave substantially improved results. Amounts as low as about 0.05% are applicable Preferably the amounts will be about 0.7 to 3%. It is important that the antidusting agent be added so that it will be dried along with the hydrolyzed protein, (i.e., so that the two are codried). However, the type of drying is not critical. Although spray drying is preferred, drying methods in general are applicable including, e.g., spray drying, roll drying, oven drying, and pan drying under widely different conditions of temperature, pressure, time, and other variables.

Caramel Flavored Hydrolysates

M.R. Sfat and B.J. Morton; U.S. Patent 3,689,277; September 5, 1972; assigned to Bio-Technical Resources, Inc. have produced a protein hydrolysate of enhanced organoleptic properties by heating a granular mixture of a protein hydrolysate and a sugar in the presence of moisture at a temperature of 75° to 100°C until a caramel flavor is imparted to the mixture. The product is useful as a flavor precursor for a fermented alcoholic beverage, particularly beer, and as a food supplement or ingredient. The product may be incorporated in a fermentation wort for producing a fermented alcoholic beverage, and when produced employing a highly fermentable sugar, provides a wort yielding a low carbohydrate beverage. Preferred products may serve as a malt flavor base in foodstuffs, particularly nonalcoholic beverages.

The properties of protein hydrolysates from various sources, both animal and vegetable, may be enhanced by this process. Vegetable seeds, especially the cereal grains and legumes, represent desirable sources of protein, and are preferred for supplying the protein hydrolysate. Enzymatic hydrolysates are preferred, as they are produced free of salts which result from other types of hydrolysis.

The protein source preferably is treated or extracted in aqueous medium with added proteolytic enzyme to produce a solution containing soluble protein hydrolysis products. Where the protein is accompanied by a substantial amount of starchy carbohydrate, as in most of the grains and legumes, a proteolytic enzyme is employed in the absence of substantial added amylolytic enzyme, to minimize the extraction of carbohydrate.

When using a cereal grain, it is preferred to employ the whole grain, for availability and production reasons. While either a hulled or dehulled grain may be employed, it has been noted in the case of barley that a beer of improved graininess is produced from the grain having the hull intact.

The grain or legume preferably is finely divided or ground for hydrolysis. Thus, a fine grind is preferred, rather than a coarse grind, as those terms are employed in the brewing industry. More particularly, it is preferred that at least about 99% of the finely divided material have a particle size below about 1 mm, i.e., passing through a U.S. Sieve Series No. 18 sieve, having a sieve opening of 1 mm. A fine grind provides greater extractability of protein, e.g., as much as twice as great as with a coarse grind.

The preferred proteolytic enzymes are substantially pure isolates, i.e., substantially free of amylolytic enzyme, and include papain, bromelain, and ficin. Commercial papain is a mixture of at least two types of proteolytic enzymes derived from papaya, and it is activated by reducing agents such as bisulfites. Bromelain is a mixture of proteolytic enzymes derived from the pineapple plant. Ficin is a mixture of proteolytic enzymes obtained from the latex of the fig tree, and it is activated by reducing agents including bisulfite. The enzyme products are free of amylolytic activity. Protein hydrolysates containing as much as 62% or more protein may be produced from whole grains employing such enzymes.

Proteolytic enzyme products having high proteolytic activity and low amylolytic activity also may be employed, but generally produce lower concentrations of protein from grain or legumes, near the desired lower limit of 40% protein. For example, Rhozyme P-11 and Rhozyme 41 may be employed to produce concentrates from barley containing up to about 45% of protein. Various other proteolytic enzymes may be employed as well.

Example: The following illustrates a preferred procedure for producing a cereal grain protein hydrolysate. A 2,000 gram quantity of hull-containing barley whole grain containing about 13.5% protein is ground at the No. 1 setting on a Labconco Laboratory mill, providing a particle size distribution, percent retained on the U.S. Sieve Series sieve number indicated: No. 18, 0.2%; No. 30, 16.7%; No. 60, 59%; No. 100, 18.5%; through No. 100, 5.6%.

The ground grain is added to 10 liters of water, made up of 9 liters of tap water and one liter of brewing water containing 240 ppm calcium sulfate, 100 ppm sodium chloride, and 55 ppm magnesium sulfate. After mixing, the pH is adjusted to about 6.6 to 6.7 by the addition of about 15 ml of 15% sodium hydroxide.

As a source of proteolytic enzyme, 4 grams of standardized papain (S.B. Penick & Co., having 0.2 Penick milk clot units per gram) is added to the mixture, with 5 grams of potassium metabisulfite enzyme activator, and the mixture is maintained at 45°C with stirring for 4 hours. The mixture is stored overnight at 0°C and decanted. The clear decanted solution is treated with a source of amylolytic enzyme, either Takamine acid fungal protease or HT proteolytic enzyme. The materials employed had the following enzyme activities:

	Amylolytic Activity, DNS units/g	Proteolytic Activity, Colorimetric Northrop units/g
Takamine acid fungal protease	42,000	21,000
HT proteolytic enzyme	97,000	1,000

DNS units are determined by the colorimetric method described in *Methods in Enzymology*, vol 1, p 149 (1955, Academic Press, Inc., N.Y.). The conversion of starch to maltose is measured colorimetrically, employing dinitrosalicylic acid as the coloring agent. A 10 mg sample of soluble starch is converted to 0.4 mg of maltose hydrate in 5 minutes at 37°C for each unit of activity. Colorimetric Northrop units are determined according to assay No. 10-313 of Marschall Division, Miles Laboratories.

The enzyme source is added in a quantity of 0.2 gram per liter, and the solution is maintained at 45°C with stirring for 2 hours. The resulting solution is starch-free by the starch-iodine test. It contains about 4.4 to 4.8% solids including about 2.2 to 2.4% protein. The protein recovery from the grain is about 75 to 85%.

Solutions produced in the above manner may be concentrated in a circulating evaporator at temperatures of 35° to 55°C and under a vacuum of 27 inches of mercury (3 inches pressure) to solids concentrations of 50 to 55% by weight, and concentrations up to 65% can be obtained if desired. The protein content of the solids in such solutions is 45 to 50% by weight of the solids.

PROCESSING WHOLE SOYBEANS

The legume known as soybean, soya bean or soja bean is the Asiatic plant Soja max. Although cultivated in Japan and China long before written records were kept, the United States now produces approximately 75% of the world production.

The raw soybean has an undesirable bean flavor and bitterness. In addition the raw beans contain certain toxic proteins and antitrypsin which is an inhibitor of the enzyme trypsin. These materials must be eliminated or inactivated to make the raw beans palatable and digestible for use by domestic animals and particularly for human consumption.

This section covers processes which treat the whole bean for use as a food product which contains approximately 40% protein.

WET DEBITTERING PROCESSES

Ammonium Salt Treatment

Raw soybeans are freed of objectionable taste and odor by *C.A. Raymond; U.S. Patent 2,795,502; June 11, 1957* using an ammonium salt. In the process selected, clean, mature soybeans, either whole or in pieces, are first immersed in a bath consisting of a mild aqueous solution of ammonium carbonate or bicarbonate, for example, 1 to 3%. The beans are soaked in this solution for about 10 hours at room temperature of 70° to 80°F, during which time a certain amount of fermentation takes place due to the natural enzymes present in the beans. In the above processing, the ammonium compound aids in the saturation of the beans with the solution. It also has a special tenderizing, softening and dissolving effect on the beans and penetrates into the oil cells, and, by its chemical action and evolution of gas, helps to break down the bean material and destroy completely the objectionable bean flavor and odor.

This process eliminates the boiling of the beans prior to use of an ammonium salt such as bicarbonate as was done in known processes. Although, the natural enzymes present in the beans will cause fermentation, active yeast is preferably added to the ammonium compound solution to the extent of 1 to 5%. By active yeast is meant yeast containing enzymic or other living organisms.

This yeast increases fermentation during the soaking period, which changes the materials into more soluble and digestible form. For example, the starches are more or less converted into sugars and the proteins into proteoses, and other well known desirable results obtained by yeast fermentation are brought about. Enzyme preparations from fungus *Aspergillus oryzae* may also be advantageously added to the processing solution to supplement the yeast and soybean enzymes, particularly preparations in which the proteolytic enzymes occur in larger proportions than the amylolytic enzymes or amylases. In this way the fermentation may be adjusted and controlled.

After the yeast is added the temperature is increased to a temperature of 110° to 130°F to decompose the ammonium compound and drive off the ammonia from the solution, as well as to complete the ammonia action on the material. After the processing with the ammonium compound and yeast, the solution is drained and the skins removed from the beans, if desired, by friction. During the processing of the beans, however, the skin or hull is rendered so tender and soluble that when the bean material is ground and homogenized, the skin becomes incorporated in the resulting paste as an unidentifiable part thereof. Whether or not the skins are removed, the beans are then placed in a kettle where they are covered with water to a depth several inches above the beans so that they will be entirely submerged. The contents of the kettle is then brought to a simmering boil at atmospheric pressure which is continued for at least half an hour.

The boiling may be continued for a further period of from one-half to one and one-half hours, depending upon the particular product being prepared. This boiling, of course destroys any further fermentative activity. It has been found that cooking the whole beans in this manner results in a lighter colored product. If the bean material is in the nature of small pieces or particles, in order to avoid waste, and after the preliminary half hour boiling, the cooking may conveniently be continued in a steam pressure cooker for about thirty minutes at ten pounds pressure. It is found that pressure cooking tends to produce a graying effect on the material, and consequently, it is not recommended in the case of whole beans where it is desirable to preserve a bright color.

In the case of bean material in the form of pieces or particles, the skin fragments may be removed by screening. After the cooking operation, the bean material in the form of whole or cracked beans is drained of the cooking water. This material is edible and palatable, and may be eaten as any other food. It may also be mixed with other foods. A desirable dish may be prepared by mixing the material with suitable sauce or other ingredients and baking in the oven, after the manner of ordinary baked beans. This bean material may also be dried or dehydrated, and ground into a powder, for incorporation in bread.

The processed bean material contains substantially all or at least the major portion of the oil originally present in the raw beans. This oil can be extracted from the processed bean material by any known method, such as pressing or dissolving out with solvents, for example, alcohol or benzene. The oil is completely free from any objectionable taste or odor, and constitutes a valuable food product. It may, for example, be used, either alone, or mixed with other edible oils, for cooking, as a shortening, or as a salad oil.

Steam Dehulling and Debittering Beans

A process by which improved soybeans free of hulls, bitter ingredients and other ingredients of unpleasant flavor is produced by *I. Kovásznay and E. Kovásznay; U.S. Patent 3,058,829; October 16, 1962* without resorting to chemical substances. Such soybeans have a rather long storage stability.

According to the process soybeans are subjected to the effect of steam but, contrary to the known processes, for a rather short period of only 2 to 5 minutes, and then suddenly cooled with cold water. The hulls so loosened are disrupted in a closed vessel by streaming, preferably swirling water. The hulls together with the embryos are separated from the cotyledons and finally the beans free of hulls and embryos are dried at elevated temperatures preferably in the range of 70° to 75°C.

The hydrodynamic separation and the removal of the embryos is effected in an apparatus comprising a horizontally disposed vessel into which the steamed and cooled beans are introduced together with a stream of water and from which the beans, the hulls and the embryos are discharged together with the water stream. The vessel is provided with a rotating shaft carrying beater blades connected with the shaft by means of spokes. The inner wall surfaces of the drum or vessel are coarsened that is rilled, grooved, hatched, wire-netted etc. in a known manner. It is not known however to introduce the goods to be treated into the drum together with a stream of water, to use the hydrodynamical force of streaming water for a separation of the hulls and embryos from the cotyledons, and to remove the separated bean components from the drum in or by means of the water stream.

The beans stripped in this manner are fully free of hulls and embryos, irrespective of their shape. The appearance and quality of the product is far better than that obtained by conventional processes, their edges being undamaged and their surface cells having not suffered any destructive effect. In consequence of the hydrodynamical peeling, the beans are not subjected to any detrimental effect. The loss in useful substances is negligible, the improved goods can be employed for any purpose and may be pressed for oil in the cold. The effect of the quick preliminary steaming, consists in a diffusion through the hulls, whereby the steam penetrating between hull and cotyledon and the water condensing there, causes the loosening of the hulls on the beans. The coefficient of expansion of the woody hull and of the cotyledon being different, the hulls separate from the beans.

The heating, effected only for a rather short period, is restricted to the superficial cells contrary to the known processes carrying out a longer steaming. The removal and conversion of the bitter ingredients and of other substances of unpleasant taste is effected, on the one hand mechanically in the course of the special peeling procedure, and on the other hand, chemically by means of the subsequent hot drying. In some special cases, inasmuch as the hulling or other operations should not yield a high-grade product, the whole process may be repeated. The peeled product is similar to split peas and are freed from the embryos, which latter generally cause trouble in processing soybeans for food. The removal of the embryos is effected simultaneously with the hulls and can be separated from the same in a known manner, for instance with separators.

An apparatus for carrying out the process is schematically illustrated in Figure 2.1 by the way of example. The horizontal cylindrical vessel or drum **1** is provided with a feeding funnel **2** and an exit conduit **3**. The rotating shaft **4** carries longitudinal beater blades **5** connected to the shaft by means of three groups of spokes **9**. The inner surface **6** of the vessel is coarse. The pretreated beans are fed through tube **7** and the water conducted through pipe **8** into the feeding funnel. It is however possible to feed the beans to the vessel together with the water, directly from the cooling apparatus.

FIGURE 2.1: STEAM DEHULLING AND DEBITTERING BEANS

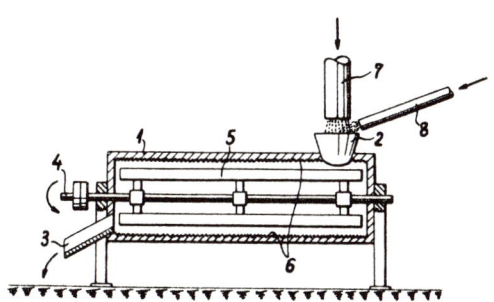

Source: I. Kovásznay and E. Kovásznay; U.S. Patent 3,058,829; October 16, 1962

According to the process soybeans are subjected for 2 to 5 minutes to steam at a temperature below 100°C and directly thereafter, suddenly cooled in cold water. The beans are then introduced with a stream of water into the closed peeling vessel in which the water is compelled to a whirling movement by means of the blades. The hulls of the beans in water are split up on effect of the swirling water by which the beans are thrown against the interior walls of the vessel. The walls can be for instance covered for this purpose with a wire net. By means of the hydrodynamic force of the streaming water the split hulls are stripped off the beans and upon this action, the embryos separate from the cotyledons which latter separate into halves. The beans are separated from the hulls and embryos and from the water in a known manner and dried then at 70° to 75°C.

The bitter ingredients and those of unpleasant taste of the stripped bean halves are decomposed in consequence of the heating, this heating is effected in all cases for such a short time that the beans cannot be denaturalized, that is to say they do not taste cooked or roasted.

Boiling Water-Saline Soaking Treatment

L.B. Rockland; U.S. Patent 3,635,728; January 18, 1972; assigned to the U.S. Secretary of Agriculture prepares quick cooking soybeans without the bitterness of raw beans. His process includes a conditioning of the beans by a brief contact with boiling water, followed by soaking in an aqueous solution containing sodium chloride, a chelating agent, and an alkaline agent. The hydrated beans are then dried, for example, by contact with air at 130° to 170°F. Alternatively, the hydrated beans may be preserved by freezing, or by partial dehydration followed by holding at refrigeration or even ambient temperatures.

In practice, raw soybeans are first contacted with boiling water for a brief time, about one-half to two minutes. This treatment has the critical effect of conditioning the beans so that they will take up the hydrating medium used in the next step, even though this medium is applied at ambient temperature and atmospheric pressure. It is important, moreover, that contact with boiling water be brief so the beans remain intact, there is no rupturing or splitting of the kernels and even the skins are retained in place. The skins exert a protective effect and it is desired that they remain intact in the final product. In a second step of the process the conditioned soybeans are soaked for about 24 hours, at room temperature and at atmospheric pressure, in a special hydration medium.

A primary consideration is that the hydration medium contains one or more tenderizing agents. These agents contribute largely to the goal of attaining a product that is quick-cooking, i.e., one that can be prepared for the table by heating in water for about 10 to 55 minutes. This is attained by having present in the hydration medium, primarily the following: sodium chloride and a chelating agent. The sodium chloride has the principal effect of tenderizing the skins, and is also believed to assist in tenderizing protein components of the cotelydons. Usually, the sodium chloride is present in the hydrating medium in a concentration of 1 to 3%. The chelating agent exerts a variety of useful effects, including the following: (1) It softens the pellicle or skin. (2) It aids in the solubilization of proteins and starchy components. (3) It acts as a buffer to maintain pH. (4) It facilitates uniform penetration of the hydrating medium into the centers of the beans, so that the final products have a uniformly smooth texture. (5) It tends to lighten the color of the product.

Various conventional chelating agents may be used such as the alkali metal salts of ethylenediaminetetraacetic acid (EDTA), alkali metal pyrophosphates, tripolyphosphates, or citrates, etc. Generally, the chelating agent is added to the hydrating medium in a concentration of 0.1 to 5%. Preferred is the combined use of sodium tripolyphosphate and tetrasodium EDTA, for example, 1% of the former; 0.5% of the latter, as providing especially good results coupled with a minimum amount of the chelating agents.

For best results, it is preferred that the hydrating medium be slightly alkaline, that is, a pH of about 9. Depending on the chelating agent selected, this may be attained directly, or it may be necessary to add an alkaline material, for example, sodium hydroxide, or preferably, sodium carbonate and/or bicarbonate. The carbonate, or bicarbonate, not only acts as an alkaline agent and buffer but also as a protein dissociating, solubilizing or tenderizing agent. Particularly good results are attained with a mixture of sodium carbonate and sodium bicarbonate and the preferred form of the hydrating medium contains these components in concentrations of 0.25% sodium carbonate and 0.75% sodium bicarbonate.

After the beans are hydrated, they may be washed with water to remove the hydrating medium from their surfaces. This washing is conveniently carried out by placing the hydrated beans on a screen and spraying them with water. However, washing is an optional step and may be omitted. Following washing, or directly after hydration, the beans are put in a condition whereby they may be shipped and stored without spoiling. Generally, dehydration is preferred as the most economical method of preservation. The dehydration can be accomplished with any of the conventional dryers used with food products. Alternatively, the hydrated beans may be preserved by freezing. This is conveniently effected by placing the beans on trays and exposing them to refrigerated air, temperature and rate of airflow selected so that the freezing takes place rapidly.

The hydrated beans may also be preserved by partial dehydration followed by holding at refrigerator temperatures. For example, the hydrated beans are dehydrated to a moisture content of 20 to 35%, then packaged, in transparent plastic bags, for instance, and held at 40°F until they are consumed. A solution was prepared having the following composition: sodium chloride, 2.5%, sodium tripolyphosphate, 0.5%, sodium bicarbonate, 0.75%, sodium carbonate, 0.25%, methyl parahydroxybenzoate, 0.025%, and propyl parahydroxybenzoate, 0.025%. The pH of the solution was 9.0.

Lots of soybeans of different varieties were each subjected to the following treatments: The beans were held in boiling water for 1 minute. The conditioned beans were then placed in the above hydrating solution and allowed to soak at room temperature and atmospheric pressure, for 25 hours. The soaked beans were removed from the solution, drained, and divided into two portions. One portion was dehydrated by exposing it on trays to a current of air at 150°F for a period long enough to reduce the moisture content of the beans to 10%. The other portion was frozen.

The dehydrated and frozen products were then tested for cooking quality. To this end, each sample was added to boiling water and simmered until the products reached a standard tenderness, typical of properly cooked beans. As controls, samples of the original (untreated) soybeans were soaked in distilled water overnight at room temperature, and tested for cooking quality. In the cooking tests applied to the products, runs 1, 2, 4, 5, 7, 8, 10, and 11, the first time figure refers to the earliest edible stage, the second when the beans were completely soft with no noticeable skin toughness. The results obtained are tabulated below:

Run	Soybean Variety	Treatment After Hydration	Cooking Time, minutes
1	Lee	Dehydrated	40 to 55
2	Lee	Frozen	20 to 30
3*	Lee	-	300
4	Kannrich	Dehydrated	25 to 35
5	Kannrich	Frozen	15 to 25
6*	Kannrich	-	270
7	Hawkeye	Dehydrated	30 to 40
8	Hawkeye	Frozen	15 to 25
9*	Hawkeye	-	240
10	Jackson yellow	Dehydrated	25 to 35
11	Jackson yellow	Frozen	10 to 15
12*	Jackson yellow	-	220

*Control

Taste tests of the cooked products (runs 1, 2, 4, 5, 7, 8, 10, and 11), indicated that they were free from bitterness; they had an excellent nutty flavor, and the beans were essentially intact with no significant mushing or sloughing. In contrast, the cooked controls (runs 3, 6, 9, and 12) had a very unpleasant bitter taste and the skins were somewhat fibrous.

HEAT PROCESSING

Infrared Heating of Raw Soybeans

Infrared heating of moisturized soybeans has been used by *E.J. Guidarelli and J.F. Lawrence; U.S. Patent 3,141,777; July 21, 1964; assigned to Cargill Incorporated* to improve the edibility and reduce the toxic factors of the beans.

The method generally comprises treating soybeans, particularly soybeans as harvested and containing their naturally occurring oil, to improve their utility as food. This is accomplished by heat treating the soybeans with infrared radiation for a controlled period of time at a specified temperature and moisture content. More particularly, the method comprises adjusting, if necessary, the moisture content of the soybeans by a tempering step, then heating the soybeans to a critical temperature in a manner which does not depreciate the dispersibility of the soybean protein. The heating is carried out using infrared radiation in combination with the controlled moisture in the soybeans. The moisture concentration in the soybeans is maintained above a critical level. Thereafter, the soybeans are cooled and may be subjected, if desired, to additional processing steps, such as cracking, dehulling, grinding or milling, flaking, and oil extraction.

The soybeans are subjected to a moisture-adjusting or tempering step, if their initial moisture concentration is not within the desired range. In this regard, the soybeans can be mixed with an appropriate quantity of water to adjust the moisture content in the soybeans to 8 to 27% by weight, preferably, between 19 and 22% by weight. To assure that the moisture has been absorbed uniformly, the beans are allowed to stand for a period of several hours until the moisture has substantially equilibrated throughout. For some uses of the soybeans, it is desirable to maintain the moisture for at least about two hours after uniform distribution of moisture to permit enzymatic action to occur within the soybeans which enhances feed efficiency in some instances.

It is very important that the soybeans be within the indicated moisture range before heat treatment to achieve the desired results. When insufficient moisture is present in the soybeans during infrared radiation, the beans have a tendency to undergo protein modification. Overheating of the soybeans may occur before destruction of toxic and inhibitory factors is accomplished. In the present method, soybeans having a moisture content within the indicated range are treated by infrared radiation at a controlled temperature for a time sufficient not only to destroy the inhibiting and toxic factors in the soybeans, but also to improve the flavor without deleteriously affecting the nutritional value of the soybeans. During the radiation, the moisture content of the beans tends to decrease. The extent and degree of infrared radiation is regulated so that the total moisture content of the beans is not reduced below 7% by weight. This is important in order to preserve the nutritional value and edibility of the beans.

Infrared radiation offers a twofold increase in the rate of drying, without overheating the nonaqueous soybean constituents. Water has a higher specific absorption of the infrared radiation than do the other soybean ingredients. Therefore, water evaporation takes place even though the soybean nutrients are not heated as much as is necessary in conventional drying processes. It has been found that gas burners designed to emit infrared radiation at wavelengths effectively absorbed by water are the most desired sources of radiation, inasmuch as they are inexpensive to operate relative to the amount of energy transferred. An infrared radiation source which peaks at between 2 and 6 microns is satisfactory, and that which peaks at 3 microns is preferred.

The radiation is preferably carried out by passing the soybeans in a layer two to four beans thick on a conveyor into contact with the radiation. Utilizing a Swank ceramic grid as a source of radiation and operating at a surface temperature of approximately 1500°F, approximately 55% of the gas energy is converted to infrared radiation in the wavelength between 2 and 6 microns. The energy is transferred directly to the soybeans without substantially heating the intervening air, and, accordingly, is very efficient. The radiant energy absorbed by the soybeans is transformed uniformly in the beans into thermal energy.

The heat treatment destroys trypsin inhibitors in the soybeans. It also has the effect of inhibiting soyin, a naturally occurring toxic protein in the soybeans. Moreover, the urease activity of the soybeans is reduced to a level whereby the Caskey Urease Test indicates a slight activity or no activity (briefly described on page 279, *Soybeans and Soybean Products,* Markley, volume 1). In addition, the infrared heat treatment has the effect of imparting to the soybean a toasted or roasted nut-like flavor. Factors which cause the soybeans to have an undesirable beany flavor are modified. These effects are achieved without substantial reduction in the protein dispersibility of the soybeans. Dispersibility of the treated soybeans is at least 55%, in contrast to a value of about 10% for conventionally treated soybean meal. A high protein dispersibility is a desirable feature, whenever the soybeans are to be used as or incorporated in livestock feed or other food products.

Example: Full fat soybeans having a moisture of about 7% are soaked in water for a sufficient time to increase the moisture content to approximately 19% by weight. The beans are allowed to temper for a period of time so that this moisture concentration is uniformly distributed. The moist beans are then passed on a conveyor in a layer not more than 2 beans thick into the path of radiation emanating from a gas burner type infrared ceramic grid operating at about 1650°F surface temperature. The beans are held in the beam path for approximately 4 minutes during which time they increase in temperature to approximately 245°F. At the end of the heat treating period, the beans are passed by conveyor between cracking rolls. They then pass to a dehulling operation. After dehulling, the beans are ground to meal and sesame oil is blended therewith in a concentration of approximately 2%, by weight, of the soybean oil content in the meal.

The soybean meal, is then incorporated in a broiler feed for chickens and subjected to poultry feeding tests, in comparison with conventional feed containing an identical concentration of soybean meal prepared by steps including dehulling, grinding, extraction of soybean oil and steam treatment of the soybean meal at elevated pressure. This soybean product of this process gave superior results in contrast to the conventional feed, in terms of increased weight per pound of feed. In addition, the present soybean product has a toasty nut-like flavor and aroma, an inactive urease level, and a substantially nil soyin concentration. The soybean product has 55% protein dispersibility. Storage stability tests on the soybean product after addition of sesame oil at a level of 2% of the soybean oil content indicate that the product can be held for extended periods of time without undergoing substantial flavor reversion or oxidation of the soybean oil content.

E.J. Guidarelli; U.S. Patent 3,407,073; October 22, 1968; assigned to Soy Food Products, Inc. has further modified the process of U.S. Patent 3,141,777 using increased moisture content in the soybeans and using infrared radiation having a peak wavelength outside the effective water absorption range.

Broadly, the method includes swelling the beans (as by soaking in hot water or other hot aqueous liquid below the boiling point to increase the moisture content of the beans uniformly, or in a high pressure steam puffing gun), and then heating the swollen moist beans with infrared radiation having a peak wavelength outside of the effective water absorption range. More particularly, the moisture content of the beans may be increased by soaking to between 35 and 75% by weight and the infrared heating is carried out with radiations having peak wavelengths between 1 to 1.7 or between 7 to 30 microns. The soaking of the beans to increase the moisture content causes the beans to swell substantially. The infrared heating of the moist beans roasts the beans to destroy their bitter taste and the enzyme inhibitors without materially adversely affecting the dispersibility of soybean protein.

The resulting beans are found to be dry and crunchy, easily edible and palatable and readily acceptable as human food. The added seasoning or flavoring may be incorporated into the soaking liquid. It may be added during roasting, as by spraying, or may be introduced into the beans after the infrared heating. In some instances it may be added as a coating to the finished roasted beans.

The beans may be swollen simply by soaking for long periods of time at room temperature. However, by soaking in hot water or other hot aqueous liquid between 180° and 210°F the same effect can be obtained in 10 to 45 minutes as that from soaking at room temperature for 20 to 24 hours. Longer soaking of 1 to 20 hours or more is not objectionable, but is usually unnecessary. By soaking under pressure, as in a pressure cooker, higher temperatures may be used and less time is required.

When the beans are expanded by puffing they are subjected to high pressure steam in the range of 90 to 200 lb/in^2 in a closed pressure vessel for 2 to 5 minutes to insure complete penetration of the bean by moisture and then the pressure vessel is opened to expand the beans as a result of expansion of the moisture contained within. The beans attain a high moisture content of 50 to 75% under high steam pressure. Upon puffing the moisture content may be reduced to 5 to 15%. It has been found that beans expanded in this manner have substantially uniform moisture content and are well adapted to infrared treatment.

The toasting is carried out by infrared heaters designed to effectively emit infrared radiation at wavelengths outside of the water absorption range. Gas burner equipment is available utilizing a replaceable double ceramic grid. The beans are preferably toasted by passing them on a conveyor in a relatively thin layer of from one up to about 4 or 5 beans thick in close proximity to the burners. Typical spacing between burner and beans is from about 6 to 18 inches. The beans are desirably subject to some agitation, such as tumbling, either as a result of movement of the belt or vibration of the conveying means. Depending on conveyor speed, a typical installation may be from about 75 to 150 feet long. The gas burner operates at a surface temperature of up to 2500° to 2600°F and preferably 1600° to 1700°F in the wavelength of between 1 to 1.7 microns.

Example: One hundred pounds of raw soybeans were soaked in two 10 gallon cans of water and sparged with steam for about 10 minutes to bring the water temperature up to about 212°F and to agitate the beans. As the result of the combined soaking and agitation, the bean hulls expanded and were separated and floated to the tops of the cans where they were skimmed off and discarded. The steam was then disconnected and the beans were permitted to soak for an additional 30 minutes at about 200°F. At this time the beans had swollen to about 3 times their normal raw size and contained about 70% water. The beans were removed from the soaking vessels and drained and then exposed for 10 minutes to infrared radiation at 1.7 microns peak wavelength. A ceramic grid having 200 holes per square inch was used. The grid temperature was 2500° to 2600°F and the grid was almost white hot.

The beans were spread in pans in a bed about ½ inch deep and placed on a vibrating conveyor passing under the gas burners. The spacing between burner and beans was approximately 8 inches. The conveyor moved at approximately 10 ft/min and the array of burners was approximately 100 feet long so that the total exposure to infrared radiation was approximately 10 minutes. During exposure the bean temperature never exceeded 220°F. At the end of the roasting treatment the moisture content of the beans was about 2.9%. The beans were then sprayed with water to cool them and to bring the moisture content back to about 10% and quickly dried. The resulting product was dry and crunchy with palatable nut-like flavor.

Two-Step Dry Heat-Hot Water Treatment

R.L. Hawley and J.T. Duren; U.S. Patent 3,594,184; July 20, 1971; assigned to Ralston Purina Company have used a combination of dry heat followed by a hot water treatment of soybeans to make them useable as food. The process comprises treating hull enclosed cotyledons, especially pea or bean legumes, particularly soybeans, to remove objectionable flavor, to remove or alter physiologically objectionable sugar constituents, to alter the density and texture to provide a desirable texture, and basically to produce a full-fat edible product retaining the desirable oils. This can be followed by roasting to obtain edible nut-like products or roasting and grinding to obtain edible spread type products.

The dry heating can be conducted with heated air, or with heated gases such as combustion gases. Preferably, the beans are dry heat treated such as in a combustion flame, i.e., in heated combustion gases during actual combustion. This not only splits the hull but also partially consumes the hulls. Experimentation with the dry heating indicates that the chief criterion to obtain the necessary results is heat input rather than specific time or temperature. The preferred temperature range is 250° to 450°F, with the respective time range being 10 minutes to 25 seconds. Although the heat input is believed the controlling criterion, the exact Btu input is extremely difficult to exactly determine for each bean. As closely as can be determined, it is believed that a mean interior bean temperature of 160° to 250°F should be achieved and maintained for a time sufficient to obtain the necessary heat input for causing the important internal as well as external changes in the complete bean.

One effect of this controlled dry heating of the complete bean is to cause the hull to split perpendicular to the long axis of the cotyledons. In the subsequent water treatment, this allows easy hull removal and also rapid hydration of the cotyledons in minutes, i.e., about 15 to 20 minutes in hot water, rather than the many hours when hydrating the cotyledons through the hypocotyl. Yet this rapid hydration will occur without disintegration of the cotyledons or significant loss of proteins or desired oils, because the heat treatment causes internal molecular changes in the bean which prevent this. The oils are apparently rendered stable to osmotic leaching. The proteins are rendered insoluble by being denatured. Further, this thorough heat treating step also causes some improvement in flavor.

After the complete beans are so heated, they are immersed in water, preferably while the beans are still hot. By gently agitating the water and beans, the hulls are readily released and fall off the cotyledons. By slightly more vigorous agitation or rubbing, the hypocotyls are also released from the cotyledons, to allow their removal with the hulls. This is desirable because the hypocotyls are a source of bitter unsaturated oils, and because their removal extends the shelf life of the product considerably. Also, the cotyledons rapidly and uniformly absorb moisture, without disintegrating, causing swelling and expansion of the high density cotyledons to a size 2 to 3 times the original, with a density of only $1/3$ to $1/2$ of the original.

The swelling and expansion causes the cell structure to expand and open. This is very significant to the texture of the final product and in facilitating osmotic extraction of constituents causing bitter flavor and objectionable tri- and tetrasaccharides such as manninotriose and stachyose which would produce flatus in the consumer. The water treatment may vary somewhat depending upon the final product characteristics desired. If the final product is to be roasted nut without other added flavor, it is desirable to remove much of the tri- and tetrasaccharides, but to retain some of the unobjectionable mono- and disaccharides to allow proper roasting to a golden color.

In this case, the water treatment may constitute water soak. This soak allows dehulling, cotyledon swelling, and significant removal of bitter flavor causing constituents. If one desires to add other sugars to the water treated items, as by soaking in a sugar solution, to obtain optimum roasting characteristics, the water treatment may comprise a water soak, followed by cooking in water to remove most of the natural sugars. This sugar leaching is very effective if the water is boiled for $1/2$ to 1 hour during the soak. Boiling also bleaches the cotyledons to a more attractive light shade. Actually, the soaking may vary from 15 minutes to 60 minutes or so, with 15 to 20 minutes being normally sufficient. The cooking time may also be 15 to 60 minutes, but in actual practice, if the cotyledons are going to be cooked, they may be placed directly into hot cooking water after heat treating since the results of water soak occur simultaneously with the results of cooking.

The separated cotyledons, when removed from the water bath, are intact, in whole form, with smooth surfaces and an appealing nature, have an excellent open texture, possess the desirable nutritional constituents, and are generally free of the undesirable internal constituents and external components.

These moist cotyledons are then roasted, either in hot oil or by dry heat, to produce appealing full-fat roasted soynuts which have good chewing characteristics, are tasty, are not bitter, and are highly nutritious. Roasting may be delayed after the water treatment, provided the product is kept moist to prevent dehydration and consequent shrinkage back to the high density product. If this roasting is done rapidly, the swollen cotyledons retain substantially all of their expanded size. The temperature range for roasting is 250° to 600°F. The time of roasting varies somewhat inversely with the temperatures, and being in the range of about 30 minutes to about 5 minutes. The nuts may have flavors such as spices or the like added before, during, or after roasting.

Example 1: Soybeans are heated with hot air at a temperature of 450°F for 2½ minutes, causing internal changes and causing the hulls to split perpendicular to the main axis of the cotyledons. The beans, while still hot, are put into a water bath previously heated to near boiling. The water is boiled for 45 minutes and agitated, causing the hulls to fall off the beans and the hypocotyls to be released from the cotyledons, causing the cotyledons to swell and expand, and causing osmotic leaching of the undesirable constituents. The cotyledons are then removed from the water, placed in hot oil at a temperature of 425°F and roasted for 8 minutes. The product is salted and eaten like nuts.

Example 2: The procedures of Example 1 are substantially repeated except that the beans are heated with hot air at a temperature of about 425°F for about 36 seconds. The resulting roasted beans have substantially the same properties as those obtained by the procedures of Example 1.

Example 3: Soybeans are treated as in Example 1, except that the soybeans are heated with a combustion flame for 2 minutes, and the water is continuously exchanged with fresh boiling water. The removed cotyledons are then mixed with an equal weight of an hydrogenated mixture of half coconut oil and half cottonseed oil, 4% by weight sugar, and ½% by weight salt. The mixture is then ground to a pasty spread.

An improvement in U.S. Patent 3,594,184 was also made by *R.L. Hawley and J.T. Duren; U.S. Patent 3,594,185; July 20, 1971; assigned to Ralston Purina Company* using a second treatment bath.

Following the water treatment and prior to washing, the cotyledons can be subjected to a second water bath containing a sugar and salt. Any sugar may be utilized, with sucrose being preferred, however, other mixtures of sugars may be incorporated in the bath to alter the flavor of the nuts. Preferably the sugar content of the bath will not exceed 50% by weight and the salt content will be 1 to 20% by weight. In addition to water, a hydrolized vegetable or animal protein or other flavoring materials may be included in the second bath. Alternatively, the second bath may include a flavor enhancer, such as monosodium glutamate or one of the prime inosates. The second bath is heated to a temperature of 150° to 212°F, preferably 210°F, to facilitate the sorbing of these ingredients by the cotyledons. The period of time the cotyledons should be immersed in the heated solution varies inversely with the temperature of the bath and is in the range of 10 minutes to 10 seconds. Preferably, when the second bath is maintained at 210°F, the cotyledons will be immersed for a period of 30 seconds.

The cotyledons when removed from the second bath have sorbed a sufficient quantity of the desirable constituents such that upon roasting a desirable flavor is developed in the nuts. The roasting should be carried out by heating the nuts to a temperature between 250° to 600°F for a period of time in the range of about 30 minutes to 5 minutes, respectively. This roasting serves to develop a desirable flavor and pleasing appearance in the nut.

Processing Whole Soybeans

As a further alternative to improving the flavor and texture of the dehulled cotyledons, the cotyledons after roasting may be treated to alter their oil content to obtain a high-oil content nut product. Soynuts normally have an oil content of approximately 25% as compared to oil contents of 50% or more for other nuts, such as peanuts or walnuts. This high-oil content contributes to the mouth feel and flavor of the nuts. It is desirable in producing a nut-like product from soynuts that the total oil content of the nuts be raised to a maximum value of approximately 75% by weight and preferably to approximately 50 to 60% to resemble other types of nuts.

Merely spraying or dipping the roasted soynuts with flavoring material results in only a small portion of the oil being sorbed by the soynuts. Also, spraying or dipping roasted soynuts with flavoring material results in only a small portion being absorbed. However, by immersing the roasted soynuts in an oil bath under a vacuum, enough oil can be added to the soynuts to alter the mouth feel and flavor of the soynut.

Example 1: Soybeans are heated with hot air at a temperature of 450°F for 2½ minutes, causing internal changes and causing the hulls to split perpendicular to the main axis of the cotyledons. The beans, while still hot, are put into a water bath previously heated to near boiling. The water is boiled for 45 minutes and agitated, causing the hulls to fall off the beans and the hypocotyls to be released from the cotyledons, causing the cotyledons to swell and expand, and causing osmotic leaching of the undesirable constituents. The cotyledons are then removed from the water, and immersed in a second bath, all percents by weight, as follows: 75.2% water, 15.0% sugar, 6.8% salt, and 3.0% hydrolyzed vegetable protein (Maggie 3H3-4), cotyledons are immersed in the second bath for 2 minutes with the second bath being maintained at 210°F. The cotyledons are then roasted by hot air at a temperature of 325°F for 15 minutes and develop a pleasant nutty flavor.

Example 2: Soybeans are heated with hot air at a temperature of 450°F for 2½ minutes, causing internal changes and causing the hulls to split perpendicular to the main axis of the cotyledons. The heat treated beans are mechanically dehulled in the dry state using a Packomatic dehuller. After dehulling, the cotyledons or beans are separated from the hulls by an aspiration system. The dehulled beans are then blanched in boiling water for forty (40) minutes and roasted in a conventional nut roasting oven at a temperature of 325°F for 15 minutes. After roasting, the soynuts are immersed in an oil bath maintained at 180°F and under a vacuum of 28 inches of mercury. The soynuts are retained in the bath for 2 minutes. The oil flavor bath, all percents by weight, is composed of the following: 92% vegetable oil (soybean oil), 3% salt, and 5% imitation flavors.

After immersion of the soynuts in the flavor bath, the soynuts were drained and the resulting product exhibited a very pleasant taste. The oil treatment also has the effect of producing a more nut-like texture to the soynuts.

Dry Heating with Infrared

H. Truax; U.S. Patent 3,343,961; September 26, 1967; assigned to Harry Truax & Sons Company, Inc. provided a machine and method for treating soybeans at point of harvest on individual farms. This uses high intensity infrared heat source without causing denaturing of the protein or case hardening of the soybean. The process also improves the organoleptic and nutritional properties of raw soybeans.

The method comprises the steps of rapidly and intermittently cascading soybeans into close proximity to an infrared energy source until the teguments of the soybeans are ruptured by vapor pressure generated internally. Immediately, at that point the soybeans are placed into a storage means without permitting any substantial decrease in temperature of the soybeans and the soybeans are stored until all of the deleterious materials contained in the beans are destroyed. The infrared energy source is maintained from 1800° to 2500°F. Generally, the soybeans are held in the storage means from one to ten hours.

In the practice the infrared energy source may be gas fired and the products of combustion emanating from the energy source may be contained in the environment surrounding the soybeans. In practice, it may be desirable to maintain the environment containing the products of combustion at about 400°F while the infrared energy source is maintained from 1800° to 2500°F.

FIGURE 2.2: INFRARED DRY HEATING APPARATUS

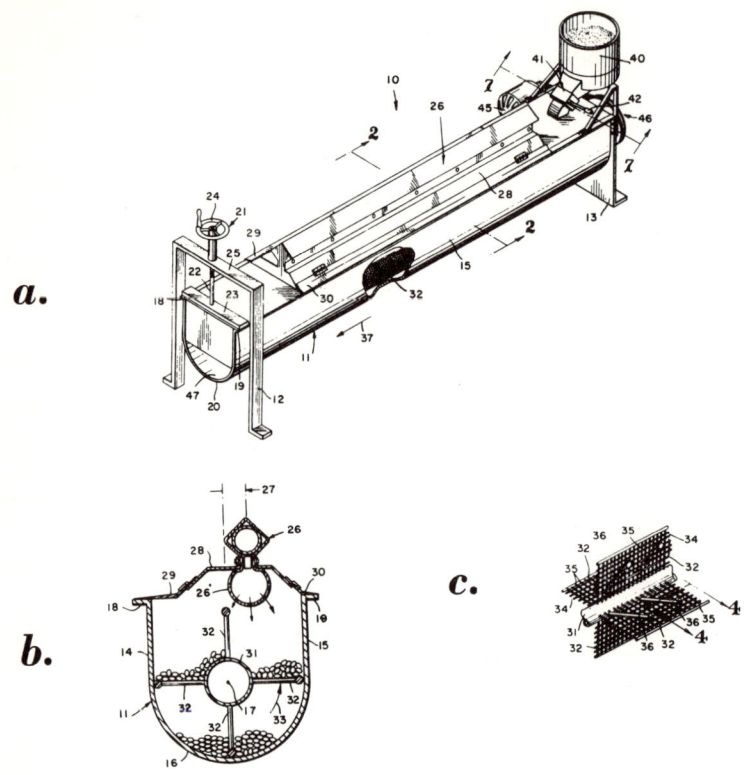

Source: H. Truax; U.S. Patent 3,343,961; September 26, 1967

Figure 2.2a is a perspective view of an apparatus for treating soybeans; Figure 2.2b is a sectional view taken from Figure 2.2a generally along the lines 2—2 and illustrating a rotary vane for cascading soybeans and the like near an infrared heat source; and Figure 2.2c is a fragmentary view of the rotary vane means.

Referring now to the drawings, a machine for treating soybeans indicated by number **10**, is comprised of a trough **11** supported at the ends by support means **12** and **13**. Referring to Figure 2.2b, it can be seen that the trough is U-shaped in cross section and has vertically upwardly extending sides **14** and **15**. The lower part of the trough, indicated by **16**, is cylindrically formed about an axis **17**. There is a pair of flanges **18** and **19** extending outwardly from the uppermost portions of the sides **15** and **16** respectively.

Referring again to Figure 2.2a, it can be seen that the support means **12** is connected to the end **20** of the trough by a vertically adjustable means **21**. The vertically adjustable means includes a threaded shaft **22** having one end swivelly connected to a member **23** secured to the trough and the other end threadedly received in an internally threaded crank **24**. The shaft extends through a hole in a cross member **25** of the support means **12** upon which the crank is supported. Therefore, by rotating the crank, the end of the trough is raised or lowered with respect to the cross member. The aforementioned member **23** is connected to the flanges **18** and **19** of the trough.

There is an infrared heat source **26** mounted above the trough as is shown in Figures 2.2a and 2.2b. The infrared heat source found most desirable is of the type having a radiant grid **26'** made of high temperature heat resisting metal which is heated to incandescence by an internal flame produced by combustion of premixed gas and air in the proper proportions. In Figure 2.2b, it can be seen that the radiant grid is cylindrical and is offset laterally with respect to the axis by distance represented by **27**. There is a reflector **28** arranged to reflect the generated radiant energy into the trough. Infrared heaters of the type discussed above are well known in the art and need not be discussed in detail.

There is a pair of adjustable ventilators **29** and **30** hinged to the shield means **28** as is shown in Figures 2.2a and 2.2b. The ventilators are provided as a ventilation means for controlling the ambient conditions within the trough. There is a shaft **31** rotatably supported in the trough on an axis substantially coincident with the axis. A plurality of radially outwardly directed vanes **32** is mounted on the shaft. In Figure 2.2b, it can be seen that the radial dimension of the vanes is such that the vanes just barely clear the trough and the radiant grid when the vanes are rotated in the direction of the arrow **33**. Therefore, when the vanes are rotated in the direction of the arrow, beans contained in the trough are picked up by the vanes and cascaded in close proximity to the radiant grid.

In Figure 2.2c, it can be seen that the vanes are preferably fabricated in the form of a grid. The grid consists of axially extending wires **34** and radially extending wires **35** arranged to form openings small enough to prevent passage of soybeans but large enough to permit transmission of radiant energy and to attenuate convection. The wires are generally round and stiff in order to utilize Fresnel diffraction and Cornu spiral pulsating effects of infrared rays emanating from the radiant grid. Calculations indicate that a rotating grid of this type will provide peak pulses up to 40,000 Btu per hour per square foot for a burner intensity of 25,000 Btu per hour per square foot of surface. Thus, the radiant flux is diffracted to increase the effective intensity of the pulse which gives deeper penetration.

In the practice of the process, soybeans are heated in an apparatus of the above type until enough vapor is accumulated beneath the tegument to provide a pressure differential which will finally rupture the tegument. Upon rupture of the tegument, the soybeans are moved without cooling into a bin having low thermal conductivity. The heat already generated inside the soybean is retained and provides strong thermal vibration within each soybean. It is necessary to retain the heat within each soybean for a predetermined period of time because the thermal reaction of some of the destructive enzymes is reversible and if the soybeans are cooled quickly, many of the destructive enzymes will not be permanently destroyed. For example, urease will not be permanently destroyed if the soybeans are cooled too quickly.

In general, soybeans which have been heat treated by this method must be stored without substantial heat loss for at least one hour and may be held without deliberate cooling for ten to twelve hours. If more than one ton of soybeans is stored and has not cooled within ten to twelve hours, the soybeans may require forced cooling. The above method of using residual heat in a post treatment bin assures that there is no overheating of the soybeans. During the tempering period in the storage bin, thermal energy is conducted to the centers of the soybeans. In addition, the entire soybean is cooked evenly to a desired color with a minimum of protein and amino acid destruction.

During the infrared heat treatment, the soybeans are intermittently exposed to a high radiant flux at a frequency which will not permit scorching of the tegument. This is achieved by cascading the soybeans in close proximity to an infrared heating source maintained at from 1800° to 2500°F. Moisture evaporated from the inside of the soybean combines with carbon dioxide to form an opaque film of the surface of the soybean, and the cascading tends to remove the film to permit efficient radiant energy transfer to the soybean.

It is necessary to control the amount of time that the soybeans are continuously exposed to the high energy infrared source. Over exposure causes denaturing of the protein, case hardening or toughening of the tegument and impairing of the organoleptic and nutritional properties of the soybeans. Using the principle of pulsating with a high energy source, the soybeans can be heat treated in four to eight minutes depending on the size of the soybeans. The time of four to eight minutes is in contrast to the thirty to sixty minutes required for treating soybeans using steam as in contemporary methods.

The exact time and temperature required for heat treating soybeans using this machine and method cannot be exactly stated because these factors depend on the moisture content of the soybean as well as the size. However, it has been found that soybeans having a moisture content of about 9 to 13% can be heat treated satisfactorily using the following method. The heat treating machine is like the machine 10 having six inch wide vanes mounted on a shaft driven at speeds of 30 to 40 rpm. The soybeans are passed through the machine and are cascaded from twenty to twenty-five times in close proximity, about one inch, to the infrared energy source. The total time in the machine is approximately 360 seconds. The soybeans are then removed from the heat treating machine and quickly placed in a post treatment storage bin, where they are held for at least one hour. To establish proper treating temperatures, the following relationship, shown graphically in Figure 2.3, is provided.

FIGURE 2.3: TIME-TEMPERATURE RELATIONSHIP

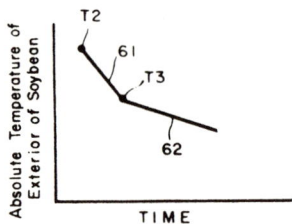

Source: H. Truax; U.S. Patent 3,343,961; September 26, 1967

Three temperatures must either be measured or extrapolated. The first temperature, T1, is the absolute temperature of soybeans prior to processing. For all practical purposes T1 will be the ambient temperature corrected to absolute. The next temperature, T2, is the absolute temperature of the soybeans taken immediately after exit from the heat treating machine. The third temperature, T3, is the absolute extrapolated maximum equilibrium temperature during storage of the soybeans in the post treatment storage bin. T3 is extrapolated from the crossing of the lines 61 and 62 shown in Figure 2.3. Line 61 represents the heat loss due to internal conduction and line 62 represents the total heat lost in the post treatment storage bin.

It has been determined from feeding tests that T2/T3 ratios must be maintained between 1.01 and 1.15 and that T1/T2 ratios may vary between 0.605 and 0.815. It has also been determined that the T1/T2 ratios are not as critical as the T2/T3 ratios. In order to maintain the T2/T3 ratios within the 1.01 and 1.15 limit, T2 can be lowered considerably below the temperature which would scorch the soybean as long as it is sufficient to rupture the teguments. Of course, T2 is determined by the size of the soybeans, moisture content, atmospheric pressure, variety of soybeans, conditions of the hull, etc. T3 can be varied by many obvious means, such as varying the thermal conductivity of the storage bin, to assist in meeting the required T2/T3 ratios.

CANNING SOYBEANS

Despite the wealth of nutritive value present, soybeans have failed in acquiring any substantial acceptance as a food product for human consumption, the primary causes being the unpalatable taste plus a tough rubbery skin which in general has defied the prior attempts to soften or otherwise render it palatable.

R.P. Baile; U.S. Patent 3,052,556; September 4, 1962 has developed a process to overcome these problems by the treatment of dried uncooked soybeans in an oil bath at a temperature of 80° to 300°F for 10 to 30 minutes. Such a bath has been found to have a marked effect upon the bean in that it permits the bean to be cooked or skinned without further processing such as a long water-cooking. This oil treatment of dried uncooked, unpeeled soybeans will of course be readily distinguished from cooking or frying or roasting processes which have previously been performed at higher temperatures and longer times on other food products such as peanuts. Rather, this treatment is a precooking treatment, which is neither long enough nor hot enough to cook the bean but, instead, substantially advantageously affects only the skin.

In the above treatment of soybeans numerous oils have been found suitable for the purpose of brittlizing the skin of the bean to make it palatable as well as removable, while also bringing out the nut-like flavor of the bean. Such oils as have been used include any edible oil such as the vegetable oils, coconut, corn, cottonseed, cocoa butter, olive, peanut and soybean. Animal oils such as lard oil and also mineral seal oil have been found to perform satisfactorily. In general, there is a wide latitude in the temperatures at which the soybeans are processed prior to any cooking. The times associated with certain temperatures may also be varied depending upon the coloration and/or use desired. For example, if the beans were cooked five minutes longer than the recommended 15 minute period at a temperature of 218°F, the beans will turn a slightly darker brown color but still all the advantages of oil processing the soybeans will be retained.

After the soybeans have received the oil treatment, they may be processed in any number of ways, depending on the final product desired. In regard to the canned product, it has been found to be most desirable to cook the oil-treated beans in the sealed can rather than in a large vat from which the cans might be filled for the reason that the increased handling involved in vat cooking and subsequent canning is not only more expensive but also may cause some of the skins to be removed, as is likely to present an unusually unsightly appearance in the final product.

Example: One pound of dried Mammoth Yellow soybeans was poured into a container of peanut oil at a temperature of 218°F and allowed to remain therein for 15 minutes. The soybeans after being removed from the oil were inspected and found not to have swelled from their original size. They were somewhat darker in color than in the original state, the color being dependent to a large extent upon the length of time in which the soybeans are left in the oil. The heated soybeans are further characterized by the skin being (1) noticeably more brittle than in the untreated state to the extent that the skin may be readily rubbed off by the fingers, for instance; and (2) substantially less tough or rubbery, therefore making it more palatable.

In addition, the hot-oil treatment brought out a desirable nut-like flavor in the soybean which was either unnoticeable or only slightly noticeable in the untreated bean, and not present in the bean soaked by the prior processes.

Other oils used in additional examples included soybean oil, cottonseed oil, lard, corn oil and the like.

PROCESSING FULL FAT SOY PRODUCTS

The simplest processing of soybeans consists of steaming the beans and removing the hulls which are more than 85% carbohydrate. This leaves a product which is ground or flaked to produce full fat flours, grits or flakes containing the full protein and oil of the whole soybean.

DEBITTERING BY HEAT OR STEAM

Steam and Pressure Debittering Process

G.C. Mustakas and E.L. Griffin, Jr.; U.S. Patent 3,290,155; December 6, 1966; assigned to U.S. Secretary of Agriculture have developed a shortened process of treating full fat soybean flakes or grits to obtain completely debittered food grade, full fat soybean flakes or flour which have superior nutritional and physical characteristics.

These superior food grade full fat soybean materials are obtained when dehulled full fat soybean flakes or grits having a normal or tempered moisture content of between 9 to 12% are subjected batchwise or preferably continuously in a jacketed container to condensed steam for from 2½ to 3 minutes to bring the flakes or grits up to a temperature of 95°C. Heated flakes or grits are immediately subjected to 5 to 10 seconds of residence in a high speed mixing apparatus having steam admission means to raise their temperature to 99° to 102°C and their moisture content to 16 to 21%, then the hot, moist flakes or grits are passed into a conventional steam-jacketed screw-type extruder having at least two chamber sections separated by an air lock die and terminated by a highly perforated extruder die.

In the extruder the temperature of the soybean material present is increased to a level of 115° to 145°C during a residence time of only 60 to 90 seconds while imposing a dynamic pressure of 375 to 400 psi by means of the extruder screw. This pressure is just sufficient to expel from the partially cooked soybean material and onto the surface thereof only the amount of oil that can be immediately resorbed by the hot material when it is abruptly relieved of pressure by passage through the extruder die (evidenced by the absence of a stream of unresorbed oil). The extruded pellet-like agglomerates are cooled with air, vacuum dried to a moisture content of 3 to 4%, and milled as substantially dry agglomerates through smooth rolls to a 100 mesh flour.

The process is operative even with screw-type extruders not equipped with internally placed pressure indicating means inasmuch as the required critical pressure can be obtained by visual inspection of the material issuing from the perforated extruder head, an only momentary

Processing Full Fat Soy Products

presence of surface oil being easily noted. Furthermore, any substantial and totally inoperative excess of pressure exerted on the substrate by the screw expels an amount of oil that forms discrete and readily observable streams issuing from the die. Although one might expect kinetic or dynamic pressure to be converted to a roughly equivalent amount of heat energy, the mechanism for the unexpected results is not that simple since the mere application of high heats even for very brief periods greatly denatures the protein yet fails to make the product bland and palatable. The following example and data of the table will more clearly illustrate the process.

Example: Dehulled full fat soybean flakes having an average thickness of 0.01 inch and a moisture content of 10.6% were introduced at the rate of 71 lb/hr into a small commercial-type steam-jacketed horizontal mixer (pre-conditioner) having steam admission and temperature indicating means. During a residence time of about 2.75 minutes (165 ± 15 seconds) sparge steam equivalent to 2 pounds of condensate per hour was admitted, giving the flakes a final temperature of 93°C and a moisture content of 13.1%. The hot flakes were then subjected to about 10 seconds of mixing in a steam-equipped high speed mixer which raised the flakes to 99°C and the moisture content to 15.6%.

The flakes were then directly transferred to a commercial steam-jacketed externally driven screw-extruder comprising an initial feeder section, a steam-lock die, two center air-lock sections each having a steam-lock die, and a terminal cone extruder section terminating in a highly perforated extruder die. The terminal pressure in the extruder was adjusted to a level that just avoided the presence of unresorbed oil at the extruder die. During a residence time of 1½ minutes the flake temperature was increased to a final temperature of 117°C, the apparent moisture content of the extrusion-exploded agglomerates being 14.8%. After both forced air and vacuum drying the agglomerates having a moisture content of 3.1% were passed through smooth rolls to provide a 100 mesh flour having the characteristics set forth in the table, which shows a comparison of extended full fat soy flour with commercial full fat soy flours.

	Extruded Flour Products Example	Specifications for General Purpose Full Fat Soy Flour[1]
Chemical:		
Moisture, percent	3.1	8[2]
Protein	44.44	40[3]
Crude fat	20.17	20[3]
Ash	5.00	6[2]
Crude fiber	3.0	3[2]
Degree of toasting:		
NSI	15.7	–
Trypsin inhibitor assay percent inhibition	95.5	–
Urease activity pH increase (Caskey-Knapp)	0	0–0.3
Physical, screen, percent through		
100 mesh screen minimum	95.7	95
Flavor and stability factors:		
Peroxide value meq/1,000 g extracted oil	0.76	–
Free fatty acid, percent in extracted oil	0.52	–
Organoleptic	(4)	–
Biological values:		
Available lysine percent of protein	5.15	–
Protein efficiency ratio, rat bioassay, g gain/g protein fed	2.53	–

[1] Soybean Council of America, *Tentative Quality and Processing Guide, Edible Soy Flour or* July 1961
[2] Maximum
[3] Minimum
[4] Debittered, bland

Alkaline Cooking at 80°C

Prior alkaline cooking of soy flour or beans has resulted in the destruction of the sulfur amino acids of the soy. *M. Rambaud; U.S. Patent 3,220,851; November 30, 1965; assigned to Societe Industrielle des Oleagineux, France* has found that cooking the soybeans in an aqueous alkaline suspension at pH 8-9 and at 80°C does not adversely affect the protein value of the product.

With a cooking temperature in the vicinity of 80°C and a pH value between 8 and 9, the duration of cooking is less than 20 minutes for a concentration of soy of 20% by weight. This period of cooking can vary with the temperature and the concentration. It is longer with a lower temperature or with a higher concentration.

The researches have shown that the cooking of a suspension at the natural pH value of the flour of whole soybeans would involve a reduction of the percentage of albumins, part of the proteins dispersible in water. This insolubilization increases with the duration of cooking and especially with the temperature. By operating on a suspension containing 20% of soy and 80% water, it has been observed that the content of albumin, which is 65% of the proteins in the raw bean, is reduced to 60%: In more than 8 hr at a temperature of 70°C, in 5 hr at a temperature of 75°C, in 2 hr at a temperature of 80°C, in 25 min at a temperature of 85°C, in 10 min at a temperature of 90°C. It thus appears that the temperature of 80°C constitutes a threshold value beyond which the speed of degradation of the albumins increases rapidly, and it is therefore essential not to exceed this value.

Some of the results obtained have been shown by means of the drawings below. Figure 3.1a indicates the cooking time in minutes at 80°C necessary to obtain a destruction of 90% of the urease (curve **U**) and of the antitrypsin (curve **A**), as a function of the pH value of the suspension. Figure 3.1b indicates the percentage of residual antitrypsin as a function of the cooking time in minutes for different values of the pH of the suspension.

FIGURE 3.1: TREATMENT OF SOY

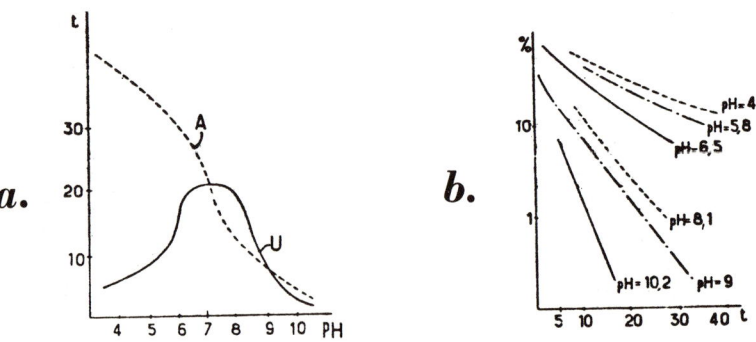

Source: M. Rambaud; U.S. Patent 3,220,851; November 30, 1965

It can thus be seen that satisfactory elimination of the undesirable factors urease and antitrypsin is obtained with a pH value comprised between 8 and 9, and that with this pH value, the elimination is satisfactory for an acceptable duration of cooking of the order of 20 minutes, which is a suitable value in order to prevent degradation of the proteins which results in a reduction of the content of albumin. The effect of the treatment is completed by the addition of a proteolytic enzyme. Such enzymes may be of vegetable origin such as papain and ficin, or of animal origin such as pancreatin, of bacterial or fungic origin obtained by biosynthesis.

When added to the soybean suspension in a proportion of 0.1 to 0.5%, the enzymatic preparation solubilizes the proteins, fluidifies the mass and permits treatment at a higher concentration. It should be observed that the treatment conditions, temperature 80°C and pH value between 8 and 9, are not the most favorable conditions for the action of the proteolytic enzyme. However, in consideration of the destruction of the undesirable compounds, urease, antitrypsin, etc., and the preservation of the proteins, the addition of the enzyme ensures a proteolysis which, although partial, is however effective.

Example: Soybeans of the so-called Yellow #2 quality, previously decorticated, are finely ground to a fineness determined by a sieve Afnor 24. The product resulting from the grinding is put into suspension while cold or at low temperature in an alkaline lye, so that the concentration of dry material is 20%. The alkaline lye utilized is such that the pH value of the suspension is equal to 9. It is preferably prepared simultaneously with lime and ammonia. A good preparation for 1,000 kg of soybean is a solution containing 4 kg of lime; to the suspension obtained, there is added 2.5 kg of ammonia (quantities expressed in CaO and NH_3).

The alkaline suspension obtained is then heated to 80°C. The heating is effected by surface or better still by injection of steam. The duration of the cooking treatment is 20 minutes. The suspension is then subjected to drying in order to eliminate the water. At the same time, the ammonia is driven off and the dried product returns to the neutral state, freed of 99% of its initial antitrypsin and entirely free of urease.

Cooking Moisturized Full Fat Meal

M.R. Gould and D.L. Swartz; U.S. Patent 3,253,930; May 31, 1966; assigned to Quaker Oats Company have developed a method of processing soybeans to obtain a highly nutritious, high energy, high fat, cooked soybean product of improved flavor having little tendency to become rancid. This can be obtained by subjecting soybeans under controlled conditions to a series of steps which comprises comminuting the beans to a relatively small particle size, adjusting the moisture content to a prescribed range, heating the comminuted, moistened beans at atmospheric pressure to a temperature not substantially above about 212°F to cook the soybeans and finally drying the cooked soybean meal to a moisture content which adapts it for use as an animal or human food product. The temperature and pressure conditions employed for heat processing the soybean meal are relatively mild as compared with prior methods.

In the present method, raw soybeans are comminuted to particle size to pass a U.S. #10 sieve. Preferably the soybeans are comminuted to a particle size so as to pass a U.S. #20 sieve. Comminution can be accomplished by grinding in conventional mills or by flaking with rolls and the like. When flaking is employed the soybeans are flaked to a thickness not substantially greater than 0.08 inch and a flake thickness of 0.005 inch is generally preferred. The moisture content is not critical and soybeans having natural moisture contents of up to 15% can be comminuted and used in the process without the necessity of predrying.

After comminution, the moisture content of the soybean material is adjusted to 15 to 30% by weight. This can be done immediately upon entry into the actual cooker but is preferably carried out as a separate preconditioning step in a hydrator or conditioner. Preferably the temperature of the ground material is raised simultaneously with adjustment of its moisture content. Commercially available high speed mixers having means to inject liquid and/or steam are satisfactory for this purpose. In a preferred procedure water is introduced into the conditioner to provide a moisture content in the ground or flaked soybeans of from 20 to 25%. Simultaneously with adjustment of the moisture content, steam is introduced into the preconditioner to raise the temperature of the soybeans to 170° to 180°F. The period of treatment in the preconditioner is relatively short, generally from 10 seconds to 60 seconds and preferably from 15 to 20 seconds.

Following the preconditioning step, the ground or flaked soybeans are fed to a cooker which preferably takes the form of a screw conveyor apparatus having means for injecting or sparging live steam into intimate contact with the soybeans. In the conveyor cooker the preconditioned soybeans are subjected to the action of steam under substantially atmospheric pressure for a period from 1.5 to 10 minutes, preferably from 2.5 to 3.5 minutes whereby cooking of the soybeans is accomplished. The temperature of the soybeans leaving the conveyor cooker ranges from 190° to 212°F. Product temperatures appreciably above 212°F have been found unnecessary to produce a satisfactory soybean food or feed product and are not employed.

Example: Raw soybeans were cracked into two to four pieces by putting them through smooth cracking rolls spaced so that the beans could not pass the rolls without being cracked. Moisture of this lot of beans was in the range of 6 to 8% and the urease activity 2+ units. Cracking in this manner permitted the hulls to be freed from the beans by aspirating the cracked mixture. The separated hulls comprised approximately 10% of the original weight of the beans. The dehulled, cracked beans were ground on conventional mill to a particle size passing a $1/16$ inch screen. The comminuted beans were subjected to a preconditioning treatment in a commercial continuous mixer employed for preparing animal feeds. The beans were fed into the preconditioner at a rate of approximately 700 pounds per hour. Water was added through a spray nozzle at a rate of 0.23 gallon per minute and steam was introduced into the preconditioner at a rate of approximately 200 pounds per hour.

The exposure time of the beans to this preconditioning treatment was approximately 15 seconds and the ground beans were discharged from the preconditioner at a temperature of about 177°F and contained 24.9% moisture. The urease activity of the preconditioned soybeans was 1.95 units, essentially the same as the raw soybeans. Urease activity is significant inasmuch as the conditions necessary to deactivate the urease are generally sufficient to effect improvement in the utilization of the protein content of the soybeans.

Following the preconditioning treatment, the soybeans were cooked in a screw conveyor into which steam at substantially atmospheric pressure was introduced through a plurality of injection points. Rotation of the worm was set so that the soybeans passed through the unit in approximately two and one half minutes. Exit temperature of the cooked material was 199°F and its moisture content 25.2%. The cooked soybeans were then dried to a final moisture content of about 10% on a pan dryer consisting of a perforated metal sheet through which heated air was passed. The urease activity of the dried soybean product was zero and the cooked material had an attractive straw color, a good flavor free of raw bean flavor, and was granular in nature. Feeding tests conducted with white rats showed that the cooked soybeans exhibited an excellent protein efficiency.

Steam Jet Dehulling and Debittering Process

A process has been disclosed by *C. Herzberg and W. Dollbaum; U.S. Patent 3,782,968; January 1, 1974; assigned to Holtz & Willemsen, Germany* for dehulling and debittering soybeans for the production of full fat flours. According to this process, raw not deoiled leguminous material is dehulled, partially crushed or disintegrated, exposed to a stream of water steam having an increased temperature, and subsequently ground or milled. The method is characterized by the steps of whirling the partially disintegrated leguminous material together with the overheated steam acting as a whirling medium having a temperature in the range of 110° to 150°C for a time period of 10 to 60 seconds in a whirl chamber containing a baffle or impacting device.

If treated material is at room temperature, it is advantageous to apply the overheated steam at a temperature between 180° to 200°C. It is possible however, to preheat the leguminous material to be treated to a temperature between 50° to 60°C and, subsequently, to expose the preheated material to water steam having a temperature in the range from 140° to 160°C. As an unexpected effect of this method, it has been found that by whirling intensively the partially disintegrated material at a relatively high temperature in the presence of water steam and, simultaneously, under the impacts of mechanical forces, the treated, not de-oiled leguminous material has become excellently refined. Without any losses in the quality, the final product is of a high biological value and digestibility, has neutral flavor and is free of bitter compounds, and undesirable enzymes. Figure 3.2 is a sectional side view of a whirl device.

FIGURE 3.2: STEAM JET DEHULLING AND DEBITTERING PROCESS

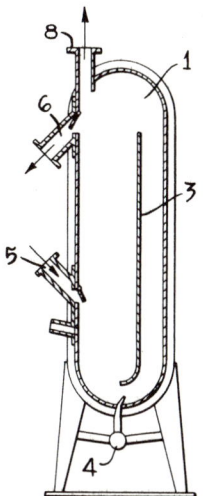

Source: C. Herzberg and W. Dollbaum; U.S. Patent 3,782,968; January 1, 1974

The operation of this apparatus is given in the following example. Water steam having the temperature of about 200°C is jetted through the nozzle assembly **4** into an isolated whirl chamber **1**, as shown in Figure 3.2. As a consequence of the expansion of steam after its passage through nozzles **4**, the temperature of steam falls to 135°C. This temperature corresponds to the working temperature of the process. Soybeans which have been halved, quartered, and preliminarily dehulled on corrugated plates, are fed through the inlet sluice **5**, at closed outlet sluice **6**, into the interior of the whirl chamber **1**. At the same time, overheated water steam at a pressure between 0.4 to 1 atmosphere is jetted into the chamber **1**. The amount of one charge of material to be treated is preferably about 3 kg. Provided that these conditions are secured, the time interval of the treatment of introduced material amounts to approximately 20 seconds.

At a rate of 4.5 kg/min treated material is discharged via sluice **6** at closed inlet sluice **5**. Without further treatment, except cooling, the material is forwarded by a conveying device (not shown) into a storing container whereby an output of 250 kg/hr of the dried soy product is attained. This soy product has been milled into flour having an average size of particles about 38μ. Resulting flour has been tested as to its chemical and biological qualities and the following average values have been ascertained:

Urease in N/g/min	0.0
Raw protein, percent	41.6
Raw oil, percent	21.6
Ashes, percent	1.9
Nitrogen-free extract stuffs, percent	26.2
Free fatty acid of the extracted oil, percent	0.9

(continued)

Peroxide number of the extracted oil	0
Iodine number of the extracted oil	129.3
Saponification value of the extracted oil	189

To illustrate the process, the values that are important for the evaluation are arranged in a comparative manner in the following table whereby: **1** means full fat soy flour from dehulled soybeans produced by the method of this process; **2** denotes a full fat soy flour produced from dehulled, but untreated soybeans; **3** is soy flour produced according to a method that is characterized by steaming the soybeans for 2 to 5 minutes, chilling the treated beans by cool water and, subsequently, by heating and drying the beans at a temperature between 70° to 75°C.

	1	2	3
Urease, N/g/min	0.0	6.6	0.87
Digestive protein, percent	40.9	33.9	36.4
Digestibility, percent	98.4	84.1	91.3

In the following table, there are disclosed data indicating the improvement of accessibility or of digestibility of amino acids that has been attained through the treatment of soybeans by this method, and those data are compared with the values of a conventional, untreated soy full fat flour as disclosed above under **2**.

Contents of Amino Acids in Enzymatic Hydrolytic Solutions

	Percent	
	1	2
Arginine	7.1	6.4
Histidine	0.9	0.8
Leucine	7.2	6.8
Isoleucine	4.6	4.4
Lysine	3.7	2.2
Methionine	0.9	0.7
Cystine	1.2	0.6
Phenylalanine	3.5	3.2
Threonine	3.6	3.3
Valine	5.6	5.4

DETOXIFICATION WITH ALKYLENE GLYCOL

Raw soybean meals used for animal feeds have been detoxified in a process developed by *A.C. Groschke; U.S. Patent 3,434,845; March 25, 1969; assigned to Agway Inc.* In place of cooking beans to detoxify them, it was found that if the ground raw soybeans are first blended with certain aliphatic diols the soybean toxins may be at least partly deactivated by subjecting them to high pressure. The process is characterized by the following steps: (a) raw soybeans are comminuted, by either flaking or grinding, to an average particle diameter (or thickness) of 10 mesh or finer; (b) the comminuted or ground raw soybeans are intimately mixed with 0.5 to 5% by weight of a diol having from 3 to 6 carbon atoms; (c) the mixture is then compressed under pressure for a time sufficient to improve the digestibility of the soybeans.

The degree of comminution of the raw soybeans is not critical so long as the soybeans are comminuted sufficiently to permit the diol readily to permeate the entire mass, and to be accommodated by the compression equipment. It is preferable that raw soybeans be ground as finely as possible. A typical soybean meal will have the following particle size distribution: 5% over U.S. Number 10; 30% over U.S. Number 20; 32% over U.S. Number 40; 11% over U.S. Number 60; 22% through U.S. Number 60.

The aliphatic diols which are appropriate for use in the process are liquids to aid in the incorporation into the raw soybeans. Moreover, the diols should be soluble in both water and fats. Propylene glycol and butanediol are the preferred diols. Substantially anhydrous diols have been found to be effective in the present method. However, the diol may contain significant amounts of water, so long as the amount of water is not so great as to vitiate the oleophilic characteristics of the diol. After blending the raw soybeans with the diol, the mixture is compressed sufficiently to detoxify partially the raw soybeans. It is estimated that compression should be carried out in an apparatus capable of subjecting the diol-soybean mixture to a pressure of several thousand psi for at least a second. Such conditions are obtained in conventional extrusion or pelletizing equipment. The presence of small amounts of moisture, combined with compression, further improves the process.

In a typical embodiment of the process, the ground, raw soybeans, after impregnating with an appropriate amount of a diol are processed in a pellet mill. In a typical pellet mill the meal is first mildly steamed in the feed supply hopper. From the hopper it falls into the mill which is in the form of a cylindrical shell having a plurality of perforations. This shell forms the die (each perforation being a separate pellet-forming die). The meal is forced outwardly through the perforations by one or more rollers acting against the internal surface of the shell, forming spaghetti-like strands which are subsequently chopped into pellets of suitable lengths. The meal being processed in an apparatus of this type may be subjected to pressures of as much as 50,000 psi, although this is only an estimate, as maximum pressures in such a die cannot be easily measured.

In practical operations, the extrusion die is operated at as high a rate as practical without choking the machine. Thus, after initially starting the machine the feed rate is increased to that point which is just short of the rate at which the machine will stall. A further limitation on the extrusion rate is the temperature of operation. Many meals employed in such equipment, including soybean meal, tend to scorch if the extrusion pressures are too high. Scorching is caused by compression and frictional heat generated when the meal is forced through the die passage, and for soybeans, occurs when the extruded pellets reach temperatures of 215° to 220°F. In a typical operation, the steamed soybean meal leaving the feed hopper of the pellet mill will be at a temperature of 180° to 185°F, while the temperature of the extruded pellets may range between about 190° to 200°F.

Example: A series of four soybean-based feeds were prepared of the following composition:

Ingredients (percent by weight)	Diet numbers			
	1	2	3	4
Ground yellow corn	51.36	51.36	51.36	51.36
Pelleted ground raw soybeans	41.00			
Pelleted ground raw soybeans with 1% propylene glycol		41.50		
Pelleted ground raw soybeans with 2% propylene glycol			41.75	
Pelleted ground raw soybeans with 3% propylene glycol				42.25
Alfalfa meal (20%)	2.0	2.0	2.0	2.0
Salt	0.25	0.25	0.25	0.25
Micronutrient Premix No. 1 [1]	0.50	0.50	0.50	0.50
Delamix [2]	0.05	0.05	0.05	0.05
d,l-Methionine	0.12	0.12	0.12	0.12
Dicalcium phosphate	1.70	1.70	1.70	1.70
Ground limestone	1.20	1.20	1.20	1.20
Miscellaneous growth factors	0.05	0.05	0.05	0.05
Cerelose 2001 [3]	1.77	1.15	0.68	
Alpha Cel [4]		0.12	0.34	0.52

[1] Micronutrient Premix No. 1 supplies the following per pound of finished feed: Vitamin A (U.S.P. units), 2,500; Vitamin D₃ (I.C. units), 500; Vitamin E (Int. units), 1.0; riboflavin (mg.), 2.5; calcium D-pantothenate (mg.), 5.0; choline chloride (mg.), 300.0; Vitamin B₁₂ (mg.), 5.0; niacin (mg.), 15.0; d,l-methionine (mg.), 227.0; santoquin (mg.), 113.5; menadione sodium bisulfite (mg.), 1.0; procaine penicillin (mg.), 2.5; zinc bacitracin (mg.), 2.5.
[2] Delamix provides the following trace elements in the finished feed: Manganese, 60.0 p.p.m.; iron, 20.0 p.p.m.; copper, 2.0 p.p.m.; cobalt, 0.02 p.p.m.; iodine, 1.2 p.p.m.; zinc, 50.0 p.p.m.
[3] Cerelose 2001 is employed as a source of dextrose.
[4] Alpha Cel is a fibrous cellulose employed as a source of inert fibers.

These foregoing formulas are designed as to provide an isonitrogenous and isocaloric diet. Accordingly, improvements in feed utilization efficiency or feed conversion, will represent improvements attributable to the use of the present method. The pelletized ground raw soybeans were processed in a standard pelletizing machine in which the ground raw soybean powder was sprayed with propylene glycol, steamed for about 15 to 30 seconds, and then compressed into pellets. In formulas 2, 3 and 4 it will be observed that the raw soybeans contain from 1 to 3% propylene glycol. These feeds were fed in a chick growth study to ascertain the effects of heat composition upon chicks' growth and feed utilization. The chicks were divided into groups of 10, and each feeding test was repeated three times. The following results were obtained.

Diet No.	Average gain (grams)[1]	Average feed conversion[2]
1	649.7	2.21
2	702.6	2.14
3	705.7	2.08
4	703.7	2.11

[1] Average gain per bird for three replications of ten birds each.
[2] Average of three replications. Each value represents pounds of feed to produce a pound of gain in live weight from day-old to six weeks of age.

APPLICATIONS OF FULL FAT SOY PRODUCTS

Soybean Beverage Powders

Full fat soybean powders useful in formation of carbonated beverages have been developed by *K.S. Lo; U.S. Patent 3,563,762; February 16, 1971; assigned to Hong Kong Soya Bean Products Co., Ltd., Hong Kong.* The soybean bottling powder is rich in protein, fat and vitamins and yet extremely economical in cost. This product may be used as a dietary complement or supplement.

Figure 3.3 is a flow diagram showing the essential steps in the production of a full fat soybean powder for use in preparing beverages.

FIGURE 3.3: PROCESS FOR PREPARING A SOYBEAN BEVERAGE

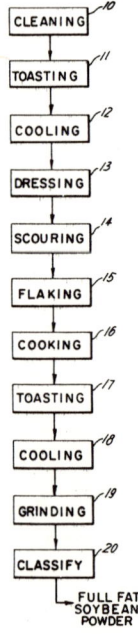

Source: K.S. Lo; U.S. Patent 3,563,762; February 16, 1971

In the process illustrated in Figure 3.3, the whole soybeans are first cleaned, **10**, to remove dirt, splits, and other foreign matter. After cleaning, the beans are heated or toasted, **11**, in a horizontal toaster, which may be a wire screen conveyor heated by a hot air flow for a period approaching five minutes. The hot air may be maintained at 300°C and is used to remove excessive moisture from the beans prior to dehulling. After heating, the beans are immediately transferred by the wire screen conveyor into a cooling chamber wherein the beans are cooled, **12**, by currents of cold air. The abrupt drop in temperature is such that the hull shrinks away from the meat and dehulling is facilitated. After cooling, the soybeans are transferred into a conventional dressing and blending machine wherein the impact action of paddles removes the hulls from the meat, **13**. The next step is to separate the beans from the hulls, **14**, and this is done in a conventional scouring apparatus where the soybeans are agitated to permit aspiration of the hulls from the top of the unit while the heavier meat falls to the bottom of the unit.

The dehulled soybeans are next fed into a hammermill in order to compress them into the form of flakes, **15**. Once flaked, the soybeans are then cooked in a Wenger expansion cooker. Within the cooker the flakes are first preconditioned with steam at a temperature of 212°F to establish a moisture content of 18 to 21%. In this condition, they are extruded through ¼ inch die openings and come out as pellets. The cooking is performed at high pressure in order to prevent boiling. It will be noted that the Wenger cooker is particularly advantageous for use in the production of full fat soy flour for use in beverages. In effect, the extrusion output renders the unit a continuous pressure cooker and the product of this process has the growth inhibitors of the soybean removed. In addition, it provides a flour that does not have the objectionable bitterness that is inherent in soybeans. Rather, the powder is bland in taste and entirely suitable for the addition of flavoring materials in order to produce any kind of beverage that will suit local markets.

The removal of growth inhibitors from the powder renders the subsequently produced beverage more digestible. This is an important factor inasmuch as the beverage is particularly valuable for its nutritional value and as a protein supplement. The pellets that are provided at the output of the Wenger cooker are again toasted, **17**, in order to reduce the moisture content to a point between 3½ and 4%. Following toasting, the beans are again cooled, **18**, rapidly by bringing the temperature to a level of approximately 90°F. The now dry pellets are ground, **19**, into a fine powder, e.g., in an Alpin pinmill, and classified, **20**, in order to obtain a final full fat soybean powder of approximately 270 to 300 mesh. The resulting powder has substantially the following composition:

	Percent
Protein	45
Fat	20
Ash	5
Fiber	3
Other carbohydrates	23
Moisture	4

The full fat soybean powder produced by the above method is further treated by cooking and high pressure homogenizing to form the beverage product. Centrifuging the homogenized product will give a beverage powder having lower carbohydrate content.

Full Fat Soy-Cereal Products

Improved particulate cereals are produced by *H.H. Kaufmann and J.F. Lawrence; U.S. Patent 3,141,776; July 21, 1964; assigned to Cargill, Incorporated* using ground full fat soybeans as the binder. It was found that highly nutritional full fat soybean when incorporated with other cereal grain or, when used by itself has highly effective binding properties suitable for use in the preparation of cereal particles which readily absorb moisture but which do not undergo substantial disintegration.

Any cereal grain that is conventionally used as an animal or human cereal feed, i.e., corn, oats, rice, wheat, sorghum, etc. can be used in the process. The grain is initially in fine particulate form, preferably in the form known as mash. To the cereal grain or mixture of grains, is added, if not already present, ground full fat soybean, either cooked or raw, in particulate form, preferably small enough to pass through a 100 mesh screen and in an amount sufficient to bind the cereal effectively together during the subsequent processing steps. The soybean concentration should be at least about 10% by weight of the cereal. It has been found necessary that all of the soybean oil naturally occurring in soybeans be present in the soybeans for use, in providing the desired binding properties. The soybeans may be raw, that is, uncooked. Instead, if desired, full fat natural soybeans which have been previously subjected to a heat treating or cooking procedure can be used. In any event, the soybeans must be subjected to a heat treating step prior to utilization of the cereal product.

The heat treating should be at a temperature below 270°F and, furthermore, at a moisture concentration initially in the soybeans of at least 8%, and the final concentration at least 7%, so as not to materially depreciate the nutritional value, particularly the protein dispersibility. In addition, the heat treatment will assure inhibition of toxic factors in the soybeans, including the soyin and trypsin inhibitors, and minimizing the raw beany flavor in soybeans. It is important that the heat treatment be carried out within the prescribed limits and that infrared radiation be used so that substantial depreciation in the nutritional value and oil content of the soybean does not occur. The following example illustrates certain features of the process. A wet feed mix having the following formula is made up in a paddle mixer: 25.8 lb corn, ground; 14.2 lb raw, ground, full fat soybeans; 3.0 lb water.

The mixing operation is carried out in a paddle mixer in which the mixture resides about 30 seconds. The mix is then passed to a premixing screen fitted into the entrance of the open ended revolving 36 inch diameter drum. The screen and drum are rotating at about 26 rpm and the screen has $3/8$ inch diameter perforations therein. The wet feed is passed through the screen, approximately 50% immediately forming particles. The length of the drum is approximately 6 feet. The granules roll along the inner surface of the drum and are rounded into smooth balls or spheres. The drum is supported with its axis at an angle of about 5 degrees from the horizontal to facilitate the granule shaping operation. The shaped granules exit from the drum onto a vibrating conveyor and are immediately passed under a bank of infrared radiation heaters utilizing Swank ceramic grids operating at about 1650°C.

As the granules on the vibrating conveyor belt pass under the infrared radiation heaters, they are dried from a moisture concentration of about 30% by weight down to a moisture concentration of between 10 and 12%, by weight, at a temperature of 246°F. The conveyor vibrates in order to minimize agglutination of the granules before the drying step. By the end of the drying step, the particles are sufficiently dried so that agglutination does not occur. The particles are then cooled and passed to storage. The finished cereal food is of distinctive appearance, has a suitable toasted flavor without a beany soybean flavor, substantially is free of toxic factors, and upon testing, exhibits a protein dispersibility of at least 55%. The particles are easy to handle, resist crumbling in the dry state, yet absorb moisture rapidly without disintengration or agglomeration so that feeding immediately after mixing with water can be carried out. They can also be eaten dry.

DEFATTED SOY PRODUCTS

Defatted soybeans are usually produced in the form of flakes, grits, meals or flour. The fat or oil has been removed by extraction or pressing leaving about 0.6% fat and a 40 to 60% protein content.

OIL EXTRACTION METHODS

Defatting Soybeans with Aqueous Alkali Metal Sulfites

Defatted soybeans, even if the various improved treatments are practiced, are inferior to the raw soybeans both in water-soluble protein content and quality (the property of the proteins, color, flavor, etc.). The process of *H. Watanabe; U.S. Patent 3,454,404; July 8, 1969; assigned to Showa Sangyo KK, Japan* for the manufacture of defatted soybeans discloses treating soybeans with an aqueous solution of alkali salts (for example, sodium or potassium salt) of sulfurous acid, hyposulfurous acid, or bisulfurous acid previous to or during the process of extraction of the soybean oil whereby the solution is absorbed in the soybeans.

By this process, the reductive, antioxidizing action of these alkali salts prevents the degeneration and coloration and other deteriorations of soybean proteins during and after the process of extraction and during the storage of defatted proteins. Suitable percentages of the above reducing agents in the soybeans which are to be subjected to extraction, are between 0.1 and 0.5% by weight.

When the percentage of reducing agent in the soybeans is less than 0.1 to 0.5%, the effects described above cannot be achieved. When the percentage of reducing agent in the soybeans is more than 0.1 to 0.5% the —S—S— bridges of the soybean proteins are decomposed by reduction and an excessive formation of —SH radicals happened and the essential properties of the soybean proteins are degenerated.

Suitable percentages of water in the aqueous solution of reducing agent, added to the soybeans, is from 1 to 2%. The addition of the aqueous solution of reducing agent to the soybeans should be done previous to or during the extraction. Crushed, or casted soybeans, are preferable because the reducing agents act effectively on the proteins contained in soybeans and the soybeans are not degenerated by heating.

The defatted soybeans produced by this process can be used extensively for the manufacture of foods, for example, breads, cakes, tofu, edible kneaded products, miso (edible fermented soybean paste), soy, adhesives for making veneer boards, coating and sizing materials for papermaking, etc. Soybean proteins made from the defatted soybeans of this process can be used as a constituent of drugs and much better products in higher yields can be obtained than the products made from defatted soybeans produced by the above described prior art methods.

Example: To crudely crushed and pared raw soybeans is sprayed sodium sulfite solution in which is contained 0.2% by weight of sodium sulfite to soybeans and 1% water. Thoroughly agitate so that the soybeans uniformly absorb the above solution, and subject them to the treatment of heating, flaking and drying etc. The product is fed to an extractor equipped with a vapor desolventizer, and the oil extracted by a solvent, i.e., hexane desolventized, and cooled. The defatted soybeans contained 7.8% of water and 91.0% of water-soluble proteins. The contents of those components are equal to those of raw soybeans, and the product is tinged with a slight yellowish green, and had a good flavor.

Contrarily, defatted soybeans made from soybeans, which were untreated with reducing agents before or during extraction, and other steps of the treatment which are equal to the treatment of this process contain 8.3% of water and 85.7% of water-soluble proteins and are tinged with brown color.

Defatting Soybeans Using Ethanol

Soybeans defatted commercially by former methods contained a residue of bound fat, principally phospholipid, amounting to about 1% or less. It is this residual bound fat that is responsible for much of the residual off-flavor or beaniness and instability in the product. While certain of the prior procedures have been partially successful in eliminating the characteristic bitter soybean taste, the defatted residue still contains an undesirable mouth-coating factor (or substance), also described as a lard-like taste which sticks in the throat. *K.H. Steinkraus; U.S. Patent 3,721,569; March 20, 1973; assigned to Cornell Research Foundation, Inc.* now has found that the residual bound fat including the mouth-coating factor along with remaining bitterness can be removed by extraction with ethyl alcohol together with or followed by chloroform.

It is essential that the alcohol extraction precede or at least accompany the chloroform extraction step. The reason for this is that the alcohol treatment results in a loosening of the bond between bound fat, principally phospholipids, and the soybean protein enabling the mouth-coating factor, phospholipids and other undesirable flavor-bearing lipids to be extracted by the chloroform. The ethyl alcohol also extracts certain bitter principles from the soybeans contributing to the resulting organoleptically bland flavor while yielding a defatted soybean with an elevated protein content. Tests have shown that extraction with ethyl alcohol alone, chloroform alone, or even chloroform followed by alcohol will not result in removal of the bound fat including the mouth-coating factor. Sequential application of ethyl alcohol followed by an application of chloroform or an ethyl alcohol-chloroform mixture results in a substantially complete removal of bound fat including the mouth-coating factor. It is not necessary to remove the ethyl alcohol between the two steps.

In a preferred method, the ground soybeans are extracted for about 2 hours with 95% ethyl alcohol. During this time not only are the alcohol solubles removed from the soybeans, but the bean materials are heated to 60° to 65°C thereby promoting the liberation of bound lipids. A second extraction step using chloroform or preferably a 1:1 (v/v) chloroform to ethyl alcohol mixture is carried out for about twenty-two hours or until residual lipid has been extracted. When using such a mixed solvent system, it is not necessary to dry or otherwise remove all of the alcohol prior to the introduction of chloroform into the extraction system. The spent solvents after extraction are recovered by conventional methods, e.g., by distillation, for reuse in the extraction procedure.

Example: The extraction of pulverized, unheated, unpressed Harasoy soybeans with 95% ethyl alcohol for 2 hours followed by extraction with a 1:1 (v/v) mixture of chloroform-

Defatted Soy Products

ethyl alcohol for twenty-two hours in a Soxhlet apparatus was compared with the method of U.S. Patent 1,297,668 to Erslev wherein extraction with a fat solvent is followed by extraction with alcohol. In this comparative example, the same apparatus and extraction times were employed in both sequences. Thus, the differences summarized below are attributed to the order in which the solvents were applied and not to the fact that warm solvents were utilized.

	Present process, ethanol/ ethanol+ chloroform, percent	Erslev process, chloroform/ ethanol, percent
Total solids extracted	26.2	24.1
Solids extracted in first step	20.6	20.2
Solids extracted in second step	5.6	3.9
Nitrogen content (defatted product)	8.79	7.92

It should be noted that this method extracts more total solids but leaves a product higher in nitrogen, i.e., protein, than that produced by the Erslev process. This method yields a bland, almost tasteless meal whereas the Erslev process yields a meal with a cereal-like flavor. Moreover, aqueous extracts of the nonfat soy solids produced by this process are nearly colorless, very bland in taste and do not coat the mouth whereas similar extracts prepared from Erslev processed soy solids are light tan in color, cereal-like in flavor and coat the mouth badly. The defatted meal produced by this process is useful in preparing soy milk products.

SOLVENT REMOVAL PROCESSES

Apparatus for Steam Treating Defatted Meal

N.F. Kruse; U.S. Patent 2,776,894; January 8, 1957; assigned to Central Soya Company, Inc. has designed an improved apparatus for treating solvent-extracted soybean meal. This apparatus provides means by which vapors and steam leaving the lower kettles could be effectively utilized in the uppermost kettle without causing meal portions to be carried into the condenser, as was the case in prior apparatus. This can best be shown in the following figure.

An elongated vertical casing or treating chamber **10** is supported upon standards **11** carried by the base **12**. A chute **13** extends into the upper portion of the casing for supplying meal. The top of the casing has an outlet **14** leading to a condenser or other suitable apparatus.

Below the topmost kettle, which will be described in detail later, there are a series of vertically-spaced kettles **15**. Each of the kettles has a side wall **16** preferably provided with a manhole **17** and a cover plate **18**, and each of the walls is preferably cut away on opposite sides to provide vapor and steam passages **19**. The side wall of the kettle is closed at the bottom by a hollow steam jacket **20**. A flow passage or discharge passage, which may be in the shape of an inverted funnel, is provided by the casing portion **21** so that meal may constantly flow from one kettle into a kettle below.

Over each jacketed bottom **20** is mounted a sweep **22** fixed by a collar **23** and a setscrew **24** to the shaft **25** which is hollow to permit the passage of steam from the supply line **32**. Following each blade of the sweep in the upper kettles is a steam line **26** having at its rear steam discharge openings **27** so that as the sweep rotates, steam is delivered at the rear of the sweep blades.

Steam may be introduced into each steam-jacketed bottom by any suitable means. As shown in Figure 4.1, a steam line **28** leads from a steam source and enters the jacket on one side while a steam outlet or condensate line **29** leads from the opposite side.

FIGURE 4.1: APPARATUS FOR STEAM TREATING DEFATTED MEAL

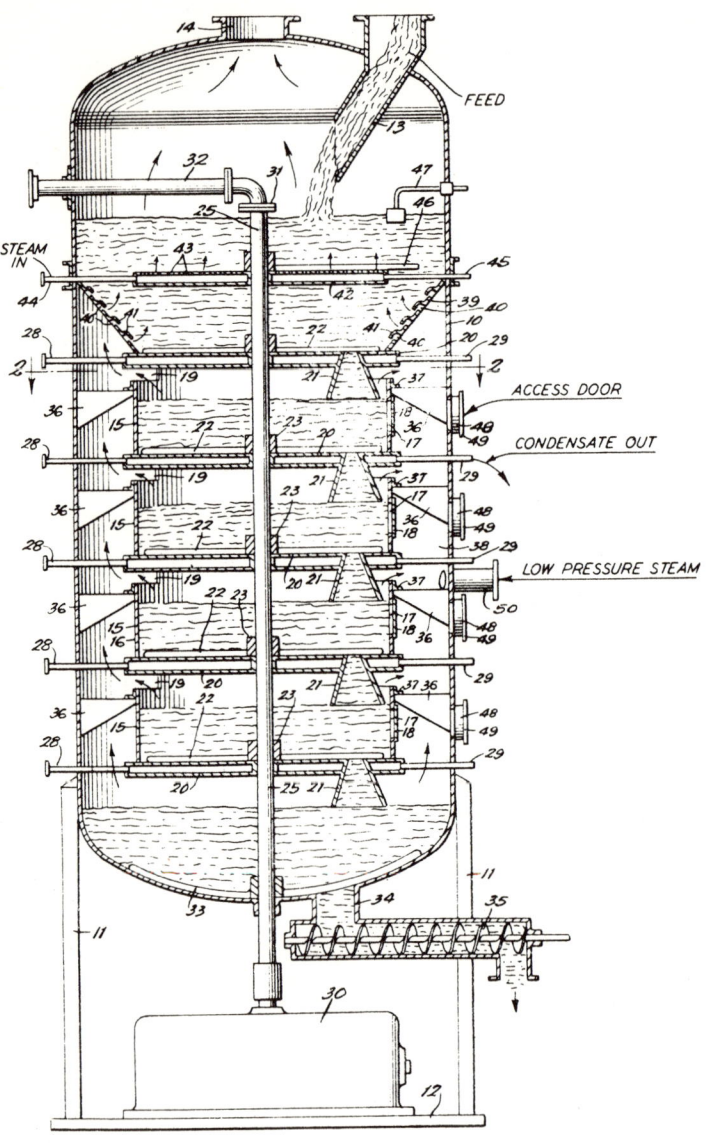

Source: N.F. Kruse; U.S. Patent 2,776,894; January 8, 1957

Defatted Soy Products

The shaft **25** is driven by a motor and reduction gears mounted within the casing **30**. A steam connection for the interior of the rotatable pipe **25** is provided at **31**, and steam enters the pipe or shaft through the fixed steam inlet pipe **32**.

At the bottom of the treating chamber, there is a curved sweep **33** which is effective in sweeping the meal into the discharge pipe **34** through which the material is moved by a variable speed screw conveyor **35** regulated to discharge the same quantity of material being fed into the top of the unit, thus maintaining proper fillage in the kettles.

Within the casing bracket supports **36** are welded or otherwise secured to the casing at their outer ends and engage flanges **37** on the kettle walls **16**, whereby the several kettles are supported in spaced relation within casing **10**. The outside of the kettles thus have an annular flow passage **38** for the passing of steam and vapors from various kettles below upwardly toward the first kettle.

The first kettle has a jacketed bottom **20** just as the other kettles, but instead of a vertical side wall, the kettle has an apertured side wall **39**, which extends outwardly and upwardly in the general shape of a funnel. The funnel wall **39** has a large number of openings **40**, each covered by an inwardly and downwardly-extending lip **41**. The lip **41** provides a shield which prevents the escape of meal into the annular passage, while at the same time directing upwardly-passing steam and vapors within the annular passage **38** inwardly within the body of meal contained in the uppermost or first kettle. By using the extensive area provided by the side walls directly above the annular passage, vast amounts of steam can be used in the kettles below and then passed through the body of meal in the first kettle without tending to carry meal particles into the condenser.

About midway of the body of meal in the first kettle, there is preferably a steam box or jacket **42**, provided with a plurality of steam openings **43**, so that a large amount of steam can be discharged into the upper layer of the meal as it enters the chamber for the early removal of solvent from the incoming flakes. The box **42** is provided with a steam inlet pipe **44**. A sweep **46** is secured to the shaft **25** that is used for moving the meal immediately above the box **42**. If desired, a float control device **47** may be used to aid in keeping the meal body at a desired level within the top kettle.

The casing **10** may have manholes **48** and doors **49**, and these are preferably aligned with the kettle openings **17** so that access can be readily had for repair and other purposes from the exterior of the casing, when this is desired. If desired, exhaust or low pressure steam or mixed steam and solvent vapors may be introduced into the annular chamber or passage **38** through the line **50** or other suitable pipe connection.

In operation, feed is fed through the chute **13** into the top kettle, where it meets immediately steam from the box **42** for the removal of solvent. At the same time, steam is delivered into the mass from the pipe following the sweep **22** over the jacketed bottom of the kettle. In addition to the steam thus introduced into kettle **1**, a large volume of steam and vapors from the kettles below is fed through the side openings **40** of the funnel-shaped sides **39**. The meal in the upper part of kettle **1** or the top kettle is at a temperature below 212°F. The incoming steam is effective in removing solvent from the flakes while at the same time steam is condensed upon the flakes to bring the moisture content up to from 14 to 30%, or above.

The vapors and steam coming up through the annular passage **38** and entering the mass of meal in the first kettle is particularly effective in supplying moisture to the individual flake surfaces in a metering type of action which distributes the moisture uniformly on the flake particles. As the meal descends through the outlets **21** from kettle **1** and successively through the kettles therebelow, the temperature of the meal is raised to a point above 212° to bring about a cooking or toasting of the meal. The heat for the meal within the lower kettles is provided by the jacketed bottoms **20**. If greater heat is desired, the side walls **16** of the kettles may be jacketed and supplied with steam in the same manner as the bottoms **20** are supplied with steam and this provision may also be made on the chamber **10**.

In the heating or cooking operation, a desirable top temperature is between 225° and 265°F, preferably 300°F. The cooking is desirably, and more conveniently, done at atmospheric pressure, but higher or lower pressures may be employed.

The number of compartments may be varied considerably, depending upon the type of oil meal or other material being treated and upon the results desired. It is possible to operate the process upon solvent extracted oil meal from which the solvent has been already removed, but in this instance it is preferred to add water or other liquid to the meal before it is subjected to the treatment described above.

Desolventizing and Toasting Apparatus

An apparatus and method for treating extracted soybeans flakes have been disclosed by G.L. Lippold; U.S. Patent 3,126,285; March 24, 1964; assigned to Dannen Mills, Inc. This permits removal of the solvent from soybean flakes from which the oil has been extracted, utilizing a continuous operation which does not require close operator control and which is designed to remove the solvent and convert the food product to palatable form in a single unit requiring a minimum of moving parts and power per unit quantity of material processed. Figure 4.2 is a verticle cross-sectional view of the apparatus.

The operation of the apparatus is as follows. Valves **129** and **125** are opened to permit steam to pass through the respective pipes and thereby into the interior of each of the wall structures **40** through header conduits **130** and into the upper tubular end of shaft **62** for conveyance into the interior of compartment **46**. Liquid formed between each of the wall structures **40** and caused by condensation of the steam, passes out through pipes **136** and drain conduit **138** and preferably to the boiler of apparatus **10** (not shown) to minimize losses of water during operation of the equipment.

Solvent extracted material such as soybean flakes are directed into compartment **46** through conduit **38** and collect on the upper surface of plate **42** of the uppermost wall structure **40** defining the bottom of compartment **46**. Upon introduction of material **74** into compartment **46**, motor **66** is actuated to cause shaft **62** to be rotated counterclockwise and cause all of the blade units **92** to be rotated within compartments **46** to **58**. The steam introduced into wall structure **40** forming the bottom of compartment **46** maintains the material therein at a specified temperature. It is to be recognized that the temperature to which material **74** is subjected within compartments **46** to **58** should be varied for different products to produce the most palatable and nutritious meal.

Steam is directed against the flakes within compartment **46** as the upper blade unit **92** is rotated by shaft **62**, and it has been found that the quantity of steam introduced into compartment **46** should be correlated with the rate of flaked material **74** passing into compartment **46**, with best results obtained if the moisture content of material **74** is raised preferably within a range of 15 to 20% with the latter percentage producing the best results when extracted soybeans are being processed.

The incoming steam removes the hexane solvent from the flakes, passes upwardly through pipe **34** and with the steam then condenses on the flakes to raise the moisture content to the stipulated value, which is necessary to produce a nutritious and palatable product.

As the material is agitated within compartment **46**, a part of the same falls through the upper chute **72** into compartment **48** whereupon the material is subjected to the heated wall structure **40** forming the lower part of compartment **48** which serves to either raise the temperature of the flaked material or maintain the product at the same temperature as that of the product at delivery into compartment **48**. The solvent remaining in the material **74** directed into compartment **48** is volatilized and passes upwardly into compartment **46** through passages defined by members **86** in the uppermost wall structure **40**. In this manner, the material **74** passes successively through compartments **46** to **58** and at a uniform rate, regardless of the quantity of material which is directed into shell **12** through conduit **38** per unit of time.

FIGURE 4.2: DESOLVENTIZING AND TOASTING APPARATUS

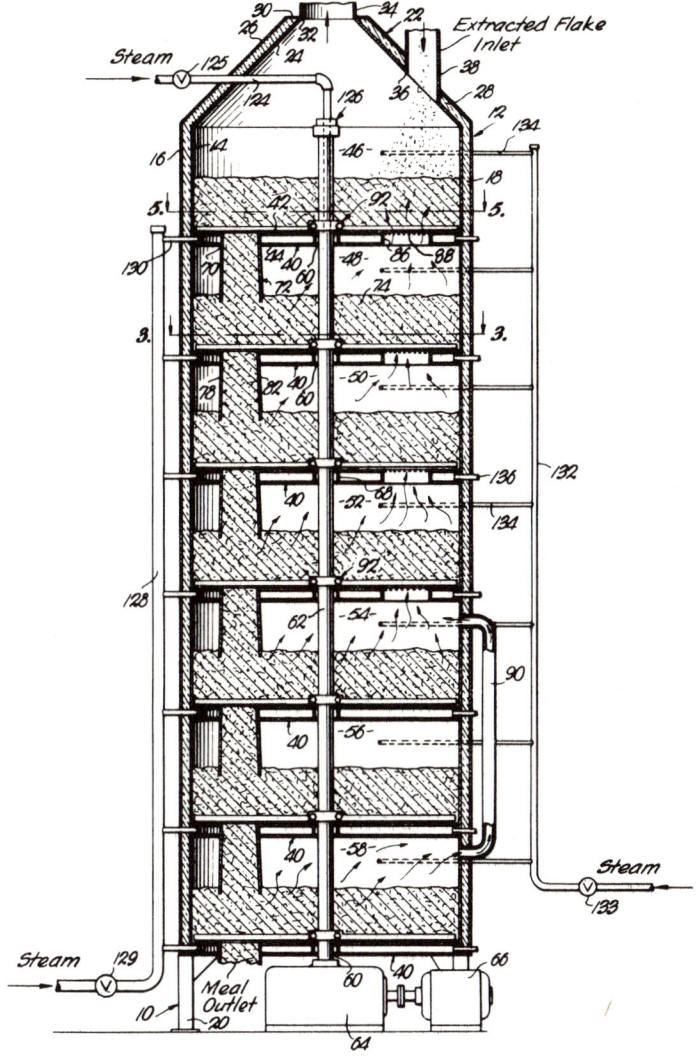

Source: G.L. Lippold; U.S. Patent 3,126,285; March 24, 1964

Raising of the temperature of the flakes in compartment **48** to the same or preferably a somewhat higher level removes substantially all of the small remaining amount of solvent in the flakes, while final desolventizing is effected within compartment **50**. The material in compartment **52** is tempered for toasting of the flakes, bringing the latter up to a predetermined heat for deodorizing and removing volatile oils and other unwanted materials. The heat is this compartment deadens the phosphatides and waxes, with some moisture remaining in the material but completely devoid of the hexane solvent. The flakes are subjected to the initial toasting temperature in compartment **54**, with this compartment being vented to compartment **52** through the vapor passages in the wall structure **40** in order to permit some moisture in the form of water vapor to be removed and permitting initial toasting of the flakes.

When the material **74** passes into compartment **56**, final toasting of the same takes place therein by virtue of the fact that this chamber is not vented to the chamber thereabove and thus, the material is subjected to a higher temperature as well as greater pressure of the order of 6 to 8 inches Hg to produce a nutritious and palatable meal by virtue of heating of the soybean meal in the presence of a limited amount of moisture.

The toasted meal **74** then gravitates into lower compartment **58** where the product is substantially dried before passing outwardly of shell **12** through meal outlet chute **72** and the conduit. The moisture removed from the meal **74** within compartment **58** is returned to compartment **54** through bypass pipe **90** which thereby maintains the moisture content of the material **74** within compartment **54** at the proper level for beginning of the toasting operation. The utilization of a substantially closed chamber such as **56**, wherein the toasting of the flakes is effected, has been found to be important in erasing the urease activity of the soybean flakes while at the same time, producing a much more palatable and nutritious product, notwithstanding heating of the same to a somewhat higher temperature than heretofore employed.

ENZYME AND YEAST DEBITTERING PROCESSES

Flavor Improvement Using Yeast

The flavor improving treatment developed by *R.W. Bradof; U.S. Patent 2,930,700; March 29, 1960; assigned to The Wander Company* involves subjecting the defatted soy flour in aqueous medium to the action of yeast. Preferably, the yeast is used together with an added oxidizing agent or a selected alkaline neutralizing agent or both. As described later in greater detail, the preferred method involves the use of all three reagents and in addition a preliminary treatment with a reducing agent, such as sodium bisulfite or sodium nitrate, and a final step to remove excess oxidizing agent. The process is carried out under controlled and relatively mild conditions of temperature and time so as to get the flavor improvement without alcoholic fermentation or protein degradation.

Although the reasons for this flavor improvement are not entirely understood, a reasonable hypothesis is that the aldehydes and ketones or other organic materials responsible for bad flavor are oxidized under the catalytic influence of the enzyme content of the yeast to form organic acids having less objectionable organoleptic properties. Such organic acids are probably neutralized by reaction with the alkaline neutralizing agent when the latter is employed. In addition, the evidence indicates that yeast also produces other beneficial effects, particularly carbohydrate degradation or hydrolysis of the polysaccharide molecules in the soy flour.

Many different strains of yeast may be employed, but ordinary household yeast and brewers yeast have been found to be suitable for this process. Although the amount of yeast used is not particularly critical, it has been found that effective results can be obtained with as little as 0.5% by weight based on the weight of soy flour and no particular advantage is gained by exceeding about 8% by weight.

For the oxidizing agent, most effective results are obtained with peroxide compounds, such as hydrogen peroxide and sodium peroxide. In addition, certain of the peracids and their salts, e.g., peracetic acid and potassium persulfate, may also be used. Oxygen (either alone or with other gases such as in air) or ozone may also be used as the oxidizing agent. However, none of the various oxidizing agents are significantly more effective than hydrogen peroxide, and in view of economy and maximum safety for food use, 30% hydrogen peroxide is preferred. The amount of peroxide employed, particularly hydrogen peroxide, may

Defatted Soy Products

be from 0.1 to 1% by weight based on the weight of soy flour, and preferably from 0.3 to 0.7% by weight.

For the alkaline neutralizing agent, the hydroxides and other compounds of sodium, potassium, ammonium, magnesium, and calcium have utility, but the ammonium, magnesium, and calcium compounds are preferred. The most suitable neutralizing agent from all considerations is ammonium hydroxide. Mixtures of calcium and magnesium compounds such as are found in commercial hydrated limestone are also useful because of their cheapness and ready availability.

Various reducing agents can be used in the preliminary debittering process including sulfur dioxide or compounds capable of releasing sulfur dioxide, particularly sulfurous acid and various bisulfites such as sodium bisulfite. Sodium nitrite has also been found to be effective.

In addition to the preliminary treatment with a reducing agent as described above, the preferred embodiment of the process also includes a final step of decomposing or removing any excess peroxide which may remain in the mixture following the debittering treatment. This is desirable to preclude the formation of oxidation products during storage or subsequent processing of the soy flour. The simplest and most effective way of removing hydrogen peroxide is by the addition of catalase which is an enzyme having specific catalytic action for the decomposition of hydrogen peroxide.

Example 1: A treating medium was prepared by adding 13.5 grams of sodium bisulfite and 900 ml of one normal ammonium hydroxide solution to 100 pounds water at 160°F. Fifty pounds of cooked, untoasted solvent extracted soy flour and 25 ounces compressed household yeast (crumbled) was thoroughly mixed with the treating solution. The temperature of the digestion mixture is kept at 140° to 145°F in a jacketed kettle for twenty minutes. A solution containing 63.0 ml of 30% hydrogen peroxide in approximately 500 ml of water is added to the digestion mixture with thorough agitation. The mixture is then maintained at 140°F for an additional 20 minutes. A solution of 13.0 ml catalase in about 500 ml of water is then added and thoroughly agitated to eliminate any excess hydrogen peroxide. The product may now be drum dried or used as such as a bland flavored soy flour ingredient for further formulation.

Example 2: A treating medium was prepared by adding two grams of hydrated dolomitic limestone, having a calculated content of 42.2% by weight of calcium oxide and 30.9% by weight of magnesium oxide, to 2 pounds of water followed by the addition of 1.25 ml of 30% hydrogen peroxide. One pound of cooked untoasted solvent extracted soy flour was thoroughly mixed with the treating solution and a slurry containing one-half ounce of compressed household yeast was then added to the mixture. The digestion mixture was heated in a jacketed kettle at 140°F with agitation and maintained at that temperature for thirty minutes. The final product was then dried.

Taste panel tests on the final treated product revealed that the product had an entirely acceptable bland flavor such that it could be readily included as a major ingredient of a food product without any objectionable taste problem.

H.-C. Chien; U.S. Patent 3,810,997; May 14, 1974; assigned to Kraftco Corporation has also developed a method for the treatment of vegetable protein with yeast to improve the flavor. An aqueous dispersion of a ground vegetable material containing protein and carbohydrates is treated to provide a substrate relatively free from microbiological contamination. A viable culture of a particular yeast or mixture of yeasts is then added to the substrate. The yeast is selected so as to be capable of utilizing at least a portion of the carbohydrate materials found in the vegetable protein source. Fermentation of the inoculated substrate is then effected under conditions which minimize production of alcohol. As used here fermentation refers to the process of effecting yeast growth under conditions which substantially inhibit production of alcohol. It is generally desirable to conduct the fermentation under aerobic conditions which suppress alcohol formation. After the fermentation is completed, the substrate is heated to inactivate the yeast. The fermented substrate may then be further treated so as to recover the vegetable protein and the yeast. During the fermentation, the yeast utilizes a portion of the carbohydrate in the vegetable protein and the yeast culture multiplies. The total protein of the system increases due to conversion of carbohydrates to yeast, which is substantially proteinaceous in composition. When this method is used to treat a soybean protein, it is a surprising result that the beany taste of the soybean is removed and a bland tasting finished product is obtained. Both the soybean vegetable protein material and the yeast have unpalatable taste. It is, therefore, a surprising and synergistic effect that a bland tasting finished product is obtained as a result of the treatment. The yeast may be any yeast which utilizes the carbohydrate of the soybean. Particularly preferred yeasts are *S. cervisiae, S. fragilis, S. carlsbergensis, Candida utilis, Candida tropicalis* or mixtures thereof.

It is preferred that the soybean protein be in the form of a soy flour for use in the process. In this connection, full-fat, low-fat and defatted soy flour may be used. Details of the process are given in the following example.

Example: Defatted soy flour, which had been produced by a solvent extraction process, was dispersed in water to provide an aqueous dispersion having 10% soy flour. The particular soy flour used is Nutrisoy 7B (Archer-Daniels-Midland Co). The aqueous soy flour dispersion was then pasteurized by heating to 160°F for five minutes. Then a viable culture of *Saccharomyces cervisiae* yeast was added to the aqueous dispersion of the soy flour. The yeast culture had 3% yeast solids in an aqueous dispersion, and 2% by weight based on the weight of the aqueous dispersion of the yeast culture was added. 0.4% of ammonium sulfate and 0.1% of potassium phosphate (dry basis), based on the weight of the aqueous dispersion, were also added.

The inoculated soy flour dispersion was then fermented for four hours at 86°F. The aqueous dispersion was air agitated during the fermentation. At the end of the fermentation, the carbohydrate and alcohol content of the dispersion on a dry basis was determined. It was found that the soy flour dispersion had 0.08% alcohol by weight.

The aqueous soy flour dispersion was then heated to 170°F for no hold time so as to pasteurize the dispersion and inactivate the yeast. The dispersion was then treated with acid to precipitate the protein and was dried to provide a dry soy flour material. The soy flour material had 64% protein of which approximately 3% was determined to have been contributed by the yeast. The treated flour is useful in the production of soy milk and cheese-type foods.

Enzymatic Debittering Methods

An improved enzymatic treatment of soybean curd or defatted soybean flour for removal of beany flavor have been developed by *M. Fujimaki, H. Kato, S. Arai, and M. Yamashita; U.S. Patent 3,585,047; June 15, 1971; assigned to The U.S. Secretary of Agriculture.* In this treatment hexane-defatted soy flour or soy curd is subjected to 2 to 6 hours of selective digestion by a microbial protease selected from the acidic aspergillopeptidase-A produced by *Aspergillus saitoi,* the neutral protease produced by *Aspergillus oryzae,* and the alkaline protease produced by *Bacillus subtilis,* or the carboxypeptidase-A isolated from bovine pancreas. The liberated or degraded astringent and beany principles and the also liberated hexanol, hexanal, saponin, and reversion-susceptible bound lipids are removed by extracting with highly aqueous ethanol.

These workers had previously found that the free amino acids and particularly the peptides liberated from soy protein by extensive hydrolysis with pepsin comprise significant amounts of leucine, isoleucine, phenylalanine, and valine. These amino acids per se are known to have a bitter flavor, which bitterness is even more intense in the concurrently formed or liberated diffusible peptides, analysis of which showed the C- or N-terminals thereof to be mostly composed of leucine. The unobviousness of using these specific enzymes is established by the following table which shows that corresponding treatments of soybean protein with the listed proteolytic enzymes or combinations were either incapable of removing both the astringent and beany tastes, or they produced an undesirable color or a different, e.g., salty taste.

Protease	Treated Hydrolysate Fraction
Coronase	Bitter
Rapidase	Bitter but free of astringent taste in 6 hours.
Pronase	Bitter and astringent (beany) taste.
Bioprase	Bitter and astringent (beany) taste.
Bromelin	Bitter and astringent (beany) taste.
Takadiastase-SS	Bitterness remained.
Mixture of Takadiastase-SS and Rapidase	Bland taste but brown discoloration.
Thermoase	Bitterness remained.
Mixture of Aspergillopeptidase-A and Thermoase	Salty taste.

Example 1: Ten grams (dry basis) of soybean curd suspended in 1,000 ml of dilute HCl (pH 2.8) was digested for 2 hours at 50°C with 0.1 gram of commercially obtained aspergillopeptidase-A, produced by *Aspergillus saitoi* (Molsin). The enzymatic digestion was then stopped by neutralizing with NaOH, and the precipitate obtained by filtration or centrifugation was treated with a 10 fold weight of 50% ethanol to remove the liberated beany and astringent factors; the wash liquid was removed by filtration, and the treated curd was lyophilized. Compared with identically treated soybean curd, excepting for the addition of the enzyme, the product was found to be free of odor and almost completely free of beany and astringent flavor (completely free when the enzymatic digestion was extended to to 4 hours).

Furthermore, the presence in an ether extract of the supernatant from the above neutralization step of greatly increased percentages compared with an untreated control extract of phospholipids, phosphatides, n-hexanol, genistein, ninhydrin positive substances, and carbonyl compounds, e.g., n-hexanal, confirms the specific enzymatic liberation of apparently implicated adverse flavor and autooxidation instability components present in the untreated soybean curd.

Example 2: Twenty grams of the defatted soy flour substrate was treated with 900 ml of dilute HCl (pH 1.5) for 2 hours at 30°C. The centrifuged supernatant was adjusted to a pH of 2.8 and a volume of 1,000 ml with dilute NaOH and then incubated for 2 hours at 50°C with 80 mg of the same commercial enzyme (Molsin) used for Example 1. The incubation was terminated by lyophilization, and the so-produced powder was treated with a 10 fold weight of 90% ethanol to remove the freed adverse flavor components, lipid materials, and carbonyl compounds. The recovered soy flour was dried at room temperature under reduced pressure. It exhibited the same organoleptic, olfactory, and oxidative stability improvements exhibited by the enzyme-treated soy curd product of Example 1.

Proteolytic Enzyme Treatment

A method of producing a controlled partial hydrolysis of defatted soy flour or meal is disclosed by *A.L. Liepa; U.S. Patent 3,687,687; August 29, 1972; assigned to Procter & Gamble Company.* The partial hydrolysis of soy flour is accomplished by mixing the soy flour, proteolytic enzyme and water for a length of time at an elevated temperature. Unexpectedly, when soy flour is reacted with water in the presence of a proteolytic enzyme and thereafter processed into a cold cereal product the product that results is more tender with less tendency to develop toughness and has greater crispness retention than does a cold cereal product containing soy that has not been subjected to a partial hydrolysis reaction.

A greater degree of hydrolysis of the soy flour over that obtained by this method results in a product having an unacceptable taste. No hydrolysis of the soy flour or only a partial hydrolysis less than that achieved by following the reaction conditions of this method gives an unacceptable tasting product as well as a poorly processable ingredient.

The proteolytic enzymes useful in the present method can be selected from several known proteolytic enzymes or their mixtures extracted from animal, plant, fungal, or microbial sources. A primary consideration in the enzyme or enzyme mixture used is that it must not contribute a significantly objectionable flavor or odor to the final product. Some examples of proteolytic enzymes found effective in the soy flour partial hydrolysis step are papain, pepsin, bromelin, ficin, alcalase, maxitase, thermoase, pronase, and mixtures thereof.

The amount of enzyme added to the reaction mixture is 25 to 2,500 ppm based on the dry weight of the soy flour. Greater amounts can be used but exert no measurable beneficial influence on the speed of the hydrolysis reaction. Lesser amounts can be used but are not because of the length of time it would take for the reaction to proceed to the desired end point. A preferred range of enzyme addition is 100 to 600 ppm based on the dry weight of the soy flour.

The temperature of the reaction is not critical provided the temperature is not so high as to kill the enzyme activity. That is, for proteolytic enzymes, temperatures in excess of 180°F have the effect of destroying the activity of that enzyme. A temperature range of 80° to 160°F is suitable for the partial hydrolysis reaction with the reaction proceeding faster at the higher temperatures. The most preferred temperature range is 120° to 130°F.

The time for the reaction to be completed depends on the temperature, water level, and enzyme level used. In general, 1 to 120 minutes are sufficient to allow the reaction to come to a completion. Times in excess of 120 minutes should be avoided in order to prevent more hydrolysis of the soy flour than is desired. Preferably, the hydrolysis reaction is allowed to occur for from 1 to 5 minutes.

Example:

Ingredients	Grams
Soy flour	92.0
Dextrose	3.0
Papain	*0.05

*543 ppm of soy flour.

The dextrose and papain are added to 130 grams of water and blended to form a dispersion. To this dispersion is added the soy flour thereby forming a mixture of 58% water. After mixing, the mixture is held at 120°F for 5 minutes. Next the mixture or dough is passed through an extruder under 750 psig and having a die outlet temperature of 170°F. The extrudant is in the form of strands having a diameter of about three-sixteenths of an inch. These strands are next cut into pellets having a length of about three-sixteenths of an inch.

These pellets are then passed through a two roll mill to produce a flake shape product having a thickness of about 0.008 inch. These flakes are partially dried to a moisture content of 12% and then puffed. A rapid heating method of puffing is used wherein the flakes are contacted with salt at a temperature of 330°F for 10 seconds. After wetting with milk the flakes are evaluated.

Flakes made by the above process and formulation with the exception of no enzyme addition are used as a control. They are tough, leathery and hard to chew after exposure to water. Flakes made by the above formulation and process are considerably less tough than the control and disintegrated fairly easily in the mouth. The flakes are also more porous, less dense, and accordingly better puffed than the control flakes.

After the soy flour has been partially hydrolyzed, various processes used for making ready-to-eat breakfast cereal are used for converting the partially hydrolyzed soy flour or dough to the final product desired.

In the preferred process, the partially hydrolyzed soy flour is extruded into strands of a relatively small cross-sectional area and then sliced into small lengths forming small pellet-like particles. These particles are next partially dried, if necessary, and formed into flakes. The flakes are then puffed to transform them into less dense and more porous or tender flakes. Toasting and/or coating may be used to enhance the color and/or flavor of the resultant high protein cereal product.

Cereals can be prepared from the soy flour above, but a cereal grain such as corn, oats, wheat, or rice can be mixed with the partially hydrolyzed soy flour in such proportions that a cold cereal having a protein content greater than 20% is produced.

OTHER FLAVOR-ENHANCING PROCESSES

Acid Digestion to Remove Flavor and Color

Bland, odorless, color-free soy flours are obtained from soy flours, preferably solvent extracted, by use of the acid digestion process disclosed by *R.J. Moshy; U.S. Patent 3,126,286; March 24, 1964 and U.S. Patent 3,168,406; February 2, 1965;* both assigned to General Foods Corporation. In the first process a bland, odorless soybean flour may be prepared by mixing 2 to 100 parts (all parts by weight) of water with one part of defatted, cooked, commercial soybean flour to form a slurry, 5 parts of water to one part of soybean flour being preferred for ease of handling. Sufficient acid is added to adjust the pH to the isoelectric point of the soy proteins, that is, pH 4 to 6, preferably pH 4.5. The product obtained at pH 4 to 6 is higher in protein and the flour easiest to handle since protein hydration and swelling are at a minimum.

The acidified slurry is then heated to 70° to 212°F, preferably 122° to 176°F for optimum color and flavor removal. The slurry is digested, i.e., maintained, at such temperatures for 1 to 30 minutes, and preferably 10 minutes when the temperature used is 176°F. The slurry is digested for a period long enough to insure complete and intimate contact of all the soybean surfaces with the liquid so that maximum beany and bitter flavor and odor removal may be effected. Periods in excess of 30 minutes at elevated temperatures cause a substantial increase in the degradation and cause an increase in the bitterness of the soy flour produced. The supernatant liquid is then removed from the digested soybean flour by filtration, centrifugation, or any other common means of separation. The digested soybean flour remains as a filter cake which is further processed. The time and temperature of digestion are in inverse relationship to each other.

The resultant filter cake is washed 1 to 4 or more times by forming a slurry with 3 to 100 parts of water per part of soybean flour originally used. Preferably 2 to 10 parts of water per part of soybean meal are employed, 5 parts to 1 being preferred for ease of handling. The temperature of the wash water is maintained above 167°F and preferably above 200°F. Temperatures above 212°F may be employed during the washing step, the maximum temperature being dependent solely upon the particular combination of time, pH and temperature employed, the only limiting factor being the degree of degradation which is acceptable in the final product. Preferably, each wash is carried out for 5 to 60 minutes. The supernatant liquid is removed from the slurry after each wash by any common means of separation. While it has been found with most wash waters that it is not necessary to adjust the pH of the resultant slurry, the pH of the slurry should be maintained between pH 4 and 6.

If a product having a neutral pH is desired, sufficient alkali may be added to the slurry just prior to the last wash to adjust the pH to 6 to 8 and preferably 6.8 to 7.2. However, satisfactory products are obtained if the pH of the product is between 4 and 9.2. After the last wash, the supernatant liquid is removed from the slurry by drum drying, spray drying, air drying, freeze drying, oven drying, and the like, and preferably spray or freeze drying, spray drying being most preferred for economic reasons. A more preferred embodiment of this process is where the slurry of washed, digested soybean flour is steam distilled by steaming, i.e., passing steam through the slurry, until the weight of vapor condensate collected equals 7 to 40 times the original weight of the soy flour. The pH of the filter cake prior to distillation is adjusted to pH 4 to 6 and preferably pH 4.5, if necessary.

The steam distilled product may then be neutralized by adjusting the pH to 4 to 9.2 and preferably pH 6 to 8 and then drying by freeze drying, drum drying, spray drying, air drying, oven drying, and the like. It has also been found that the bitter and slightly beany flavor and odor which may result when steam distillation is employed can be avoided by steam leaching a slurry of soybean flour and acidified water.

Example 1: One pound of cooked, defatted, commercial soybean flour was mixed with five pounds of water and to this mixture was added sufficient concentrated hydrochloric acid to adjust the pH to 4.5. The resulting slurry was heated to a temperature of 176°F and maintained at that temperature for ten minutes. The digested slurry was filtered in a vacuum filter and the resultant filter cake was washed five times with five pounds of water at 176°F. Each wash was carried out for 10 to 15 minutes and the wash water removed from the filter cake by filtration before the last washing. After the last wash the slurry was drum dried on 6" diameter stainless steel rolls internally heated with 25 psig steam and rotating at 2 rpm. The resulting product was a cream-colored, bland, odorless soybean flour.

Example 2: Three hundred grams of 120 U.S. Standard mesh cooked, defatted, commercial soybean flour were mixed with 500 ml of water having a pH of 6.6. The pH of the mixture was adjusted to pH 4.5 with 28.0 ml of 6 N hydrochloric acid. The slurry was placed in a steam leaching apparatus comprised of a cylinder having a perforated glass plate at the bottom covered with a filter cloth upon which was placed the acidified water-soybean flour slurry. Steam was passed through the acidified water-soybean flour slurry from beneath and the vapors produced by such steaming carried off to a condenser. The steam vapor which condensed in the acidified water-soybean flour slurry environment was permitted to drain through the filter cloth and perforated glass plate, being aided to some extent by the application of a vacuum.

The steaming was carried out for two hours during which time 2 liters of leach water (the water draining through the perforated glass plate) and 1 liter of condensate were collected. The digested soybean flour having a pH of 5.1 was adjusted to pH 6.0, the water being filtered from the soybean flour and then dried. The product was bland, odorless and light tan in color. Two hundred ten grams of the product were obtained.

In the second process the water washing of the digested product is replaced, at least in part, with a water-miscible solvent. Such treatment is particularly effective when the soybean flour is not readily filterable or separable such as when a very fine mesh meal is being used or when an uncooked, defatted, soybean flour which swells during the treatment is being used. In such cases the soybean flour slurry treated at about pH 4 to 6 is diluted with at least an equal volume of a water-miscible alcohol and then the supernatant liquid is removed from the slurry by any common separation means.

The digested soybean flour which is in the form of a filter cake is then mixed with a water-miscible solvent which is volatile, polar and organic in nature. While any water-miscible solvent such as ethers like 1,4-dioxane; ether alcohols like propylene glycol monoethyl ether; water-miscible esters like ethyl lactate may be used to remove the undesirable bitter and beany flavor and odor, it is preferred to use water-miscible alcohols such as ethyl, isopropyl, methyl, tertiary butyl, etc., since these do not impart any flavor of their own to the product. While generally water-miscible solvents may be employed during solvent extraction or leaching at a concentration of 60 to 100% solvent, it is preferred when employing water-miscible alcohols to use them at a concentration of 80 to 100%. In the case of the preferred alcohol, ethyl alcohol, it has been found that alcohol concentrations of 80 to 100% are eminently satisfactory although it is preferred to use the azeotrope for economic reasons.

Example 3: Two hundred pounds of cooked, defatted, commercial soybean flour having a moisture content of 4 to 10% was mixed with 800 pounds of hot water and seven pounds of concentrated hydrochloric acid was added to obtain a pH of 4.5. The resulting slurry was treated for one hour at 180°F. The treated slurry was filtered in a vacuum filter and 400 pounds of filter cake containing 60% moisture was obtained. The filter cake was reslurried and washed in 600 pounds of water and held for 30 minutes at 180°F. The slurry was again filtered through a vacuum filter and 400 pounds of filter cake was obtained. The filter cake was mixed with 360 pounds of 99% isopropyl alcohol and the slurry treated for one-half hour at 160°F. The slurry was again filtered through a vacuum filter and 350 pounds of filter cake was obtained. The filter cake was again treated with 360 pounds of 99% isopropyl alcohol and maintained at a temperature of 160°F for one-half hour. Sufficient 50% sodium hydroxide solution was added to the hot slurry to adjust the pH to 6.0 (when measured as a 10% aqueous slurry of the soy flour).

The slurry was refiltered and the resulting filter cake slurried with 450 pounds of 99% isopropyl alcoyol. The slurry was heated to 100°F and maintained at such temperature for one-half hour. The treated slurry was again filtered and 260 pounds of filter cake was obtained. The filter cake was dried in a vacuum pan dryer at 28" of vacuum, the temperature within the dryer being maintained at 110°F. One hundred and forty pounds of product was obtained. The product was bland and odorless.

Acid and/or Peroxide Treatment of Soy Flour

Three processes have been disclosed by *T.M. Paulsen; U.S. Patent 3,100,709; August 13, 1963; U.S. Patent 3,361,574; January 2, 1968; and U.S. Patent 3,361,575; January 2, 1968;* all *assigned to Archer-Daniels-Midland Company* for improving the flavor and odor of soy flours. In the first process, the flour is prepared by simultaneous or stepwise treatment of relatively dry particulate soybean material with water-soluble protonic acids and/or their water-soluble ionizable salts in combination with hydrogen peroxide and then drying. Drying is effected in any conventional or suitable manner to provide modification of the treated soybean material without necessitating separation of the constituents of the particulate soybean material. This chemical treatment surprisingly removes or improves odor, flavor and mixing properties without impairment of baking and baked qualities of the finished product.

In a practical plant application the treatment is applied to conventionally prepared defatted and desolventized soybean flakes as they come from the processing unit at a temperature of about 190° to 210°F.

The acids used are water-soluble and may be either organic or inorganic and are protonic as defined by the Bronstad-Lowry theory. As indicated, the salts of these water-soluble organic or inorganic acids must also be ionizable. The acids and salts are preferably suitable for human consumption. For preparation of edible chemically treated soybean products, such protonic acids and their salts are exemplified as: sodium chloride, calcium chloride, acetic acid, citric acid, phosphoric acid, hydrochloric acid, and the like, one or more of which is used in combination with the hydrogen peroxide. The concentration of the acid and/or its salt, and the hydrogen peroxide, may be from 0.5 to 5 parts and preferably 1 to 2 parts of each based on 100 parts of the protein-containing vegetable material. The monovalent metal salts, if used alone, are in concentrations of over 0.5% and preferably in combination with a divalent metal salt.

In addition, this process includes the treatment of particulate soybean material, and mixtures thereof with other particulates of protein containing vegetable material, with other nutritionally required or essential minerals, such as iron, copper, manganese, cobalt, zinc, and the like which are applied inclusively as water-soluble salts, of protonic acids in combination with the hydrogen peroxide treatment. The overall result produces a similar effect of change in water dispersibility of the protein and a change in other physical factors.

Example 1: A solution of calcium chloride and hydrogen peroxide was sprayed onto hexane extracted soybean flakes at 70° to 80°C while blending in such a manner, i.e., in a baffled rotating drum or high-speed mechanical blender, so as to obtain uniform distribution of the solution. The treated soybean flakes were air dried to equilibrium moisture and ground to pass 100% through a 200 mesh screen. The proportions, hereinafter expressed in parts by weight, for chemical treatment were:

	Parts
Soybean flakes	100
$CaCl_2 \cdot 2H_2O$	3
H_2O_2 (50%)	1
H_2O	15

Example 2: A solution of citric acid and H_2O_2 in water was sprayed onto hexane-extracted soybean flakes at 70° to 80°C with thorough agitation such that the solution had maximum opportunity for even distribution on the dry soybean product. Increasing the moisture content to about 20% established more ideal conditions for migration of the aqueous solution throughout the flakes. The chemically treated soybean flakes were allowed to air dry to equilibrium moisture, ground and classified to 100% through a 200 mesh screen. The components for treatment were:

	Parts
Soybean flakes	100
Citric acid	3
H_2O_2 (50%)	1
H_2O	15

The high content soybean products are surprisingly edible without the usual bitter soy taste and odor and the cooking qualities have not been altered in the manner characteristic of toasted soybean material.

It has also been found that the chemically treated soybean products, as described, can be used at normal and high levels in the bread baking industry. The bread is prepared in a conventional formulation with from 2 to 20% of the combined chemically treated soybean material added or substituted for the usual bread flour. In addition, the preferably treated soybean material can be used in making soybean cereal and other foods with a suitable edible binder.

In the second process, the protonic acid alone is used to treat the flour. Soybeans, such as in flakes, at high solids content, are contacted with either solutions or protonic acids, solutions of salts of the protonic acids or solution mixtures of the acids and salts, under temperatures of 65°C to not over 100°C, for a period of a few second to not over 15 minutes, with controlled moisture content at least sufficient to moisten, e.g., about 7% but not in excess of 20% in addition to equilibrium moisture, and dried in a manner to effect removal of soybean taste and soybean odor without detrimental effect to normal physical and functional characteristics and without detrimental color change to produce a final light-colored food product for human consumption.

Example 3: Full-fat soybean flakes (7% moisture) were extracted to remove the oil using a treating solution of glacial acetic acid in hexane. The defatted flakes were desolventized by steam stripping under vacuum, i.e., 90° to 100°C flake temperature and 10 inches of Hg vacuum, air dried to equilibrium moisture and ground to pass 100% through a 200 mesh screen. Proportions for chemical treatment were:

	Parts
Full-fat soybean flakes	100
Glacial acetic acid	3
Hexane	1000

In the third process, only the peroxide is used to treat the flour. One hundred parts of soybean particulates are treated with 0.25 to 5 parts of a water-soluble peroxide, such as hydrogen peroxide, sodium peroxide or urea peroxide added to the soybean particulates which contain not over 20% moisture in excess of equilibrium moisture; the treated particulates are held at a temperature of 65° to 100°C for a period of uniform treatment of 30 sec to 15 min; and the particulates are dried by cooling until equilibrium moisture is achieved.

Example 4: A solution of H_2O_2 in water was sprayed onto dehulled, hexane-extracted soybean flakes at 80° to 100°C while blending in a manner, i.e., in an insulated, baffled, rotating drum or a high-speed mechanical blendor, so as to obtain uniform distribution of the solution on the flakes. The treating temperature was held at temperature for five minutes and the treated soybean flakes were air dried to equilibrium moisture and ground to pass 100% through a 200 mesh screen. The proportions for chemical treatment were:

	Parts
Soybean flakes	100
H_2O_2 (50%)	2
H_2O	8

Example 5: Same as Example 4 but dehulled, full-fat soybean flakes were used instead of hexane-extracted soybean flakes.

Example 6: Same as Example 4 down to proportions, except time for heat treatment was 2½ minutes instead of 5 minutes.

	Parts
Soybean flakes	100
H_2O_2 (25%)	4
H_2O	4

Countercurrent Alcohol Extraction

G.C. Mustakas and E.L. Griffin, Jr.; U.S. Patent 3,023,107; February 27, 1962; *assigned to U.S. Secretary of Agriculture* have found that commercial hexane extracted soybean meal can be completely debittered with very little further denaturation by extracting the meal countercurrently with a solvent selected from 95% ethanol and 91% isopropanol for 18 to 36 minutes at 24° to 38°C, draining the meal for at least 5 minutes, and desolventizing the drained meal by subjecting it for not more than 5 seconds to fluidization in a superheated (149° to 158°C) closed system vapor stream, the vapor stream comprising solvent, air, moisture, and added carbon dioxide being superheated by indirect steam at 115 to 120 psi. Fractional removal of the vapor stream and almost instantaneous recovery of the meal particles for the necessary rapid cooling may be conveniently accomplished with any suitable cyclone separator. A convenient apparatus for desolventizing the meal particles in a vapor stream comprising superheated solvent is shown by Brekke et al, *Journal American Oil Chemists' Society* 36, 256 (1959).

If absolute ethanol is used in this process, only partial removal of the objectionable bitter beany flavor is obtained, while with more dilute alcohols than 95% ethanol or 91% isopropanol denaturation is greatly increased.

Example 1: Commercial hexane-defatted soybean flakes having a nitrogen solubility index [percent dissolved total nitrogen after agitating 2½ parts of ground defatted soybean for two hours in 100 parts of water at 25°C (NSI)] of 81.1 and a moisture content of 9.72% were fed to a commercial horizontal extractor approximately 20 feet long with a plurality of slotted 12 inch diameter paddle wheels to provide twenty mixing and separating stages. The soybean flakes were fed into the extractor at a rate of 25 pounds per hour along with 95% ethanol fed countercurrent to the flakes at a rate of 37.5 pounds per hour. The temperature in the extractor was maintained at 38°C and the residence time of the flakes was 18 minutes.

The wet flakes, which now had an NSI of 75.8 and contained about 6% moisture and 37% ethanol, were drained on a drag-link conveyor for 5 minutes before being fluidized by introduction into the 52.2 ft/sec velocity recycle vapor stream of a closed system flash desolventizer, the vapor stream comprising air, moisture, ethanol, and diluent carbon dioxide (which was added to retard denaturation) the vapor stream being heated by indirect steam at 115 to 120 psi pressure to a temperature of 156.7°C. After a critically limited residence time not exceeding 5 seconds in the desolventizer tubes, the flakes were removed from the vapor stream by a cyclone separator and cooled in a stream of cool air. The bland final product had an NSI of 69.71 and contained 1.8% of residual ethanol.

Example 2: Commercial defatted soybean flakes having an NSI of 78.8 and a moisture content of 7.85% were treated as in Example 1 with the exception that 91% isopropanol was substituted for the 95% ethanol at a temperature of 24°C during a residence time of 36 minutes. After draining for 6.5 minutes the flakes had an NSI of 76.6 and contained approximately 37% isopropanol. The flakes were then fluidized in a 149°C vapor stream comprising isopropanol, air, moisture, and added carbon dioxide, the velocity of the vapor stream being 49.8 ft/sec and the average residence times of the particles being 2 to 5 sec. The particles recovered in a cyclone separator had a temperature of 82°C and were then quickly further cooled in a stream of air. The bland final product had an NSI of 68.9 and contained 1.95% of residual isopropanol.

Use of dilute alcohol solutions for debittering soybeans is also disclosed by *G.C. Mustakas and E.L. Griffin, Jr.; U.S. Patent 3,268,503; August 23, 1966; assigned to U.S. Secretary of Agriculture.* This is a process which employs more dilute alcohols as well as higher temperatures and increased fluidization times than those employed in U.S. Patent 3,023,107 to produce substantially denatured soybean protein having a water absorption capacity at 88°C of about 350 to 400 ml per 100 grams of the protein product.

Previously defatted aqueous alcohol-moist soybean flakes that have been drained following slurrying in a 50 to 70% solution of methanol, ethanol, or isopropanol in water to remove the bitter factors and the extensive amount of soluble carbohydrate material, may be freed of their absorbed aqueous alcohol without inducing degradation and other adverse processing side effects such as balling-up, scorching, and poor moisture absorbing ability apparently resulting from the joint action of the large amount of water and the high temperature required for volatilizing the latter, by fluidizing the damp flakes in a suitable desolventizer for 7 to 9 seconds at about 370° to 375°F (about 190°C) to obtain soybean flakes that will absorb about 3.5 to 4 times their own weight of water at 80°C.

The critical desolventization phase of this process is conducted at 190°C in an indirectly heated desolventizer resembling that of Brekke et al, *Jour. Am. Oil Chemists' Society* 36, 256 (1959) excepting for the insertion of an inverted U-section in the horizontal duct which adds four elbows, thus functionally practically doubling the overall length of the steam-jacketed desolventizing duct so that the blower-induced critical fluidization time of the aqueous alcohol-wet soybean flakes prior to their discharge at a conventional cyclone separator is 7 to 9 seconds.

Whereas 100 grams of the original defatted soybean flakes absorbed only 250 ml of water at 80°C and had a bulk density of 55 lbs/cu ft, 100 grams of the product obtained by the instant process absorbed about 350 to 400 ml of water under the same conditions and had a bulk density of about 46 lb/cu ft. When flakes drained from 70% aqueous ethanol were slowly desolventized in a heated vessel comprising internal agitation means, the resulting soybean flakes were partly caked and scorched. They had a bulk density of 64 lbs/cu ft and 100 grams absorbed only 275 ml of water at 80°C.

Example 1: Defatted soybean flakes were slurried in 50% methanol at room temperature for 30 minutes. The drained wet marc containing about 31.8% moisture was introduced at the rate of 10 lb/hr into the 61 ft/sec velocity recycle vapor stream of an inverted U tube-containing closed system flash desolventizer, the recycle vapor stream comprising air, moisture, methanol, and diluent carbon dioxide being heated by indirect steam at 255 psi pressure at a temperature of 373°F. After an average residence time of 9 seconds in the desolventizer tube, the essentially dry flakes were removed from the vapor stream by a cyclone separator and cooled in a stream of air. One hundred grams of the thusly treated soybeans were found to absorb 368 ml of water at 80°C. The bland flakes assayed 74.2% protein and had a nitrogen solubility index [(water-soluble protein/total protein) x 100] of 4.1%.

Example 2: Defatted soybean flakes were slurried in 50% ethanol at room temperature for 30 minutes. The drained wet marc containing 32.6% moisture and 16.9% ethanol was introduced at the rate of 10 lb/hr into the resolventization apparatus of Example 1 excepting that the recycle vapor stream was indirectly heated to 376°F and had a velocity of 72.4 ft per sec. After a residence time of 7 seconds, the cyclone-recovered air-cooled flakes having a nitrogen solubility index of 5.07 and a protein content of 77.0% were found to absorb 3.68 times their weight of water at 80°C.

Example 3: Defatted soybean flakes were slurried in 50% aqueous isopropanol at room temperature for 30 minutes. The drained wet marc containing 35.6% moisture and 21.5% isopropanol was introduced at the rate of 20 lb/hr into the desolventization apparatus of Example 1 excepting that the recycle vapor stream was indirectly heated to 380.2°F and the stream had a velocity of 68.4 ft/sec. After a residence time of 8 seconds, the cyclone-recovered air-cooled flakes having a nitrogen solubility index of 4.96 and a protein content of 72.3% were found to absorb 3.48 times their weight of water at 80°C.

MODIFICATION OF DEFATTED SOY FLOURS

Heat and Humidity Conditioning for Soy Food Supplements

E.J. Salbego; U.S. Patent 3,775,542; November 27, 1973 has developed a treatment process for soybean flour which includes a dry-heat period which either kills or renders satisfactorily dormant undesired organisms within the stored-food product without raising the temperature to such a level as to denature desired enzymes and proteins.

This improved soybean feed for livestock is obtained by subjecting raw (uncooked) soybean flour, with or without soybean oil present, to a temperature of 35° to 50°F, preferably 45°F, for at least three weeks and preferably at least four months, and exposed to an atmosphere having a relative humidity of at least 60% and preferably 95%. During or after the above processing the raw soybean flour may also be mixed with other feed material such as soybean meal, shelled corn, oats, baled alfalfa and the like.

While raw soybean flour, made by grinding soybeans as they come from the field, may be used it is preferred to obtain meal after the oil has been extracted by a process where the soybeans are not heated to an extent to produce substantial changes in their composition (i.e., substantially uncooked), and this meal is then ground without substantial heat to a flour. The oil in such raw soybean flour generally ranges from about two to five percent by weight of the flour.

In a preferred embodiment, the raw soybean flour either alone or mixed with soybean meal or other food products is first maintained at a temperature of 70° to 104°F, preferably 75° to 100°F, in a surrounding atmosphere of air or inert gas of approximately the same temperature and at a relative humidity of less than 20% and preferably only 5% for at least 3 weeks so as to bring about substantial drying of the flour. This dry heat treatment also appears to kill or render dormant unwanted and undesirable living organisms. While this first step is not essential it is desirable and greatly preferred.

The next step is the essential incubation period. In this incubation period the relative humidity of the surrounding atmosphere is maintained at at least 60% and preferably 95% while the temperature of the atmosphere and the flour is held at 35° to 50°F, preferably 45°F. This incubation period is at least three weeks and preferably for about 4 months or longer to obtain maximum growth of the desired living organisms which are believed to cause the growth-promoting results obtained with feed containing soybean flour so treated. The incubation period, in the preferred treatment, is followed by drying at, for example, 45°F and 55% relative humidity for several days (three or more days, preferably twenty-five days or more).

Following the drying period, a storage period with the relative humidity being 50 to 60% and the temperature about 35°F is preferred. The treated material can be stored indefinitely under those temperature and humidity conditions.

A further feature is that a relatively small amount (such as a few grams) of a food-supplement composition suffices to show a beneficial effect when consumed by an adult human or by an animal of about the size or weight of an adult human. The treatment is applied to bagged soy flour in storage rooms or chambers where temperature and humidity can be controlled by any means. Use of the treated soy product in feed mixtures is shown in the following table.

	Mixture														
	1	2	3	4	5	6	7	8	9	10	11	12	13	14	15
	Gr.	Gr.	Gr.	Gr.	Gr.	Gr.	Gr.	Gr.	Gr.	Gr.	Gr.	Gr.	Gr.	Gr.	Gr.
	Bd.	Gd.	Pr.	Gd.	Gd.	Gd.	Gd.	Pr.	Pr.	Pr.	Pr.	Pr.	Gd.	Pr.	Pr.
Treated product	25	84	25	25	17	24	85	75	80	87	83	81	88	86	84
Lecithin	50	42	20	10	8	11	40	36	40	46	41	39	46	44	40
Oatmeal	200	24	15	10	5	9	25	21	20	28	22	20	28	26	25
Cornmeal	50								10						
Nonfat dry milk										18	20	14	26		
Dry Yeast										7		7	7		
Gelatin										9	18		18	18	9
Nicotinic acid							1	1	1	1		1	1	1	1
Vitamin A								X	X	X					X
Total	325	150	60	45	30	44	151	133+	151+	196+	184	162	214	199	159+

Note.—X=300,000 U.S.P. units.

In the above table, the numbers opposite the listed components, appearing in the columns 1 to 15, represent grams as indicated by Gr. at the head of each column. The parts, of course, could be given in ounces, pounds, and the like, so long as indicated proportions are maintained. Extensive tests have shown which mixtures are good Gd., which are poor Pr., and which is bad Bd., as regards serving its intended purpose. The columns are labeled accordingly.

One way of utilizing any useful mixture, listed in the above table, for example, is to give it directly to the human or animal recipient, a level tablespoon (for example) to an adult human or to an animal about the size of an adult human. Alternatively, the composition in the desired amount may be mixed with and eaten with other food. If desired, the composition may be wetted to an adhesive state and made into small, medium, or large balls or pills, glazed with sugar if desired, or even chocolate coated.

Increased Water Solubility of Soy Flour Using SO_2

The water solubility of soy flour has been increased by a process developed by *R.R. Cooke, J.E. Hunter and R.W. Mitchell; U.S. Patent 3,542,562; November 24, 1970; assigned to The Procter & Gamble Company*. It has been found that by exposing soy flour to SO_2 for a period of time the soy flour becomes appreciably more water-soluble. Solubility of soy flour so treated can be increased to a point of substantially equalling that of milk solids in prepared culinary mixes, (i.e., cake mixes) specifically, the process of treating soy flour involves: (a) providing a moisture content of the soy flour of from 5 to 20% by weight of soy flour; (b) treating, at temperatures of from 60° to 150°F, the moisturized soy flour of step (a) with from 0.5 to 4.0 grams of SO_2 per pound of soy flour.

The soy flour used as a starting material is prepared by grinding soybeans into a meal-like substance and then extracting the fatty oils with an appropriate solvent such as hexane. The resulting residue is run through a heating and steaming process to remove any remaining

Defatted Soy Products

traces of hexane. The temperature during this heating and steaming process is from 230° to 260°F. This is continued for from one-half to one hour.

Prior to the SO₂ treatment the soy flour is preferably pulverized or ground, preferably to a particle size that will pass through a 100 mesh U.S. standard screen. The ground soy flour is then placed in a conventional blender. Any blender is suitable, for example, a double cone blender gives satisfactory performance. The blender should be of such a nature that it can be freely agitated and preferably have some means of allowing injection of gaseous substances.

Prior to the SO₂ treatment and while the soy flour is being continuously agitated, steam is injected to increase the moisture content of the flour, preferably within the range of 6 to 12% by weight of soy flour. This moisture content is necessary to obtain the resulting (after the SO₂ treatment) increase in protein solubiluty. If less than 5% by weight of moisture is present in the soy flour, the SO₂ treatment will not increase the solubility to levels comparable to that of milk solids; on the other hand, if over 20% by weight of soy flour of moisture was added, undesirable side reactions are increased.

The table shown below shows the effect of varying the moisture content of the soy flour prior to the SO₂ treatment. The process used was in accord with that described in detail below. In all runs the temperature of the SO₂ treatment was 78°F; the time or treatment was ten minutes. Compressed SO₂ gas was employed. The percents as shown in the table are all weight percents. As can be seen, all else being constant, a change in the moisture level prior to the SO₂ treatment significantly affected the amount of water soluble soy protein obtained after the SO₂ treatment. Generally, the lower levels of moisture content resulted in higher percentages of water-soluble protein. The percent of water soluble protein was calculated by the well-known Kjeldahl nitrogen determination.

Water-Soluble Protein as a Function of Moisture Content

Moisture content, percent:	SO₂, gm./lb. of soya flour	Percent water soluble protein	
		Run I	Run II
7	1.5	35.1	35.6
7	2.0	33.4	33.1
7	2.5	30.6	32.4
11	1.5	33.4	32.2
11	2.0	28.7	29.1
11	2.5	26.9	28.4
13	1.5	32.2	32.0
13	2.0	31.0	29.7
13	2.5	28.2	26.1

During the process the temperature should be kept within the range of 60° to 150°F and preferably within the range of 60° to 120°F. If the temperature is much below 60°F, the kinetics of the reaction are such that it will proceed very slowly. High temperatures (over 150°F) cause a competing reaction which renders the protein insoluble in water and thus unsatisfactory for use in culinary mixes. It is possible that the competing reaction at higher temperatures is a protein heat denaturizing reaction in which the protein is placed in such a structural form that its use in a culinary mix is not satisfactory.

While maintaining the temperature within the above referred to 60° to 150°F range, SO₂ gas is passed into the reaction system, preferably while continuously stirring or agitating so that the SO₂ will be exposed to all of the flour in the reaction vessel. The amount of SO₂ added should be from 0.5 grams per pound of soy flour to 4.0 grams per pound of soy flour; however, the optimum amount is between 1.0 and 2.0 grams of SO₂/lb of soy flour. When 1.0 to 2.0 grams of SO₂ per pound of soy flour is added to the reaction mixture the resulting percent of water-soluble protein is from about 34 to 35% or more which is at a level comparable to that of milk solids.

The soy flour is subjected to the SO₂ treatment for from about 5 to 20 minutes. Five minutes is about the minimum amount of time at which the increased solubility effect becomes significant and after about 20 minutes there is no appreciable increase in protein solubility. Thus the SO_2 and soy flour may continue in contact for several hours without any significant adverse effect, however, there is no real need to extend the time of contact beyond the twenty minute level as to do so would merely increase the amount of expended time without a corresponding increase in performance. The flour so produced is shown to be quite useful in preparing dry cake mixes.

Increasing Water Absorption of Soy Meal

Commercially satisfactory soybean meal generally has a moisture content of 12%. It has been difficult to maintain this moisture content due to the drying of the meal during shipment or while on inventory. These difficulties can be overcome by treatment of soybean meal according to the process disclosed by *F.A. Norris and D.C. Johnson; U.S. Patent 3,155,524; November 3, 1964; assigned to Swift & Company.* This treatment causes the meal to have an increased moisture absorption, and lose moisture less readily, thereby enabling moisture contents in general to be kept nearer the optimum and eliminating short weights arising from moisture loss during shipment.

The process comprises the steps of increasing the pH of spent soybean flakes before desolventizing and toasting, and heating the treated flakes for a relatively short period of time. Flakes which have been treated by this process have greatly improved water absorption properties, and also exhibit a color which is commercially desirable.

The process is carried out on soybean flakes which have already been extracted to remove the oil. Such soybean flakes are known as spent flakes. This process cannot be satisfactorily applied to fat containing soybean flakes. The principal reason is that soap is formed by the reaction of the alkali with the free fatty acids in the soybean flakes, raising problems of solvent recovery and soap removal before the oil can be sold.

The use of this process to unextracted flakes is unsatisfactory because cooked soybean flakes do not extract well unless dried considerably below the moisture used in cooking and carefully conditioned. In some cases extraction may be impossible if it is preceded by cooking. Further, the cost of processing is increased since cooking must later be repeated in order to remove the solvent from the extracted flakes. The effect of increasing the pH on the percent moisture absorption of the soybean product is shown by the following table. The flakes used in these tests were treated at 20% moisture for 30 minutes under 25 pounds jacket steam pressure.

pH	Percent Moisture Absorbed
6.4 (control)	177
7.6	208
8.4	222
9.4	214
10.7	257

Generally any basic substance can be used to increase the pH of the soybean materials to improve the moisture absorption of the product. Particularly useful basic materials are aqueous solutions of sodium or other alkali metal hydroxides, sodium or other alkali metal carbonates, tri-sodium phosphate, ammonia gas, or lime. Increasing the pH by the addition of an alkaline material is critical in the production of soybean meal having an increased moisture absorption. The addition of an acid (lowering the pH) results in a decrease in moisture absorption. For example if the pH is lowered to 4.2 by the addition of acid the moisture absorption is reduced to 132%, under the same processing conditions used in the above table.

The addition of the alkaline or basic material has the further effect of eliminating urease activity in the soybean material. This is also beneficial since there is an increasing commercial demand for soybean meal having a low urease activity. The use of very short heating times has been found to increase the moisture absorption properties of the soybean meal. The use of short heating times has formerly been impossible since the product produced using short heating times had a commercially unacceptable light color. However, the use of short heating times in this process, coupled with the alkali treatment of the spent soybean flakes produces a commercially acceptable product. The following tables illustrate the effect of shorter heating times on the moisture absorption of the soybean meal produced by the instant process.

Raw Material and Conditions	Percent Moisture Absorption		
	10 Minutes	30 Minutes	50 Minutes
Soybean spent flakes at 173°F	220	213	196
Soybean spent flakes at 188°F	212	193	175
Soybean spent flakes at 220°F	200	192	174

This increase in absorption as the heating time is decreased also occurs as the pH is varied over the alkaline range, as is shown in the following table.

Raw Material and Conditions	Percent Moisture Absorption		
	10 Minutes	30 Minutes	50 Minutes
Soybean spent flakes pH 7.5, 220°F	212	207	—
Soybean spent flakes pH 8.5, 220°F	226	216	—
Soybean spent flakes pH 10.0, 220°F	265+	235	—

Example: 1,000 grams of soybean flakes (pH 6.4) were placed in a Hobart laboratory mixer, and 32 ml of 20% sodium hydroxide (mixed with enough water to bring the moisture content up to 20%) were added. The pH was thereby raised by 7.6. The mixed product was placed in a steam jacketed reactor at 220°F and kept under agitation for 30 minutes. At the end of this time, the temperature had reached 225°F and the moisture content had dropped to 17.7%, pH 7.5. This product on drying down to 8% moisture, had a water absorption of 208% compared to 177% for a control run which was made under exactly the same conditions except that no alkali was used. In the control run, the moisture out of the reactor was 16.2%, which indicates that flakes treated with alkali retained moisture more strongly, and that the retention increases as the pH is raised.

Noncaking Soybean Meals

Kaolin or similar materials have been used by *M.A. Williams; U.S. Patent 3,469,994; Sept. 30, 1969; assigned to Central Soya Company, Inc.* for treating soybean meal to inhibit caking. The process is particularly useful in preventing soybean meal from "setting up" during storage in large blocks which resist removal from the storage vessel.

By toasting soybean meal above 212°F, it is possible to produce a large fraction of particles which will pass through a No. 8-10 mesh U.S. standard screen, and such particles which may constitute between 30 and 50% of the meal do not require treatment. Such particles have hard, glazed, and often glossy exteriors and do not have substantial amounts of surface water-soluble protein. The oversize particles are passed to a hammermill or other grinding equipment to be reduced in size so that they will pass through No. 10-12 mesh screen or a screen of larger or smaller mesh as may be desired.

It was found that it is the oversize and recycled ground material which is responsible for the caking or setting up of the meal. These larger particles have interior portions in which the protein remains substantially more water-soluble, and when ground, the interior water-soluble protein material is exposed. To solve the problem, it is only necessary to treat such exposed portions, using an inert powder or dust-like material. Best results have been obtained by using kaolin preferably having 87 to 98% in a particle size less than two microns.

Instead of kaolin, beneficial results can be obtained with other inert powders or fine material, such as, limestone, clays, starch, etc. None of such inert, dust-coating agents, however, have been found to be as effective as kaolin.

It was surprising to find that so little kaolin was needed to obtain a meal product in which caking was inhibited. With a coating of kaolin as low as 0.1%, beneficial results were obtained, while with a coating of from 0.25 to 0.5% by weight, caking was overcome and without increasing the dustiness of the product. While the addition of fat to soybean meal does not in itself prevent caking, the addition of small amounts of fat with the inert dusting agent is effective not only in inhibiting caking but actually causes the soybean meal to become free-flowing.

In one embodiment of the process, the soybean meal is toasted at a temperature above 212°F as in a desolventizer toaster to form small particles which will pass through a screen such as, for example, a No. 10 mesh screen, and such particles may then be separated and recovered as product. The oversize particles which do not pass through the screen may be passed to grinding apparatus for reduction to smaller size. Such larger particles have interior protein portions which are water-soluble and which are responsible for the caking or setting up of the soybean meal. The coating dust material, such as kaolin, is applied to such interior portions exposed by the grinding and the treated product may be withdrawn or, if desired, merged with the product from the first screen. The kaolin, or other coating dust material employed, may be added at a number of places in the process.

The coating material (preferably kaolin) may be advantageously added where the soybean meal is withdrawn from the desolventizer toaster and where it is hot and wet. As the steaming soybean meal leaves the toaster, condensation of the steam on the withdrawn particles gives the particles a surface moisture of about 17 to 18%, and such moisture under the temperature conditions of the meal (190° to 200°F) causes the kaolin or other coating dust to adhere tightly to the particles, and particularly to the oversize particles which later are subjected to grinding operation. Thus, in the later grinding operations, the adhering kaolin, etc. is mixed thoroughly and bound to the interior portions of such particles.

Example 1: 0.25 to 0.5% kaolin was added to the hot, wet soybean meal as it was leaving a desolventizer toaster. The addition of the kaolin at this point helped to reduce the dustiness of the coated meal. Further, it was found that the addition of the kaolin in this location improved the resistance to bin set-up as follows. While it required 15 lb to break down the pack test before the kaolin was added, after the kaolin was added it required 0 to 5 lb, the results of the test being set out in the following table:

Sample No.	Percent kaolin added	Meal temp. when on pack	Weight on pack (lb.)	Weight to break pack (lb.)	Moisture of meal	Protein as is
1	None	95	100	15	12.31	49.92
2	.25	95	100	0	11.99	50.07
3	.50	72	100	0	11.94	49.70
4	.50	72	200	0	11.94	49.70

Example 2: The process was carried out as described in Example 1, 0.25% kaolin being added to the hot, wet meal from the desolventizer toaster, but in addition 0.125% of feed fat was added to the meal recycle system after grinding. Before adding the kaolin, 13 lb were required to break down the pack. After the kaolin and fat were added, the finished meal flowed. Many other tests with kaolin alone (no fat) have never produced a product which flowed directly on removing the 2" plug as the combination coating did in this test.

SOY PROTEIN CONCENTRATES

Soy protein concentrates are produced from soybean flakes by various washing processes using such solvents as water, aqueous alcohol or dilute acid. Soluble carbohydrates are removed by this washing process, particularly raffinose and stachyose which cause flatulence. The concentrates have improved flavor (i.e., less "beany" flavor) and the protein content is raised to 60 to 70%.

EXTRACTION PROCESSES FOR PRODUCING CONCENTRATES

Countercurrent Extraction Yielding Concentrated Wheys

A process has been developed by *S.J. Circle and R.W. Whitney; U.S. Patent 3,365,440; January 23, 1968; assigned to Central Soya Company* for extraction of soybeans to produce the insoluble protein plus a concentrated extract or whey. This concentrated extract requires little further concentration to form a syrup and overcomes the problem of finding an ecological and economic method for disposal of large volumes of waste extract fluids from soybean processing. In this process the soybeans (preferably defatted) are extracted to provide a soy concentrate that can then undergo further extraction and precipitation for the formation of soy isolates.

In one embodiment of this process, a stream of soybean material (full fat or defatted) containing as part of its ingredients a water-soluble portion, is passed in countercurrent relation with a stream of solvent for the water-soluble portion. The ratio of the streams is adjusted to provide a concentration of the water-soluble fraction in the solvent effluent stream of at least 2%. In a specific application, the source of material of the part to be dissolved may be soybean flakes. Illustrative of other particle forms, the soybeans may be cracked, granulated, etc.

The solvent for the water-soluble portion may take a variety of forms, including aqueous solutions which are effective to immobilize the alkali-soluble acid-precipitable protein while separating out all of the other hydrophilic constituents. Among the protein-immobilizing solvents available are various aqueous acidic media, preferably having a pH in the isoelectric range, that is, a pH in the range 4 to 5. The acids selected may include sulfurous, sulfuric, hydrochloric, lactic, acetic, carbonic, and phosphoric, although other acids may be conveniently employed depending upon the ultimate usage of the protein product. Another protein-immobilizing solvent is chilled water, with or without the presence of alkaline earth cations such as calcium and magnesium ions. Still another group of protein-immobilizing solvents includes water-miscible organic solvents such as acetone, the lower alkanols, dioxane, dimethyl sulfoxide, etc. The most effective range is 20 to 80% of the solvent mixed with water, with maximum effectiveness being reached at about 60%. Organic solvent concentrations less than 20% result in undesirable swelling of the soybean source material due to water imbibition. Organic solvent concentrations in excess of 80% do not remove as high an amount of the water-soluble solids, in particular the bitter components.

Aqueous organic solvents are advantageous in that they reduce swelling of the flakes, achieve more efficient removal of the bitter constituents and other solubles, and provide a bacteriostatic action. In this latter aspect, the aqueous organic solvents protect both the protein and the solubles against degradation. Further, the aqueous organic solvents operate to remove the least amount of alkali-soluble, acid-precipitable protein and albuminous protein when used for leaching out the aqueous organic solvent-soluble portion. Only about 2 to 4% of the total soybean nitrogen is taken up into the solubles, this low quantity being achieved since most of the nitrogenous components are generally insoluble in the organic aqueous solvents.

Ordinarily, the weight ratio of the solvent stream in contact with the source stream should be kept below 15, based on the source stream being defatted flakes, to achieve a substantial nonevaporative concentration of the solubles portion in the solvent effluent so as to make further concentrating of this stream practicable. In general, the variation of solids concentration in the clarified supernatant liquor with different solvent-flake ratios would be represented by the data in the table below. The concentrations of solubles in the clarified supernatant liquor as shown below represent those theoretically obtainable in a single stage batch extraction. Ordinarily, in a single stage batch aqueous acid-leach extraction the dry soybean source material takes up at least about four times its own weight of water in a nonfluid or nonmobile form, similar to water of hydration. If the extraction is made in a countercurrent manner, the concentration of solubles in the clarified supernatant liquor is essentially doubled for the same solvent-flake ratio.

Ratio, Solvent:Flakes	Percent Solids in Clarified Supernatant Leach Liquor
27	1.2
15	2.2
10	3.2
5	6.7
4	8.3
3	11.1
2	16.7

The passing of the two streams in countercurrent relation may be achieved through the use of a variety of mass transfer contactors. This includes vertical towers, basket extractors of various types, batch vessels interconnected for countercurrent operations, and the like. The process also may use either continuous or discontinuous flow apparatus.

Example: This example demonstrates the operation of a continuous countercurrent extraction system employing a small-scale plateless column with an input rate of 22 lb per hour of dry soybean flakes at the top. A constant flake-bed height of about 4 feet from the bottom was maintained. A few hours were required to obtain stratification of solvent-soluble solids and steady state operation. Sampling ports, placed at various distances from the bottom of the column, were used to obtain the stratification of solids data. The flakes at a pH of 6.8 (reaction in water) were introduced at the top of the column, along with hydrochloric acid, and acidified water was introduced at the bottom at a rate of 320 lb of water per hour. Under these conditions, an overflow of liquid of about 80 to 90 lb per hour and an underflow of about 230 lb per hour were obtained. The residence time of the flakes in the column was about 1 hour. The following table shows the data obtained.

Commercial soybean flakes usually have a pH in water of 6.5 to 7.0. Thus to immobilize the acid-precipitable protein, the flakes were contacted with an aqueous acid in the proper proportions to bring the pH reaction to 4.0 to 5.0. Where a column is used in the continuous countercurrent extraction system, acid may be introduced at both ends of the column. At the top and where the flakes (usually dry) are introduced, a small volume of highly concentrated acid, which is readily diluted by the large volume of exiting leach liquor prior to its withdrawal from the column is added whereas at the other end a large volume of very dilute acid to contact the exiting spent flakes at a point prior to their withdrawal from the column is introduced.

Sample Location	Number of Feet from Bottom	Temperature °F (Aqueous Solvent)	pH	Clarified Supernatant Leach Liquor % Solids	Clarified Supernatant Leach Liquor % N x 6.25
Top	6.7	100	4.4	9.09	*1.79
6	5.8	-	4.4	9.07	1.79
5	4.8	-	4.3	9.09	1.80
4	4.1	-	4.8	8.14	1.59
3	3.3	-	5.0	2.74	0.54
2	2.4	-	5.2	0.55	0.18
1	1.4	-	5.2	0.05	0.04
0	0.75	-	4.9	0.05	0.02
Bottom	0	110	4.9	0.05	0.02

*Overflow

In this manner, the optimum pH condition for immobilizing the protein throughout the column was maintained while at the same time achieving an optimum gradient of concentration of solvent-soluble solids. There is a high concentration of solubles in the leach liquor supernatant, at the point where the flakes are added to the system and leach liquor is withdrawn, and an extremely low concentration of solvent-soluble solids in the leach liquor at the end where the spent flakes are withdrawn and the aqueous leach solvent is introduced. The clarified supernatant leach liquor containing 9.09% solids was thereupon evaporated to a syrupy consistency.

Two-Phase Extraction of Full Fat Soy Products

Bland, light colored soy concentrates are obtained by *R.G. Schweiger and S.A. Muller; U.S. Patent 3,714,210; January 30, 1973; assigned to Grain Processing Corporation* using a two-phase solvent extraction. The full fat soybean flakes or meal are extracted with a two-phase liquid solvent system; one phase consisting essentially of one or more lipophilic solvents and the other phase consisting essentially of a mixture of water and one or more water-miscible solvents. Oil and nonproteinaceous materials are simultaneously extracted providing a soy protein concentrate product which is light in color and bland in taste.

The one phase of the solvent system consists of one or more lipophilic solvents, such as hexane, cyclohexane and the like, and the other phase of the solvent system consists essentially of a mixture of water and one or more water-miscible solvents, such as methanol, ethanol, isopropanol, acetone and the like. In the aqueous phase of the solvent system the ratio of water to water-miscible solvent ranges from 2:8 to 4:6, preferably 3:7, on a volume to volume basis. The volume ratio of the nonaqueous to the aqueous phase ranges from about 1:2 to 2:1 and most preferably is 1:1.

Typically, the process is carried out by mixing full fat soybean flakes or meal (ground flakes) at room temperature or above for 0.5 to 2 hours with 5 to 20 parts by weight of a mixture of three types of solvents, i.e., a lipophilic solvent, such as hexane, a water-miscible solvent, such as methanol, and water, at a ratio of, for example, 10:7:3 (v/v) This solvent mixture separates into two liquid phases, and, after extraction, the lipophilic solvent phase contains essentially the soy oil while the other phase contains low molecular weight compounds present in the soybean flakes, such as carbohydrates, flavor components, pigments, etc. The temperature during extraction is not critical and can be below or above room temperature but, for practical purposes, should be below the boiling point of the solvents employed. After extraction and solvent removal, the protein-containing product is dried, preferably in vacuo, to avoid excessively high temperatures which may denature the protein. Temperatures below about 50°C can be satisfactorily employed for drying the extracted material. The extraction can be repeated a number of times depending upon the content of protein desired in the product. The extraction can be conducted in accordance with known solvent extraction procedures but countercurrent extraction procedures are preferred.

Example 1: Full fat soybean flakes (20 g) were suspended in a two-phase liquid solvent mixture consisting of 100 ml of hexane and 100 ml of 70% aqueous methanol (v/v). After mixing at 26°C on a reciprocal shaker for 1 hour, the mixture was filtered and the solids washed on the filter with 40 ml of the same solvent mixture. The extraction and washing was once repeated and the material thus obtained dried in vacuo at 40°C. The average protein content of the resulting soy protein concentrate from seven runs carried out in this manner was 67% (dry basis) with a protein recovery of 92% and a fat content of below 1%. The products were lighter in color and considerably blander in taste than those obtained by conventional two step process.

Example 2: Full fat soybean flakes (150 g) were stirred one hour at room temperature with 3 liters of a two-phase liquid solvent mixture consisting of hexane, methanol, and water in a ratio of 10:7:3 (v/v), which had been used twice previously for extracting soybean flakes. The solids then were filtered off on a medium fritted glass funnel and extracted similarly with a similar solvent mixture which had been previously used once for extracting soybean flakes and finally were extracted again with a fresh solvent mixture of the same composition. The product thus obtained was dried in vacuo below 45°C. The protein content of the product was 72.7%, fat content 0.9%, and the weight recovery 80%, all on dry basis. Tastewise, the product compared favorably with high quality commercial soy protein products.

Two Step Extraction of Defatted Soy Products

L.P. Hayes and R.P. Simms; U.S. Patent 3,734,901; May 22, 1973 assigned to A.E. Staley Manufacturing Company have prepared soy protein concentrates by removing residual lipid and water-soluble constituents from defatted soybean flakes. The residual lipids are initially extracted from the soybean flakes with a hydrocarbon/monohydric alcohol solvent followed by aqueous extraction of the water-soluble constituents. In addition to the soy concentrate, high lecithin-containing oil is obtained by admixing the resultant lipid miscella and aqueous miscella and then separating an oil phase from the mixture. In more detail, the process for separating residual lipids and water-soluble constituents from solvent-extracted seed materials and providing seed protein concentrates comprises the following steps.

(a) Extracting residual lipid constituents from a solvent-extracted seed material which contains on a dry weight basis from 0.5 to 10% lipids, from 25% to 40% water-soluble constituents with the remaining portions being primarily protein and a minor amount of fiber, the lipid constituents being extracted from the solvent-extracted seed material by subjecting 100 parts by weight of the material to a lipid extraction medium containing a hydrocarbon solvent and at least 2 parts by weight to about 30 parts by weight monohydric alcohol for each 80 parts by weight hydrocarbon solvent, the extraction of lipids being conducted under conditions whereby the amount of alcohol is maintained at a level of at least 2% to about 40% of the seed material dry weight;

(b) Separating the lipid miscella from the residual seed material;

(c) Extracting at least a major portion of the water-soluble constituents contained in the residual seed material by subjecting 100 parts by weight of the residual composition to 150 to 300 parts by weight aqueous extraction medium containing at least 40% but less than 70% by weight aliphatic monohydric alcohol;

(d) Separating the resultant aqueous miscella from the seed material providing a seed protein concentrate;

For maximum advantage of the process the residual lipids are subjected to the following additional recovery steps.

(e) Mixing the lipid miscella and aqueous miscella and effecting phase separation of the resultant mixture to provide a nonpolar phase containing substantially all of the lipid extracts and polar phase containing substantially all of the water-soluble extracts, with the total amount of monohydric alcohol contained within the resultant mixture being sufficient to provide a polar phase containing at least 40% but less than 70% by weight of the total polar phase weight;

(f) Separating the nonpolar phase from the polar phase and recovering the respective lipid concentrates and water-soluble concentrates therefrom.

This process is useful for recovery of residual lipids, water-soluble constituents and protein concentrates of materials obtained by solvent extraction of oil seeds such as cottonseed, safflower, sunflower, peanuts, sesame, soybeans and the like which have a protein concentration of at least 30% on a dry weight basis. The process is particularly adapted to soy compositions. The following example illustrates the processing of soy compositions although it should be understood that other solvent-extracted oil seed protein materials may be used.

Example: Using a six-stage countercurrent treatment, a bland, soy protein concentrate having a protein content of 74% by dry solids weight was prepared. The first three countercurrent stages were used to extract residual lipids from defatted soybean flakes saturated with hexane. The latter three stages were used to extract the water-soluble constituents. To the first countercurrent stage, there were fed soybean flakes saturated with hexane where excess hexane solvent used in a conventional solvent extraction process had been drained from the flakes. The flakes contained 56% by weight dry solids, 8.5% by weight water and 32.5% by weight hexane. On a dry solids weight basis, the soybean flakes were comprised of about 0.6% residual lipids and 57% protein.

The three countercurrent equilibrated slurries of stages 1, 2 and 3 were maintained at 120°F with a total residual lipid extraction solvent medium to dry solids weight ratio at about 5:1 (i.e., about 4:1 on full fat flake weight). Including the hexane and water from the saturated flakes plus additional water of the 180 proof ethanol used each of the first three equilibrated countercurrent stages had a residual lipid solvent extraction medium comprised on a weight basis of 64% hexane, 26% ethanol and 10% water. The first three stages were slurried for 20 minutes with the flakes from stages 1 and 2 being separated from the lipid extraction solvent by means of a perforated basket centrifuge operated at 2,000 g. A portion of the full miscella containing the extracted residual lipids was recovered from the first stage and placed in decanter flasks for further processing with the aqueous miscella from stage 4.

The intermediate miscella of the second and third stages were recycled for use as a residual lipid extraction solvent in the first and second stages and to maintain an appropriate solvent level for each stage. The centrifuged cakes from stages 1 and 2 were transferred respectively to equilibrated stages 2 and 3. The net solvent usage in extracting the residual lipids was 1.8 parts by weight of residual lipid extraction solvent for each part by weight of dry solids admitted to the first stage. Thus, for each part by weight of dry solids admitted to the first stage, 1.8 parts by weight of fresh solvent comprised (on a weight basis) of a 64% hexane, 26% ethanol and 10% water was introduced to the third stage. After completion of the third stage, excess residual lipid solvent extraction medium was drained from the solid soy composition. The drained soy material (approximately 50% dry solids, 27% hexane, 14% ethanol and 9% water) was then subjected to a desolventization whereby all of the hexane was removed. The soy material, free of hexane, contained 74% dry solids, 16% ethanol and 10% water.

The water-soluble constituents were removed by forwarding the soy composition (free of hexane) to the next three equilibrated countercurrent aqueous ethanol extraction stages. Each slurry stage was maintained at about 6.75 parts by weight solvent for each part by weight dry solids and at a temperature of 120°F. After twenty minutes of slurrying for each stage, the equilibrated soy compositions were separated by means of a perforated basket centrifuge operated at 2,000 g. A portion of the full aqueous ethanol miscella from the fourth equilibrated stage was recovered for processing with the residual lipid miscella from stage 1. The intermediate miscella from stages 5 and 6 were forwarded and recycled for use in stages 4 and 5 and to maintain the appropriate solvent level for each stage. Recovered centrifuged cakes from stages 4 and 5 were respectively forwarded to stages 5 and 6. Fresh solvent having a 1:1 weight ratio of ethanol and alcohol was introduced into the sixth stage at a rate of about 2.7 parts by weight fresh solvent for each part by weight of dry solids admitted to the first stage. The centrifuged cake recovered from the sixth stage was comprised (on a weight basis) of 40% dry solids, 30% ethanol and 30% water. The water and ethanol were removed from the cake by steam stripping. The resultant desolventized product was ground into a meal. Its assay was 91% by weight dry solids of which 74% by weight was protein.

About 6.2 parts by weight of the full residual lipid miscella recovered from stage 1 containing on a weight basis of 2.75% dry solids, 70.50% hexane, 23.25% ethanol and 3.50% water was combined with 4.1 parts by weight of the full aqueous ethanol miscella from stage 4. The full aqueous ethanol miscella from stage 4 was comprised on a total weight basis of about 46% ethanol, 43% water and 11% dry solids. The combined miscella were then vigorously agitated to provide a homogeneous mixture which was then allowed to stand for five minutes in a decanter flask where it separated into two distinct phases. The upper nonpolar phase was separated and upon analysis found to be comprised on a total weight basis of 93% hexane, 2.85% lipids, 2.42% ethanol and 1.73% water.

An oil of a high lecithin content, 40% acetone insoluble, was recovered by evaporation steam stripping. Based upon the total amount of dry solids submitted in the first stage, the recovered oil represents 2.6% of its weight. The dry solids were recovered from the lower polar phase by evaporating off the excess solvent. Recovered solids represented 13.6% by weight of the total dry weight of soybean flakes submitted to the first stage with the recovered solids being primarily comprised of sugars.

Aqueous Polysaccharide Extraction

The protein loss usually encountered when preparing soy concentrates is avoided in the process disclosed by *D.F. DeLapp; U.S. Patent 3,762,929; October 2, 1973; assigned to American Cyanamid Company*. The soy flake is contacted with an excess of water in the presence of 0.5 to 4.0%, by weight, based on the weight of the soybean flake, of a polysaccharide such as carboxymethylcellulose, guar gum, carrageenan, sodium alginate, gum karaya, alginic acid, and agar fractions. The polysaccharide somehow chemically combines with the soy flake and substantially prevents and minimizes the dissolution of the protein in the water at the pH employed.

The extraction is conducted at a temperature from 20° to 50°C for 30 minutes to 2 hours. A pH in the range of 4.2 to 4.6 is employed by the use of a sufficient amount of any food grade acid such as hydrochloric acid, acetic acid and the like. After treatment the resultant high protein containing concentrate is recovered. The concentrate generally comprises from 65 to 70% protein, the remainder being carbohydrates and crude fiber. Recovery can be effected by filtration, centrifugation etc. The concentrate produced by this process can be used as a protein supplement in that from 6 to 8% used as an additive to food products increases their nutritional value. In the following examples, all parts and percentages are by weight unless otherwise specified.

Example 1: To a suitable vessel are added 2,500 ml of water containing 5 ml of glacial acetic acid to give a pH of 4.4. 100 parts of ground defatted soybean flake are added along with 2.0 parts of carrageenan. The resultant mixture is stirred for one hour at ambient temperature. The water is then centrifuged off and the insoluble fraction is dried. The resultant product is in excess of 65% protein as calculated from nitrogen and represents a commercially attractive protein supplement, i.e., concentrate.

Examples 2 through 5: The procedure of Example 1 is again followed except that an equivalent amount of (2) carboxymethyl cellulose, (3) guar gum, (4) sodium alginate and

(5) gum karaya are substituted for the carrageenan. In each instance, a high protein containing concentrate is recovered.

Undenatured Soy and Peanut Concentrates

A controlled method of extracting defatted soy or peanut meals is disclosed by *L. Sair and I. Melcer; U.S. Patent 3,809,767; May 7, 1974; assigned to The Griffith Laboratories, Inc.* Their method produces an undenatured protein concentrate described as having natural structure or natural texture and having a nitrogen solubility index above 15% by weight (preferably above 40% by weight). In the process, first, vegetable protein material, a substantial portion of which will not pass through a 40-mesh screen, which has good natural structure and texture, a nitrogen solubility index (NSI) of more than 15% by weight (preferably, more than about 40% by weight) and which, preferably, has been defatted or deoiled, including defatted or deoiled soybean or peanut meal, is made up into an aqueous slurry (e.g., having about 5 to 20% by weight solids) in which the pH of the vegetable protein material is adjusted to the vicinity of the isoelectric point.

The duration of the isoelectric wash is dependent upon the particular process conditions used. When the isoelectric extraction treatment has been concluded, the resulting liquid extract, which includes soluble, vegetable protein material and undesired, characteristic taste- or flavor- and odor- or color-conferring ingredients found in the starting vegetable protein, is separated and discarded by a centrifuging, screening or filtering operation. The recovered insoluble, vegetable protein material having natural structure and texture may be further washed with water which may have a somewhat acid pH for bacterial purposes, followed by being filtered, screened or centrifuged. The resulting insoluble, vegetable protein slurry has a pH in the vicinity of the isoelectric point of the protein content thereof, and has, for example, about 65 to 75% by weight protein determined on a dry basis.

Following the extraction procedure, the defatted vegetable protein material must be at least partly dried under controlled conditions and then neutralized. Neutralization is accomplished by adding an edible, inorganic alkali or inorganic buffering agent to the insoluble, vegetable protein material in the presence of water and with mixing to raise or adjust the pH of the vegetable protein material to 5.5 to 10.5, preferably to a pH of at least 6. The neutralization step raises the pH of the defatted protein material, thus rendering the undenatured, vegetable protein material soluble and thereby enhancing the water-binding and emulsifying characteristics of the protein material, which characteristics are important when the concentrate is used in meat and other food products.

It has been found that one can facilitate the retention of natural structure and texture by partly drying (e.g., flash drying) the vegetable protein slurry prior to neutralization (e.g., reducing the moisture content from about 60 to 80% by weight water to within the range of about 5 to 55% and, preferably, to within the range of about 20 to 45% by weight water), whereby the subsequent neutralization step does not break down the natural structural and textural coarseness or integrity of the partly dried, vegetable protein material. The drying step, must be conducted under sufficiently low, controlled, drying-temperature-time conditions which enable one to obtain a soluble, defatted vegetable protein concentrate having a nitrogen solubility index of at least 15% or well above 15% by weight, and, preferably, should be conducted under conditions which do not substantially denature the vegetable protein concentrate as evidenced by the concentrate having a nitrogen solubility index of at least about 40% by weight.

Example: An aqueous, isoelectric, soy protein slurry was prepared with the following materials: 10 lb of cargill grits, 6.8 gal of water, 0.05 lb of sodium bisulfite and 0.568 lb of 22°Bé hydrochloric acid. The slurry was agitated for 30 minutes and was found to have a pH of 4.15. The slurry was centrifuged in a Tolhurst basket centrifuge having a 12" diameter basket. The insoluble, soy protein cake was reslurried with 4.7 gal of water (a washing step), again centrifuged in the Tolhurst basket centrifuge, and was washed with an additional 2.1 gal of water while being centrifuged. The wet, insoluble, washed, soy protein concentrate contained 71.3% by weight moisture and constituted a yield, on a dry basis, of 67.4% by weight. The wet, soy protein concentrate was partly dried in a vertical, pilot plant, fluid bed drier. The operating conditions in this first drying step were as shown in Table 1 below.

TABLE 1: FIRST STAGE DRYING

Length of time fluid bed drier was operated	Air inlet temperature, °F.	Air in lower chamber, °F.	Air outlet temperature, °F.	Temperature partly dried product, °F.
0 (start of run)	222	202	209	
5 min	231	113	104	
10 min	221	104	98	92
15 min	224	101	98	92
20 min	225	100	99	93
23 min. (air heater cut off)			85	
25 min. (stopped collecting product)			80	80

The recovered, partly dried, soy protein concentrate which weighed 4.1 lb and had 35.8% by weight moisture, was placed in a ribbon blender with 6% by weight sodium bicarbonate based on the solids. The sodium bicarbonate was readily mixed with the concentrate without forming a viscous or gummy mass. The neutralized, soy protein concentrate had a pH of 7.4. The neutralized concentrate was dried in the vertical, pilot plant, fluid bed drier under the conditions shown in Table 2 below.

TABLE 2: SECOND STAGE DRYING

Length of time fluid bed drier was operated	Air inlet temperature, °F.	Air outlet temperature, °F.	Temperature of dried product, °F.
0 (start of run)	256	239	
2 min	260	206	130
3 min	257	200	130
4 min	257	195	130
5 min	257	198	135
6 min. (air heater cut off)	170	167	156
9 min		127	98

The neutralized concentrate from the second drying step was blended and was found to have 10.8% by weight moisture. The second drying of the concentrate was continued from 5 minutes in the vertical, pilot plant, fluid bed dried under the conditions shown in Table 3, below.

TABLE 3: CONTINUATION OF SECOND STAGE DRYING

Length of time fluid bed drier was operated	Air inlet temperature, °F.	Air outlet temperature, °F.	Temperature of dried product, °F
5 additional min	120–198	115–143	102–126

The moisture content of the neutralized concentrate from the continuation of the second drying step was 7.9% by weight, and a portion of it was ground in a laboratory Fitz Mill so as to pass through a 103-mesh screen. The analysis of the ground, dried, neutralized soy protein concentrate, pH 7.7 (after the continuation of the second drying step) was moisture, 7.2%; protein content based on dried neutralized concentrate, 64.3%; protein content based on 100% dried neutralized concentrate, 69.3%; and nitrogen solubility index (NSI), 43.5%. The above product was shown to have improved water absorption and

SHEARING FORCES TO LIBERATE PROTEIN

Urschel Mill Rupture of Protein Cells

M.D. Wilding and A.C.-Y. Peng; U.S. Patent 3,583,872; June 8, 1971; assigned to Swift & Company have developed a process of isolating vegetable protein by subjecting defatted protein to high intensity shearing forces. The process produces a suspension of protein which is not defined as to actual composition.

The procedure comprises contacting raw vegetable protein material with an aqueous solvent under neutral pH conditions and temperature not substantially in excess of 100°F, (preferably 40°F to 90°F) and passing the resulting slurry through a narrow opening where high intensity rupturing forces disrupt the natural cell structure freeing the desirable proteinaceous material. The entire product exiting through the narrow opening is centrifuged to separate cellular material from extracted protein. The residue can then be washed and recentrifuged to separate additional protein from the fibrous residue. Finally, the soluble protein may be concentrated and kept in liquid form or precipitated by treatment with acid to form a solid product. The nutritious protein may be used per se as a food substance or may be incorporated into other food material.

In carrying out the process, sufficient water is provided to hydrate the protein cells to their maximum and to provide a slurry-type mixture. If insufficient water is present, the material will tend to form a paste and the efficiency of the extraction process will be diminished. On a weight basis, the water should be present in a ratio of between 5 to 14 parts, preferably 8 to 10 parts for one part of soybean protein source material. When using a water to soybean material ratio of 9:1 it is possible to recover 65 to 75% of the available protein without adjusting the solution to an alkaline pH. On the other hand, a water to soybean ratio of about 4:1 will enable one to recover only 40 to 50% of the available protein. On the other side, a water to soybean ratio of 14 to 1 or greater will result in decreased protein recovery somewhere in the order of 35 to 45% of the available protein.

The temperature of the water used to hydrate the soybean material may vary, however, temperatures in excess of 100°F are not required. Normally, tap water is used and hydrolysis of the water-soluble protein does not take place. It is preferred to use water between 40° and 90°F. After the soybean protein source material has been hydrated, it is subjected to a severe mechanical working to rupture the cell wall and membrane of the protein bodies to increase the speed and yield of extraction of the protein bodies. This is accomplished by passing a slurry of soybean protein source material at high centrifugal speeds through a narrow opening (about 0.003" to 0.012", preferable 0.005" to 0.008").

In order to rupture the protein cell, high intensity forces are required to cause both shearing and differential pressure effects on the subcellular protein structures. Other severe mechanical workings such as ballmilling, hammermilling, pebblemilling, crushing, flaking, etc., are insufficient to recover protein in yields of 65 to 75% (without alkaline pH adjustment of solution) of the available protein. In order to rupture the protein cell, it has been found that a tangential speed of the shearing blade should exceed 2,500 in/sec and that the discharge orifice should range between 0.003" and 0.012". A typical shear-type solubilizer for this purpose is an Urschel Mill (MG model) with a microhead attachment. Using a 6" diameter, 9,000 rpm will give a tangential speed of about 2,830 in/sec while 36,000 rpm produces a tangential speed of around 11,290 in/sec.

By using high intensity shear-type apparatus, it was possible to increase the percent of protein in the supernatant by 12% or more over the best conditions of conventional water extraction procedures of 2 hours in length. This process not only substantially improves the yield of protein, but shortens the time of extraction and reduces the equipment requirements thus reducing the over-all cost of production per unit of protein.

Example: One part of undenatured, defatted soy flakes was mixed with 9 parts of cold tap water and agitated for 5 minutes before being processed through an Urschel Mill. A microhead attachment was affixed wherein the orifice size range was in the range of about 0.004" to 0.008". The soybean flake-water slurry was then put through the same to rupture the protein cells. A portion of the initial mixture before passing through the microhead was taken out as a control sample. Both the control mixture and the Urschelized sample were centrifuged at 5,000 rpm for 20 minutes. The supernatant was decanted and saved. The residue was washed once with a small amount of cold tap water and recentrifuged. The total protein extracted into the supernatant by the Urschel process was 69.29% compared to the control of 41.96%, thus creating a difference of 27.33% higher protein content by the process. Protein left in the residue of the Urschelized sample was 30.7% and that of the control 58.3%. Two other experiments using the identical procedure, as shown above, show the total protein extracted in accordance with the process of 69.1% and 73.1% respectively.

Use of High Intensity Shearing Forces

Protein concentrates having markedly different viscosities are obtained in the process of *D.M. Miller and M.D. Wilding; U.S. Patent 3,723,407; March 27, 1973; assigned to Swift & Company*. These high viscosity concentrates are obtained by contacting undenatured, defatted, vegetable protein with an aqueous system, acidifying the system and subjecting the material to centrifugal speeds and differential pressure effects while passing it through a shearing orifice so as to disrupt the natural cell structure of the protein bodies.

The procedure comprises contacting defatted vegetable protein source material with an aqueous solvent under approximately mild acidic conditions, i.e., pH range of about 3 to about 6, usually 3.5 to 5.5 and quite often between 4 and 5, the aqueous solvent having a temperature not substantially in excess of 100°F, but preferably between 40° and 90°F and passing the resultant slurry through a narrow opening whereby high intensity rupturing forces disrupt the natural cell structure thereby freeing the desirable proteinaceous material. The entire product exiting through the narrow opening is centrifuged to separate soluble material from insoluble protein solids. The solids may be washed with tap water and recentrifuged.

The resuspended solids are normally heated to about 40° and 80°C and the pH adjusted to about 6.5 to about 8.0 with an alkaline food grade substance. The adjusted mixture can then be heated and spray dried. The nutritious protein concentrate may be used per se as a food substance or may be incorporated into other food material. While the process is typically applied to the recovery of soy protein, other defatted vegetable materials, especially oilseed materials, may be treated by this process to recover the respective protein therein. Typical raw material which may be utilized, following conventional oil extraction procedures, comprises soybeans, peanuts, castor beans, cottonseeds, sesame seeds, and sunflower seeds.

Example 1: Prior Art — 300 grams defatted soy flour of 65% protein solubility was suspended in 3 liters of tap water and the pH adjusted from the ambient pH of 6.6 to pH 4.1 with six normal hydrochloric acid. After stirring for 30 minutes, the suspension of flour and water was centrifuged at a relative centrifugal force of 4,000 times gravity for 20 minutes. This force was sufficient to yield a transparent, slightly colored supernatant and a firmly compacted residue. The supernatant containing soluble carbohydrates, ash, nonprotein nitrogen constituents and acid soluble protein was decanted and discarded and the residue resuspended in 1.5 liters of tap water.

The pH was again carefully checked to insure that it remained at pH 4.1. After stirring for 30 minutes the wash water was separated from the solids by centrifuging as above. The supernatant was again decanted and discarded. This time the washed residue was resuspended in about 2 liters of tap water, the pH adjusted to 7.5 with sodium hydroxide and the suspension, not quite a true solution, heated to 40°F and held at that temperature for 23 minutes. Following the heating step, the suspension was dried using a Nichols type spray dryer.

Example 2: This Process — 1.5 lb of a defatted, relatively undenatured soy flakes and 15 lb of tap water were stirred for 30 minutes and then acidified to pH 4.1, the suspension was passed twice through an Urschel Laboratories microhead mill with 0.006" slit openings. Following this procedure which resulted in a disruption of the soy flake at the cellular level, the pH was checked, the entire suspension centrifuged, and then washed in the manner indicated in Example 1 using a wash volume of 7½ lb of tap water. After the wash step, the washed solids were resuspended in about 10 lb of tap water, the pH adjusted to 7.5, the suspension heated to 40°C and held there for 30 minutes. Spray drying of the suspension followed. The comparative viscosities (expressed in cm g) of these two soy concentrates are given in Table 1 in which:

Region 1 represents the change in viscosity during the period of temperature transition from 50° to 71°C;

Region 2, the change in viscosity during the 6 minutes the suspension is held at 71°C;

Region 3, the viscosity change during the cooling cycle from 71° to 50°C;

Region 4, the change in the viscosity during the final cooling phase from 50° to 25°C;

High represents the highest viscosity reached during the heating and cooling cycle;

Low, the lowest viscosity; and

Rise, the total change in viscosity from low to high.

TABLE 1

Example	Region 1	2	3	4	High	Low	Rise
1	-20	10	60	80	220	70	150
2	-100	200	20	100	980	650	330

Table 2 indicates the comparative emulsion stabilities for soy concentrates.

TABLE 2

Example	% Emulsion Remaining After	
	Heating	Centrifuging
1	90.6	13.3
2	100.0	44.2

RECONSTITUTED SOY CONCENTRATES

Materials having the general properties of soy protein concentrates are prepared from soy isolates and fibrous materials.

Soy Concentrates from Isolates and Fibrous Residue

Soy concentrates having improved flavor have been prepared by *L. Sair; U.S. Patent 3,635,726; January 18, 1972; assigned to The Griffith Laboratories, Inc.* This method involves combining (a) highly proteinaceous, soy protein isolate recovered from a liquid extract from an extraction of soybean material at a pH above the vicinity of the isoelectric pH of the glycinin content, with (b) fibrous residue recovered from the above extraction, to produce an improved, soy protein concentrate. This blend concentrate has better flavor than concentrates produced by conventional methods of water extraction at the isoelectric point of the protein. A very small residue of beany taste or flavor remains in the concentrate by this former process. The extraction of defatted soybean flour or flake at a pH above the vicinity of the isoelectric pH of the glycinin content effectively removes undesired beany taste or flavor bodies, and in addition extracts desired, soluble protein. In this extraction, for example, about 35 to 38% by weight of the original soybean material is in a soluble form which can be recovered as soy protein isolate which is free of undesired beany taste or flavor bodies, and about 33 to 37% by weight of the original soybean material is insoluble and can be recovered as fibrous residue having polysaccharides and insoluble protein and which is likewise free of undesired beany taste or flavor bodies. The soluble protein may be precipitated and ultimately recovered from the extract as soy protein isolate which is free of such undesired beany taste or flavor bodies, by reducing the pH of the extract to a pH in the vicinity of the isoelectric pH of the glycinin content.

Since the recovered fibrous residue and protein isolate, particularly when washed, are both free of the undesired, characteristic beany taste or flavor, they can be combined effectively and economically to produce an improved, soy protein concentrate. Thus, the insoluble, fibrous residue which previously has been used in animal feed can be combined effectively, in desired proportions, with the more proteinaceous, insoluble, soy protein isolate, and the combined product may be neutralized to a pH within the range of from about 5.5 to 10.5, and dried, to provide an economical, highly proteinaceous, soy protein concentrate suitable for human consumption that is free of undesired, characteristic beany or chalky taste or flavor, and beany odor and color.

Briefly, in this process soybean protein materials (e.g., defatted or deoiled soybean material), including flakes or flour, is extracted with water or water having alkaline material so that the pH of the soy protein material is above the isoelectric pH of the glycinin content (i.e., pH of 4 to 4.8), preferably at a pH of at least 7. The use of an aqueous alkaline solution having sodium sulfite is desired because the sulfite serves as an alkaline buffer and protein preservative. The liquid extract, which includes soluble protein and undesired, characteristic beany taste, and odor and color conferring ingredients, is separated from the insoluble, fibrous residue. Preferably the insoluble, precipitated, fibrous residue is subjected to additional extractions or washing with aqueous alkaline solution and/or water. At least part of any remaining liquid extract is removed from the fibrous residue by a centrifuging, screening, or filtering. The resulting washed, insoluble, fibrous residue has 50 to 65% by weight protein on a dry basis.

The separated liquid extract from the initial extraction step at a pH above the isoelectric pH of the glycinin, which has soluble protein, and, if desired, the liquid extract from the additional extractions or washing, is acidified to a pH in the vicinity of the isoelectric pH of the glycinin (i.e., pH of 4 to 4.8). The characteristic beany taste, and beany odor and color conferring ingredients and soluble sugars and nitrogen-containing bodies are present in the resulting liquid whey, and the whey is removed from the resulting precipitated, soy isolate curd which contains glycinin protein. If desired, the precipitated isolate may be subjected to additional extractions or washing with an aqueous acidified solution and/or water. The separation and recovery of the protein isolate may be accomplished by centrifuging, screening or filtering. The recovered, wet, soy protein isolate curd contains, about 90 to 70% by weight protein on a dry basis.

The resulting soybean protein isolate is then combined, in desired proportions, with the washed, insoluble, fibrous residue from the extraction of the soybean material to produce a wet, protein concentrate. An edible, inorganic alkali or inorganic buffering agent or admixture is added to the soy protein concentrate material to raise (neutralize) the pH of the concentrate to within the range of from 5.5 to 10.5.

The resulting wet, neutralized, soy protein concentrate product, which may be in the form of dough, is then dried; by spray drying the wet concentrate, or by drying a thin layer of the wet concentrate under vacuum. If desired, the fibrous residue and/or soy protein isolate curd may be subdivided or comminuted in a wet state before they are combined, or the concentrate may be subdivided before or after it is dried. Although soybean flour and the like can be used in this process, there are economic advantages in using soybean flakes.

Example: 100 grams of defatted soybean flakes were suspended in 500 ml of water to which was added 0.5 gram of sodium sulfite. The pH of the slurry was adjusted with 50% by weight aqueous caustic soda to a pH of 8 and was agitated for 10 minutes, followed by the screening of the material through an 8 mesh screen. The flakes were washed on the screen with water and were then suspended in 500 ml of water and again agitated for a few minutes, followed by screening and washing with water.

The solids content of the washed fibrous flakes or residue was 10.6% by weight and the flakes had a protein content of 6.2% by weight which on a dry basis was calculated as 58.3% by weight protein. The weight of the dry flakes on dry solids basis was 30.4 grams which was calculated as 33.8% by weight of the initial soybean flakes. The liquor extracts from the screening steps were combined. Hydrochloric acid was added to adjust its pH to 4.2. The soy protein isolate curd was allowed to settle and the supernatant whey decanted off. This procedure was repeated two times to purify the soy protein isolate curd. The solids content of the washed isolate curd was 6.4% by weight and the protein content was 5.7% by weight which is equivalent to 90% by weight protein. The weight of the protein curd was 500 grams which was calculated as 32 grams of solids, indicating 35.5% by weight of the starting soybean material was recovered in the protein curd fraction.

The discarded liquid whey from the process was combined to a total volume of 1,530 cc. The solids content was 1.41% by weight which was calculated as 21.4 grams of solids or 24% by weight of the starting soybean material. The soy protein isolate curd and the washed soy protein flakes or residue were combined to form a soy protein concentrate. The concentrate was centrifuged, and the supernatant liquid was discarded. The solids content of the recovered centrifuged concentrate material was 21.7% by weight and the protein content was 16.0% by weight which was calculated as 73.5% by weight protein on a dry basis in this fraction. The weight of the centrifuged concentrate material was 302.5 grams which was calculated as 65.5 grams, indicating that in the combined fraction of the washed soy protein flakes and the washed soy protein isolate curd, 73% by weight of the starting soybean material was recovered in the concentrate.

The soy protein concentrate material from the centrifuge was then neutralized with a 10% aqueous sodium hydroxide solution to a pH of 6.5. The concentrate product had the appearance of a heavy sticky batter. It was placed on trays and dried at 130°F under vacuum. As soon as the vacuum was pulled, the viscous material expanded to approximately fivefold its original volume. This expansion of the concentrate permitted rapid drying similar to foam drying. The dried soy protein concentrate of this example was lighter and had less color than a dried soy protein concentrate prepared by conventional methods.

Water was added to separate portions of the dried soy protein concentrate material produced in this example and a dry soy protein concentrate made by conventional methods to provide 5% by weight concentrate samples that were tasted. The concentrate of this example had less of a beany or chalky taste or flavor. In addition, it was noted that when the dried soy protein concentrate product of this example was suspended in water, the resulting product was more translucent and there was less of a tendency of the insoluble hemicellulose to settle on standing, than when the dried, neutralized soy protein concentrate prepared by conventional methods, was suspended in water.

Soy Protein Deposited on Cellular Material

The process developed by *R.A. Hoer and F.E. Calvert; U.S. Patent 3,649,293; March 14, 1972; assigned to Ralston Purina Company* produces what might be considered a reconstituted soy concentrate. In this process the soy protein is precipitated from solution onto the insoluble cellular material of defatted soybean meal or flakes and then heated. Unlike concentrates prepared by alcohol extraction, the protein is not denatured and has excellent emulsifying, binding and water absorbing properties. The process is best accomplished by forming a slurry with defatted soybean flakes and water. The slurry is agitated for a sufficient time to extract the soluble protein and carbohydrate materials contained within the physical cellular structure of the original flakes. The slurry consists of the suspended solid, cellular seed material with a major portion of the protein and carbohydrate materials in solution. The aqueous slurry may be formed at a pH of from 6.5 to 9 and preferably from 6.8 to 8 to speed up solution formation. As an alternative, the protein and carbohydrate materials may be dissolved out of the flakes in an acid solution having a pH of from 1.2 to 2.5. The pH of the slurry is adjusted to the isoelectric point of the protein to cause the protein to precipitate out of solution on and into the surfaces of the liquid swollen solid thereby depositing a rich protein coating on the cellular seed material. The slurry is adjusted to a pH of from 4 to 5 and preferably to 4.5. The soluble extracted carbohydrates remain in solution.

The solid protein coated cellular material is separated from the liquid as completely as practicable and washed to remove substantially all of the soluble carbohydrates that may remain in the protein coated solid material. The solid material is then suspended in water to obtain an appropriate solids content and the slurry is adjusted to a pH of from 5.5 to 7.5 (preferably 6.8) to obtain a dispersible product. If an acid product is desired, the pH is adjusted to 2.0 to 3.5 (preferably 3). Where it is desirable not to adjust the pH, the slurry may be maintained at the isoelectric point, e.g., pH 4.5, for further processing. The slurry in any one of the pH ranges may be subjected to dynamic instantaneous heating at an elevated temperature range while also being dynamically and physically worked, preferably almost simultaneously. A most satisfactory way of achieving this is to pass the slurry through a device commonly known as a jet cooker.

For example, it includes adjacent jet nozzle orifices, normally concentric, through which the slurry and the pressurized steam used as a heating agent are ejected at high velocities in intersecting flow patterns, so that each tiny bit of slurry is instantly and dynamically heated by the steam while simultaneously being subjected to severe physical forces at the nozzle. The physical working of each tiny portion is believed to expose the undesirable flavor components associated therein to physical action, and this physical action with the elevated temperature heat treatment is believed to weaken and/or loosen the tenacious attraction between the undesirable components and the complex protein molecules, to an extent where these components can be removed with the flashed off vapors in a subsequent vacuum chamber treatment.

The temperature range to which the slurry is heated for the desired results is 220° to 400°F, although the temperature should not be in the lower region unless the product is subsequently vented into a high-vacuum chamber after being held under pressure in a special holding chamber. Normally, the temperature should be 285° to 320°F for best results. The slurry is held at such temperatures for a period of from 5 to 30 seconds. The heated slurry is then immediately released to atmospheric pressure or to a vacuum chamber to flash off the volatile, undesirable flavor components and subsequently the evaporatively cooled material is transferred, preferably, to a spray drier where the product is dried to the desired particle size. It is important to note that the above described treatment produces a product which is nearly sterile and, therefore, highly desirable for use in producing food products.

Example: A slurry of soybean flakes, and water was prepared with a water to flakes ratio of 14:1 by weight. The water had a temperature of 90°F prior to formation of the slurry and the slurry was agitated for about 30 minutes. The slurry was passed through a wet grinder where the solid particles were reduced to a particle size that would pass through a 0.010 inch perforated grinder screen. The ground slurry was adjusted to a pH of about 4.5 with phosphoric acid to deposit the protein on and into the surfaces of the solid cellular seed material. The protein coated slurry was concentrated in a centrifuge to about 25% solids by weight, diluted with 90°F water back to the original volume and again concentrated to 25% solids by weight.

The slurry containing the protein coated solids was diluted to 15% by weight solids and the resulting slurry adjusted to a pH of about 6.8 with sodium hydroxide. The pH adjusted suspended slurrry was fed to a jet cooker and heated instantaneously to 310°F, held at that temperature for about 5 seconds and, thereafter, injected into a vacuum chamber to flash off the excess vapors and volatile undesirable flavor components. The deflavored, essentially sterile protein product was fed to a spray drier under pressure of about 5,000 psi and exhaust temperature of 180°F to obtain a dry, particulate product. The properties of this product (A) were compared with a product (B) which is commercially available. Product B is produced by contacting soybean flakes with an acid solution at the isoelectric point of the protein (about pH 4.5) to dissolve out only the carbohydrate materials. The protein remains within the cellular structure. The solids are separated from the liquid and the resulting solid product is processed in a known manner to obtain a dry, neutral, protein product. Two tests were employed (Test No. 1 and Test No. 2) on products A and B to determine their properties of forming good emulsions (Consistency) and absorbing liquids (Fat Separation). Results are shown in the following table.

Product	Test number 1		Test Number 2			
			Before heating		After heating	
	Consistency	Fat separation	Consistency	Fat separation	Consistency	Fat separation
A	2.0	2.0	2.0	2.0	1.5	1.5
B	4.0	4.0	4.0	3.0	4.0	4.0

Consistency: 1=Thick, non-pourable; 2=Thick, pours slowly; 3=Thick, pours readily; 4=Thin, pours easily.
Fat separation: 1=no traces of free fat; 2=slight traces of free fat; 3=considerable traces of free fat; 4=substantial traces of free fat.

From the results shown in the table, it is readily seen that the product (A) has substantially better properties, e.g., emulsifying and absorbing.

MODIFICATION OF SOY CONCENTRATES

Reducing Microorganism Count

Soy protein concentrates having a reduced bacterial activity and a desirable viscosity have been produced by *L. Sair and I. Melcer; U.S. Patent 3,669,677; June 13, 1972; assigned to The Griffith Laboratories, Inc.* They found that the application of heat to effect bacterial control of the soy protein material can be used effectively only after the extraction and removal of soluble material at a lowered acid pH or in the vicinity of the isoelectric pH and before any neutralization step which involves adjusting or raising the pH of the soy protein material or drying step.

Briefly, the process covers the steps wherein (1) defatted soybean material is extracted with a liquid extracting medium at the isoelectric pH, (2) the liquid extract having an undesired soluble beany flavor-conferring material is removed and insoluble soy protein material is recovered, (3) subjecting this insoluble material to bactericidal heating at about 160°F at an acid pH to reduce the bacterial activity, (4) raising the pH of the heated soy protein material to 6 to 10.5, and (5) drying the material.

This process provides excellent control of bacteria and produces improved, edible soybean protein products by raising the temperature of a slurry of extracted soy protein material up to at least about 160°F, preferably about 180° to 212°F or 190° to 212°F, and maintaining the slurry at such temperatures for an effective period of time prior to adjusting or raising (neutralizing) the pH of the protein material in the range of about 6 to 10.5. By pinpointing the destruction of bacteria at this stage of processing, a soybean protein product can be produced having a thermophile count of less than 500/gram, or even less than 300/gram, and a total bacterial count of less than 50,000/gram, or less than 25,000/gram.

By going to a higher temperature for the bactericidal heating step, such as 175°F or higher, the apparent result is to extract the protein material enclosed in the insoluble gum cell network. Upon neutralization to a pH of 6 to 10.5, the extracted protein becomes available and the neutralized slurry has a higher viscosity. More importantly, a very viscous soy protein product is obtained on cooling. The cold (about room temperature) viscosity is an index of the ability of the protein to hold water and the more viscous the cold paste, the more desirable it is from the standpoint of emulsification and water retention. By raising the temperature of the bactericidal heating step up to 190°F, this effect is even more pronounced. The slurry of extracted soy protein material, heated to about 190°F, followed by neutralization and cooling, gives a paste with a high cold viscosity and with no flowability. By way of comparison, an unheated slurry, when neutralized, produces a paste which is almost fluid.

It is important that the soy protein material used in this process be extracted of soluble materials by being subjected to water at a pH in the vicinity of the isoelectric pH of the protein or glycinin content of the soybean protein material, and that the resulting solution be separated from the insoluble solids, followed by the washing of the insoluble solids to remove soluble beany taste conferring material. This process does not remove the desired protein material, but rather extracts the undesirable soluble material from the protein. The resulting extract is discarded. Only after extraction should the proteinaceous slurry, having a pH of not more than 4.8, be heated to at least 160°F, preferably higher, for times ranging between 10 minutes and 2 hours, control the bacteria concentration without producing the undesired side effects mentioned above.

Example 1: A soy protein slurry was prepared with 100 parts by weight of deoiled or defatted soy flour and 835 parts by weight of water. The slurry was held at 55°F and one part by weight of sodium bisulfite was added and the pH was adjusted to 4.2 using diluted hydrochloric acid (1:1). This formed a slurry having about 11% by weight solids. The microorganism count for the various steps indicated in the table below was determined by the method set forth in *Recommended Methods for the Microbiological Examination of Food,* second edition, edited by J.M. Shaif, chapter VI, pages 55 to 64, published by the American Public Health Association, 1966. The table shows that the treatment with acid and sodium bisulfite, as described above, caused about a 68% drop in the mesophile count and about a 20% drop in the thermophile count.

Time of count	Mesophiles/ml. of slurry	Thermophiles/ml. of slurry
After bisulfite addition but before acid addition	50,000	16,000
After acid addition	23,000	12,000
One hour after bisulfite and acid treatment	18,000	13,000

A portion of the original 11% by weight slurry was agitated for 30 minutes and then permitted to settle for 6 hours. The supernatant liquid was decanted, leaving an insoluble soy protein slurry which was then washed with a three-fold volume of water. The wash water was discarded and the insoluble protein material was again suspended in water to produce a slurry having about 14% by weight solids. This slurry was divided into 2 parts, 1 of which parts was heated in the acid condition (a pH of 4.2) at 160°F for 1 hour. The pH of the second part was raised to a pH of 6.5 using an aqueous solution of sodium hydroxide (the term sodium hydroxide as used herein includes commercial sodium hydroxide such as commercial caustic soda) and was heated at 160°F for 1 hour. Using the same bacteria assay procedure referred to above, the results shown in the table below were obtained.

Duration of heating at 160°F., min.	Heating of extracted protein material		Heating of neutralized protein material	
	Mesophiles/ml. of slurry	Thermophiles/ml. of slurry	Mesophiles/ml. of slurry	Thermophiles/ml. of slurry
0 [1]	24,000	20,000	48,000	24,000
15	27,000	22,000	38,000	22,000
30	23,000	15,000	32,000	21,000
60	19,000	10,000	31,000	19,000

[1] The "0" time was the point at which 160°F. was reached.

The above table shows that a one hour heat treatment of the soy protein material obtained after extraction killed about 50% of the thermophiles; whereas a similar heat treatment after neutralization to a pH of about 6.5, killed only about 21% of the thermophiles.

Example 2: Using the procedure outlined above in Example 1, the bisulfite addition, acid addition, and washing steps were repeated. The recovered slurry having about 14% by weight solids was then heated to about 180°F. Approximately 15 minutes were required to to raise the slurry temperature from about 55° to about 180°F. The following table shows the bacterial count, as determined by the procedure referred to above, after maintaining the about 180°F temperature for the periods of the time indicated therein.

Duration of heating of extracted protein at 180° F., min.	Mesophiles/ ml. of slurry	Thermophiles/ml. of slurry
0 [1]	22,000	61,000
15	890	900
30	190	60
60	<10	30

[1] Point at which 180° F. was reached.

Partially Hydrolyzed Concentrates

Soy containing cereals having improved tenderness, i.e., remain crispy and tender on exposure to milk, have been produced by *W.T. Bedenk; U.S. Patent 3,687,686; August 29, 1972; assigned to the Procter & Gamble Company.* Soy concentrate having a protein content of 70 to 90% on a dry basis is partially hydrolyzed in the presence of a proteolytic enzyme. It can then be combined with a cereal grain such as corn, rice, oats or wheat. The resultant mixture can be processed to produce a cold cereal product having a high nutritive value, i.e., a protein content greater than 20%.

The partial hydrolysis of soy concentrate in the process is accomplished by mixing the soy concentrate, a proteolytic enzyme and water for a length of time at an elevated temperature. Unexpectedly, when soy concentrate is reacted with water in the presence of a proteolytic enzyme and thereafter processed into a cold cereal product the product that results is more tender with less tendency to develop toughness and has greater crispness retention than does a cold cereal product containing soy that has not been subjected to a partial hydrolysis.

The enzymes useful in the process are selected from known proteolytic enzymes or mixtures thereof extracted from animal, plant, fungal, or microbial sources. A primary consideration is that it must not contribute a significantly objectionable flavor or odor to the final product. Some examples of proteolytic enzymes effective in the soy concentrate partial hydrolysis are papain, pepsin, bromelin, ficin, alcalase, maxitase, thermoase, pronase and mixtures thereof. The preferred amount of enzymes added to the reaction mixture is 25 ppm to 2,500 ppm by dry weight of the soy concentrate. Greater amounts can be used but exert no measurable beneficial influence on the speed of the hydrolysis. Lesser amounts as a practical matter are not used because of the length of time it would take for the reaction to proceed to the desired end point. The most preferred range is 100 ppm to 600 ppm by dry weight of the soy concentrate.

The amount of water needed for the partial hydrolysis reaction is basically determined by apparatus limitations. That is, the lower limit of water is determined by the capability of the mixing equipment. The lower the level of water the more viscous will be the resultant mixture. On the other hand, an excessive amount of water in the partial hydrolysis reaction would necessitate additional work in reducing the water level in subsequent processing steps. The preferred level of water is 50 to 80% based on the total weight of the mixture. The most preferred level is 55 to 60% based on the total weight of the mixture.

The temperature during the reaction is not critical provided it is not so high as to kill the enzyme activity. A temperature range of 80° to 160°F is suitable for the hydrolysis reaction with the reaction proceeding faster at the higher temperatures. The most preferred temperature range is 120° to 130°F. The time for the reaction to be completed depends on the particular temperature, water level, and enzyme level used. In general 1 to 120 min are sufficient to allow the reaction to come to completion. Times in excess of 120 min at a reaction temperature should be avoided to prevent more hydrolysis of the soy concentrate than is desired. Most preferably 1 to 5 min allows for a degree of hydrolysis sufficient to make an acceptable product. The following example illustrates production of the cold cereal products. Unless otherwise indicated all percentages given are on a weight basis.

Example: A formulation containing 100 grams soy concentrate, 0.05 grams bromelin (500 ppm of soy concentrate), 300 grams brewer's grits, 30 grams sucrose and 10 grams salt, was prepared. 125 grams of water and the proteolytic enzyme, bromelin are blended in a vessel until the bromelin is dissolved. Soy concentrate having a protein content of 75% is then added slowly to the water-bromelin mixture to form a mixture having a total water content of 56%. A 180°F water jacket is maintained around the vessel so that after completion of the soy concentrate addition the mixture temperature is 125°F. This mixture is held at 125°F for approximately 3 minutes.

Brewer's grits (corn), sucrose, and salt are blended with 230 grams water to make a mixture having 40% water. This mixture is added to a rotary cooker and cooked for 1 hour at 250°F under a pressure of 18 psig. At the end of this time the corn is thoroughly gelatinized. The mixture is then placed in a forced air dryer for 16 hours at 160°F after having been passed through a hammermill to break up any lumps. The moisture content at this point is 2.5%.

The partially dried cooked corn mixture is then blended with the soy concentrate mixture and extruded under 500 psig and a die temperature of 200°F. The mixture is extruded into strands having a circular cross section of 3/16 inch. These strands are then sliced into pellets of 3/16 inch length. These pellets are next passed between a 2-roll mill to produce a flake having a thickness of 0.010 inch. The flakes are then placed in a rotary drier having an outlet temperature of 200°F and a hold time of 4 minutes. The flakes after this operation have a moisture content of 10 to 12%. A jet zone hot air oven operated at 310°F and with a hold time of 3 minutes puffs the flakes. The resultant flakes having a protein content of 23% are more crisp and more tender after exposure to milk and have a more pleasant taste than flakes made by the same process with no bromelin addition. The flakes made by this example also are easier to puff than are the control flakes.

SOY PROTEIN ISOLATES

The purest soy protein form is the isolates from which the bulk of the oil and carbohydrates have been removed leaving less than 10% of nonprotein material such as ash and minor constituents. The isolates have a protein content of 90 to 95% and have most of the beany or bitter flavor constituents removed. The isolates are produced by dissolving the soy product above or below the isoelectric point of the soy protein and then adjusting the pH of the solution to the isoelectric point to precipitate the purified protein.

PREPARATION OF ISOLATES

Aqueous Extraction of Heat Treated Soy Protein at pH 6 to 8

A process for the isolation of soy protein was developed by *S. Circle, P. Julian, and R. Whitney; U.S. Patent 2,881,159; April 7, 1959; assigned to The Glidden Company* which related to the isolation of soy protein in a substantially unhydrolyzed, relatively undenatured state from heat treated oil-free protein source material. Instead of alkaline extraction at pH of 9 to 11 or higher, use of heat treated or debittered flakes as the protein source material and extraction with an aqueous solution at pH 6 to 8 at 100° to 180°F resulted in an excellent yield of bland flavored, light colored, substantially unhydrolyzed protein isolate.

The protein thus obtained is substantially unhydrolyzed since it has not been subjected to the deleterious effects of strong alkalies. The recovery of total protein from the heat treated source material was increased not only by limiting the losses due to hydrolysis but also by recovering from the protein source material a substantial amount of the albuminous protein usually lost in the whey. The albuminous protein is, in the presence of caustic alkali, solubilized and not precipitated at the isoelectric point of the globular protein. However, it is coagulated by heat and rendered recoverable with the spent residual flakes.

Temperatures above 130°F effect the recovery of increasing amounts of the albuminous protein; at temperatures above 180°F the heat denaturation of the globular protein increases inordinately. Within the pH range of 6 to 8, it is practical to carry out the extraction between 130° and 160°F. The process is preferably carried out at 140°F and at a pH of 7 in order to obtain the maximum yield of high quality vegetable protein in substantially unhydrolyzed condition. The process, however, is not limited to the extraction of heat treated, i.e., debittered soy protein source material with water at between 100° and 180°F. Other aqueous extractants such as dilute lime, alkali metal and ammonium hydroxides and the like can also be used within the pH range of 6 to 8. In each instance the yield and quality of protein isolated from heat treated protein source material is excellent and within the requirements of a practical commercial procedure. Further, the yield of protein can be increased by subsequent treatment of the extracted protein source material with aqueous caustic alkali solutions.

Example: A mixture of 100 parts of solvent-extracted soybean flakes containing 11% moisture and which have been prepared by a process including the step of removing the last traces of extraction solvent from the flakes by treatment with superheated steam and 1,400 parts of water was agitated for 1.55 hours at 140°F. The slurry was filtered over screens to give 1,200 parts of aqueous liquor. The residual flakes were washed by being reslurried for 10 minutes at 80°F in 1,200 parts of water. The slurry was filtered over screens to yield 1,200 parts of reslurry liquor and the spent flakes were pressed to yield 100 parts of press liquor. The pressed spent flakes were dried and amounted to 38 parts at 12% moisture.

The combined reslurry and press liquors (1,300 parts) were added to the aqueous liquor (1,200 parts) and the entire mass, after being clarified by centrifuging, was acidified to pH 4.6 by addition of aqueous hydrochloric acid, whereupon a proteinaceous curd was precipitated. The curd was permitted to settle and the whey (1,700 parts) was decanted therefrom. The wet curd was washed by reslurrying it in 1,700 parts of hot (150°F) water after which the wet curd was separated by filtration. The curd was dried to give 32 parts of substantially unhydrolyzed bland-tasting isolated protein of light color at about 10% moisture. Comparable results were obtained when the soybean flakes were extracted with aqueous ammonium hydroxide at pH 7, with aqueous caustic soda at pH 7.5, or with aqueous lime in sufficient amount to give a pH of 8 in the extracting liquor.

Alcohol Treated Defatted Meals for Preparing Isolates

Bland protein isolates are produced by *N.J. Beaber and J.H. Obey; U.S. Patent 3,043,826; July 10, 1962; assigned to J.R. Short Milling Company* from defatted soybean flours. This process is based on the finding that desirable protein products can be recovered from soybean material if deoiled soybean material which contains alcohol-soluble, nonoil solubles is treated before recovering the protein with a lower aliphatic alcohol at a temperature below 58°C and under conditions that the alcohol does not remove any substantial amount of solubles from the soybean material. The protein is then extracted from the alcohol-treated soybean material by an aqueous medium at a pH substantially different from pH 4.8 to 5.2. The protein is then precipitated from the aqueous extract at a pH of 4.8 to 5.2, and the resulting protein curd is then recovered and dried at a temperature not exceeding 58°C.

This differs from prior art processes in that no attempt is made to remove before recovery of the protein the alcohol-soluble components described by those skilled in soybean protein recovery as responsible for the bitter or beany taste and responsible for inhibiting gelation of the recovered protein. Yet the protein products obtained by the present process are both completely bland in taste and capable of forming thermally irreversible gels. Other advantages also result from the present process. The protein curd precipitated at pH 4.8 to 5.2 has a high solids content, in excess of 40% when extraction is carried out at a pH not exceeding 10.5, and often more than 50% when extraction is at a lower pH. The protein curd is glutenous in nature; it tends to exude water. It can be extruded preparatory to the drying step without the usual necessity for recycling dried product to make it extrudable. Finally, because of its high solids content, such curd can be dried much more quickly and under less severe conditions than the curds obtained by conventional procedures.

Defatted soybean materials are contacted intimately with a monohydric aliphatic alcohol such as methanol, ethanol, propanol or isopropanol. The time of contact is not critical, being as short as one minute or as long as several hours. The temperature of the soybean material during alcohol treatment must not exceed 58°C; successful results have been obtained at temperatures of 0° to 5°C. The alcohol treatment step is carried out so that the alcohol does not remove alcohol-soluble components of the soybean material. This is accomplished by first using the alcohol to extract an initial quantity of soybean material until the alcohol becomes saturated, at the temperature used, with the nonoil soybean solubles, this initial quantity of soybean material being discarded. The saturated alcohol is then used to contact fresh soybean material; and since the alcohol cannot take up any more of the solubles present in the soybean material, no extraction occurs.

Enough alcohol is used to thoroughly wet the soybean material. At the lower limit, as to alcohol, the alcohol to soybean material weight ratio can be 1:2. Under most operating conditions, it is desirable to employ an alcohol to soybean weight ratio on the order of 6:1, to allow ease of handling in a continuous procedure as hereinafter described. The upper limit as to alcohol is not critical, except for practical economic considerations, and the weight ratio may be as high as 20:1.

The oil-extracted, alcohol-treated soybean material is combined with water amounting to at least several times the weight of the soybean material and extracted with continual agitation for at least 30 minutes at a pH substantially different from 4.8 to 5.2. Thus, the extraction may advantageously be on the alkaline side of the isoelectric point, at pH 6.0 to 11.0, with the pH being adjusted by addition of sodium hydroxide. On the other hand, the extraction may be at a pH of 2.0 to 4.0, with that pH obtained by addition of sulfur dioxide or a mineral acid.

The aqueous extract is then adjusted to a pH of 4.8 to 5.4 by addition of an acid or alkali. A light colored, glutenous curd precipitates promptly and after a short settling period in the precipitator is easily removed by centrifuging. If the soybean material was freed of alcohol at 50° to 58°C prior to extraction of the protein, the curd recovered will have a solids content of 35 to 50% by weight or more. The glutenous curd recovered is unlike the curd recovered from soybean material treated with hexane alone in that it is glutenous in nature and exudes water. Thus, if the protein curd of this process is merely allowed to stand at room temperature for 15 to 30 minutes, it will be observed that a substantial amount of water has separated from the curd. The resulting protein product is a light colored, substantially all soybean material having the two characteristics desirable for food applications, namely, bland taste and the ability to form a thermally irreversible gel.

Example: The starting material was 400 pounds of commercially prepared hexane-extracted soybean flakes having an oil content less than 1% by weight and a moisture content of 4%, the hexane extraction having been carried out at temperatures below 58°C. The hexane-extracted flakes were delivered to a continuous screw type, countercurrent extraction apparatus operating with a hold-up time of 35 minutes. Ninety-five percent ethanol was continually circulated through the apparatus at 22°C countercurrent to the flakes, the liquid treating medium having been previously contacted with similar soybean flakes until saturated with nonoil soybean solubles. The ratio of liquid to flakes by weight was 6:1.

After the extraction and draining off of solvent, the drained flakes from the extractor were delivered to a desolventizer, the remaining liquid being removed thermally from the flakes at a temperature of 55°C, the hold-up time in the desolventizer being 55 minutes. The full amount of treated flakes was delivered from the desolventizer to a mixer with a slow speed rotary agitator and there extracted with 1,330 gallons of water, adjusted to pH 7.2 with 30% NaOH at 22°C, the extraction period being 75 minutes. The aqueous extract was recovered by centrifuging. By means of 8:1 acetic acid, the pH of the extract was then adjusted to 5.0. A light colored curd precipitated promptly and after a 30 minute settling period, the curd, totalling 222 pounds, was recovered by centrifuging. The solids content of the curd was 43.2%.

The curd was then extruded through a meat grinder without addition of dried material and dried in a Roto-louvre dryer. The air entry temperature in the dryer was 60°C. The wet bulb temperature was held below 43°C. The dried product was ground in an attrition mill and classified to +30 mesh. The total yield of protein amounted to approximately 24% of the weight of the starting material. A 10% water suspension of a portion of the product was made at pH 4.9 with the addition of 2.5 grams ammonium hydroxide (28% NH_3). After 2 hours the dispersion formed a thermally irreversible gel at 89°C and after

24 hours at 91°C. The dried product was a tan, friable material completely bland in taste.

Reverse Osmosis for Isolating Proteins

A soy isolate containing proteins which are both soluble and insoluble at the protein isoelectric point has been obtained by *D.R. Frazeur and R.B. Huston; U.S. Patent 3,728,327; April 17, 1973; assigned to Grain Processing Corporation.* In this process, an aqueous homogenized slurry of soybean particles is subjected to a separation so as to separate insoluble materials from liquid and the liquid is subjected to reverse osmosis to obtain as a retentate soy proteins including both those which are insoluble and soluble in water at the protein isoelectric point.

Suitable reverse osmosis equipment is commercially available from such manufacturers as Dorr-Oliver Corporation and Havens OSMOTIK Corporation. The choice of membrane is routine, being governed by the above considerations. Suitable representative membranes are Dorr-Oliver Corporation XPA and APA membranes and the Havens 215 membrane.

Concentration is achieved during the reverse osmosis procedure by simply permitting the retentate volume to decrease as diffusable materials and water pass from the retentate to the diffusate without the admission of water. Washing the retained proteins with about 0.5 to 4.0 volumes of water is generally sufficient to obtain a high quality soy protein isolate upon subsequent drying. Concentration provides both purification and economies in processing. Concentration of up to threefold or more may be practical.

The retentate from the reverse osmosis step is removed from the reverse osmosis equipment and dried to provide a high quality soy protein isolate. The retentate which is retained by the membrane in the reverse osmosis procedure possesses a number of very important characteristics. The retentate generally contains about 9 to 14% more of the total soybean protein content than is the case of soy protein isolates obtained by conventional protein precipitation procedures. Not only is this increased protein recovery important from an economic standpoint, but in addition the product possesses improved nutritional advantages because the recovered protein contains high levels of sulfur-bearing amino acids. The process provides definite economic advantages with respect to costs of equipment, equipment operation and in the overall cost of producing the soy protein isolate.

Example: A 5% weight/volume slurry of white soybean flakes is made with cold tap water. After complete mixing of the soy flakes and water, the resulting slurry is continuously introduced into a triplex pump homogenizer and homogenized at 3,000 psig. The homogenized soy slurry is then separated into supernatant liquid and sediment by passing the slurry through one or more conventional centrifuges. The sediment is discarded and the supernatant liquid is concentrated fourfold by reverse osmosis through a Dorr-Oliver XPA membrane in Dorr-Oliver reverse osmosis support modules. The retentate is then washed with 2 to 3 volumes of water during reverse osmosis.

The retentate is then removed from the reverse osmosis equipment and dried in a conventional spray drier to give a soy protein isolate characterized by high nutritive value, high protein solubility (>75% NSI), excellent flavor, light color (white or light tan), superior water binding properties (viscosity of 2% solution 12 cp at 30°C), essentially all (>75%) protein in a natural (nondenatured) form, improved stability to heat (<20% denatured at 99°C for 5 minutes), improved fat emulsifying properties (>200 ml soy oil emulsified per gram of isolate) and controllable heat gelation characteristics.

A distinguishing feature of this soy protein isolate which is of considerable importance is its gelling characteristics. Soy protein isolates made by known prior art precipitation procedures generally form gels in 7% aqueous solutions at 65°C in a period of 10 to 15 minutes. However, soy protein isolates made by this process do not gel under these conditions even if the time is greatly extended. The absorption of calcium by the soy isolate of this process is required to permit heat gelation. Typically, approximately 0.2 mg of soluble calcium per gram of these isolates is required to permit heat gelation under the above specified conditions. Generally, addition of approximately 0.3 mg of soluble calcium per gram of soy isolate will cause gelation of a 7% aqueous solution of the soy isolates at room temperature (25°C) in 30 minutes. This property is of great value when the soy protein isolate is used in foods such as in processed meats where both calcium binding and heat gelation are desired.

Low Viscosity Soy Isolates

The process of *P. Melnychyn and J.M. Wolcott; U.S. Patent 3,630,753; December 28, 1971; assigned to Carnation Company* discloses a method of controllably decreasing the viscosity of soy protein in aqueous dispersion. The method includes treating a soy protein dispersion with a treating agent at selected temperature and for a time sufficient to effect a desired viscosity reduction without substantial hydrolysis of the soy protein. The protein is then separated from the treating agent.

Such treating agent may comprise halogens, inorganic salts containing both halogen and oxygen atoms, soluble inorganic persulfates, azodiamides or thiol-containing reducing agents. As one example of the process, soy protein after treatment with the treating agent is contacted with an odor- and flavor-improving agent selected from aliphatic mono- and polyhydroxy alcohols, ethers and ketones or mixtures thereof at a time, temperature and concentration sufficient to improve the odor and flavor. Such contact is effected by first precipitating the soy protein at about its isoelectric point, i.e., about pH 4.5 and then washing the precipitate with the improving agent. In another example, an aqueous dispersion containing the treated soy protein is mixed with the improving agent.

Referring more particularly to the steps of the present method, the aqueous soy protein dispersion is obtained by alkaline extraction of soy protein from a suitable soy protein source such as defatted soy flakes. Defatted soybean meal or the like can also be used as the starting material. Alkaline extraction conditions are mild and rapid with low temperature and short time being employed to avoid substantial hydrolysis of the soy protein.

There is no necessity of using a high temperature to carry out the aqueous alkaline extraction. Instead, the temperature is usually kept below 100°F and preferably at about room temperature (70° to 75°F). During the initial stages of alkaline extraction, the pH of the slurry is 9.5 to 12.5 but preferably not above 11.7 to 12.2. Stirring is advocated to increase contact of the protein with the extracting agent and reduce required treating time.

After the extraction is under way and preferably substantially completed, for example, in about 1 minute, the treating agent may be added directly to the slurry. There is no necessity to separate the extraction medium from the protein source before treating with the treating agent, although this protein source-separating step can be carried out if desired.

The treating agent comprises an agent capable of effectively decreasing the viscosity of soy protein when dispersed in an aqueous medium. It can be an inorganic salt containing both halogen and oxygen atoms, for example, potassium iodate, sodium chlorate, sodium chlorite, potassium bromate, etc. or can be a halogen, e.g., chlorine or bromine. It can also be a suitable inorganic persulfate, such as sodium persulfate, ammonium persulfate or the like. The treating agent can be an azodiamide, for example, azodicarbonamide, or an active sulfhydryl-bearing reagent, such as dithioerythritol, mercaptoethanol or cysteine, capable of reacting with disulfide linkages in soy protein.

For most purposes, the concentration of treating agent can be very small, for example, about 1.4×10^{-3} M. A typical concentration when chlorine is used is 0.1 to 0.3%, preferably 0.1% by weight of the protein source. The most desirable concentration for the treating agent will depend on the concentration of the protein source, the particular treating agent selected, and the temperature and time used in the viscosity-reducing reaction. The upper limit of treating agent concentration should be below that which could impart an undesirable flavor and odor to the soy protein.

The time, pH and temperature for the alkaline extraction of the soy protein are as previously specified. However, when the treating step using the treating agent of the present method is carried out, another set of conditions preferably is employed. Accordingly, the treating step is carried out at a temperature of 160° to 180°F, preferably 170°F. This high temperature necessitates maintaining the pH of the aqueous medium at not substantially in excess of pH 10.5 to avoid hydrolysis of the soy protein and development of sulfurous odors in the extract. Optimum pH is about 8.5. The treating step can be short, for example, 1 minute, preferably with stirring, or can be extended up to 10 minutes or more. The time-temperature relationship is balanced to avoid substantial hydrolysis of the soy protein while still permitting complete reaction between the treating agent and soy protein. The following example will illustrate the process in greater detail.

Example: Commercially available defatted soybean flakes ground to about 60 mesh were immersed in water at a 1:10 ratio of the flake weight-to-water volume. The water contained sufficient sodium hydroxide to provide a pH of 8.5 and also contained except in controls a suitable concentration of treating agent previously added to the alkaline water. Alkaline extraction of the soy protein and simultaneous viscosity reduction of the extracted soy protein were carried out over a period of 5 minutes at 170°F, after which the resulting dispersion was centrifuged at 4,000 x gravity for 5 minutes, the temperature dropping to 100°F.

The clarified supernatant was decanted and the soy protein in the decanted supernatant precipitated at 100°F by addition of hydrochloric acid until the pH of 4.5 was reached and a protein precipitate formed. The precipitate was then separated and recovered by centrifuging it at 4,000 x gravity for 5 minutes. This precipitate was then washed in water, centrifuged, and then dissolved in water at pH 7.0, frozen and lyophilized to obtain a dry powder for viscosity determinations.

In those instances in which the soy protein was subjected to subsequent treatment to improve the odor and flavor, the protein precipitate (at pH 4.5) was dispersed in a quantity of improving agent at 70°F with stirring to provide a concentration of the improving agent of 25 to 90% by volume. The protein remained in contact with the improving agent for 5 minutes, after which a 5 minute centrifugation at 4,000 x gravity was carried out, the supernatant was decanted, the resulting soy protein precipitate was water washed, recentrifuged, separated and then dissolved in water at pH 7.0, frozen and freeze dried.

The lyophilized product, whether or not it had been subjected to the treating agent and/or the improving agent, was pulverized in a Waring blender to a uniform particle size and then dissolved in a predetermined amount of water to provide a specific protein concentration for viscosity measurements. Solution of the protein was accomplished by mechanical stirring for 15 minutes, followed by homogenization at 1,500 psi and 75°F. Immediately thereafter, viscosity measurements were made on each aliquot at 75°F using a Brookfield viscometer. After initial readings, dilutions with measured amounts of water were carried out to obtain viscosity readings at more dilute protein concentrations. Typical test results utilizing the above indicated procedure are set forth in the table on the following page.

In all of the tests reported in the table, the soy protein obtained after the treating step and also that product obtained after the odor and flavor improving (where both steps were used) had a substantially reduced viscosity in contrast to a control which was merely alkaline extracted, this viscosity being evident over a wide range of concentrations. Although the treating agents varied somewhat in their relative effects on viscosity reduction, they were all effective even when used in very small concentrations.

In those tests where the soy protein was only subjected to the flavor and odor improving agent, the viscosity was slightly increased over that of the control. Parallel tests were performed using sodium persulfate and mercaptoethanol as treating agents, with and without subsequent treatment with odor and flavor improving agents. Viscosity results were obtained which were comparable to those set forth in the following table.

Treating agent	Flavor and odor improving agent	Protein concentration, percent	Viscosity in centipoises
Alkaline extracted control (no treating agent)	None	2 4 6 8	20 90 560 1,820
Alkaline extracted (no treating agent)	86% ethanol	2 4 6	35 250 2,300+
Chlorine 1.4×10^{-3} M	None	2 4 6 8 10	5 10 20 70 350
Do	86% ethanol	2 4 6 8	10 20 40 480
Do	80% isopropanol	2 4 6 8	10 20 125 890
Chlorine 7×10^{-3} M	86% ethanol	2 4 6 8	5 15 100 730
Chlorine 1.4×10^{-2} M	do	2 4 6 8	5 15 85 650
Ammonium persulfate 1.4×10^{-3} M	do	2 4 6 8	10 20 75 570
Sodium iodate 1.4×10^{-3} M	None	2 4 6 8 10	5 10 30 110 840
Potassium bromate 1.4×10^{-3} M	do	2 4 6 8 10	5 10 30 85 375
Sodium chlorate 1.4×10^{-3} M	do	8	265
Bromine 1.4×10^{-3} M	do	8	195
Azodicarbonamide 1.4×10^{-3} M	do	≃7	18
Dithioerythritol 1.4×10^{-3} M	do	≃7	7.3
Sodium chlorite 1.4×10^{-3} M	do	≃7	22.6

ISOLATES HAVING IMPROVED SOLUBILITY OR DISPERSIBILITY

Water-Soluble Soy Concentrates, Isolates and Whey

A process for forming soy protein concentrates and soy protein isolates having high solubility in water and related soy products is disclosed by *P.U. de Paolis; U.S. Patent 3,682,646; August 8, 1972; assigned to Hunt-Wesson Foods, Inc.* The process involves mixing defatted, dehulled soybean material in the form of flakes or meal in water at a temperature between 175° and 212°F (to destroy the trypsin inhibitors, urease, etc.) and maintaining a pH of preferably between 6.8 and 9 throughout the processing.

The aqueous soybean mixture is cooled to between 100° and 140°F. Peroxidase enzymes may be added to the aqueous soybean mixture to decompose peroxides contained therein or antioxidants may be added. The insoluble materials are then separated from the water-soluble soybean solution and mixed with an alcohol-water solution such as 70% methanol to yield insoluble soy protein and semisoluble soy protein. The water-soluble portion of the aqueous soybean mixture is mixed with an alcohol-water solution such as 70% methanol to yield soy protein concentrates and/or soy protein isolates and soy whey solids and sugars. The process is illustrated by the flow diagram in Figure 6.1 and the following example.

Example: 1,000 pounds of defatted, dehulled soybean flakes were mixed with 9,000 pounds of water at 185°F for 20 minutes. The pH was raised to 7.0 by the addition of a buffer such as sodium hydroxide. The mixture was allowed to cool to 130°F and then

Fermcolase (a peroxidase enzyme) was added to provide 1,000 Baker units per 100 grams of flakes and reaction was allowed to proceed for 20 minutes. Separation of the insoluble material (solids line **16**) from filtrate **18** takes place by means of rotary vacuum filter **14**.

FIGURE 6.1: FLOW SHEET FOR MANUFACTURE OF SOY CONCENTRATES, ISOLATES AND WHEY

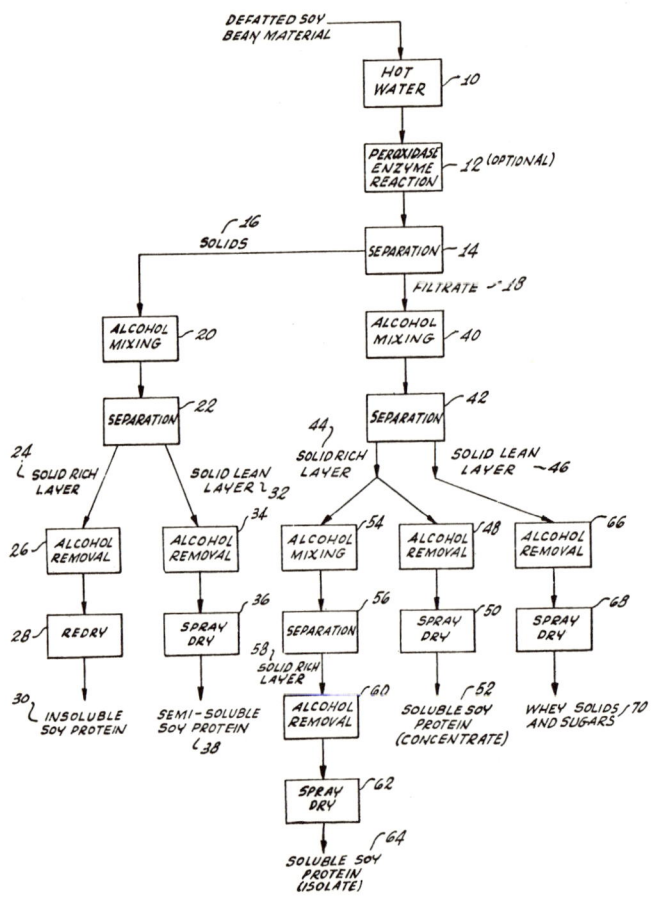

Source: P.U. de Paolis; U.S. Patent 3,682,646; August 8, 1972

A 70% methanol solution is then added to the insoluble material (step **20**), the insolubles line containing 667 pounds solids and 2,612 pounds water. About 7,600 pounds of 70% methanol is added. After about 20 minutes the supernatant (solid-lean) layer is separated from the sludge layer in separation step **22**. The supernatant layer is then stripped of alcohol (step **34**) and dried. The resultant product weighs 112 pounds, contains about 6% moisture and 80.50% protein of about 52.1% NSI solubility. Alcohol is also removed from the sludge (solid-rich) layer and dried. The resultant product weights 538 pounds and contains about 3.4% moisture and 63.10% protein of 11.46% NSI solubility.

Soy Protein Isolates

Turning to the filtrate **18**, it contains about 333 pounds solids and 6,476 pounds water making a total of 6,809 pounds entering a methanol reactor or alcohol mixing step **40**. 11,720 pounds of 70% methanol is added to the filtrate. Mixing continues for about 20 minutes in step **40** and thereafter separation (step **42**) as by centrifuging, takes place into two phases, the solid-lean layer **46** and the solid-rich layer **44**. The solid-rich layer has alcohol removed therefrom by a flask desolventizer (step **48**) and is thereafter spray dried (step **50**). The resultant product weighs 185.6 pounds, contains about 5.2% moisture and 72.50% protein (dry basis) having an NSI solubility of about 98.8%.

In order to obtain the isolate, the solid-rich layer **44** rather than being processed as just set forth, is mixed with 230% of 70% methanol by weight and separated by centrifugation into a further solid-rich layer **58** and a solid-lean stream. The layer **58** is then stripped of alcohol and dried. The resultant product weighs 236.4 pounds and contains 5.9% moisture and 90% protein (dry basis) of 93.1% NSI solubility. Finally, the solid-lean layer **46** separated from step **40** containing about 157 pounds solids, 6,500 pounds water and 12,500 pounds methanol is passed through a methanol stripping column and thereafter the product is spray dried (step **68**). The resultant product **70** weighs 166.4 pounds and contains 5.4% moisture, 82.6% polysaccharides and about 12% proteins. The results are tabulated in the table below.

	Concentrate, 52	Isolate, 64	Semi-soluble, 38	In-soluble, 30	Whey solids, 70
Moisture and volatile matter, percent	5.2	5.6	7.6	3.4	5.4
Crude protein, percent (N=X6.25)	72.50	90.0	80.50	63.10	12
Nitrogen solubility index, percent	98.8	93.1	52.1	11.46	
Polysaccharides					82.6

The soluble soy protein concentrates obtained by this process are substantially unhydrolyzed, are cream colored and free flowing. They are very bland products which have little flavor and odor and possess high protein low sugar content, low sodium high calcium content, excellent nutrition, good water and fat absorption, good extension properties, good thickening ability and good texture control. The soluble soy protein concentrates obtained by this process have a solubility of the order of 92 to 99%, whereas the best soy concentrates available on the commercial market have a Nitrogen Solubility Index of the order of about 5%.

The high solubility of the soluble soy protein concentrate makes it readily suitable for instant beverages of high nutritional value, nondairy whole milk, nondairy coffee whitener, frozen desserts, aerosol topping, instant breakfasts, dry mix foods, baby and junior foods, dietetic and ethnic foods, emulsifiers, meat industry uses, soy uses, confections, margarine, soy coffee and the like.

Catalase Improvement of Dispersibility and Flavor in Concentrates and Isolates

A catalase treatment of soybean protein which both increases dispersibility and improves flavor has also been developed by *P.U. de Paolis; U.S. Patent 3,716,372; February 13, 1973*. He found that after destruction of the trypsin inhibitors in defatted, dehulled soybean material by boiling or near boiling in water for a short period of time, the resulting solids are rendered almost 100% soluble or dispersible in water by introducing from 0.1 to 2 milliliters of a solution containing the enzyme catalase, each milliliter of solution containing 1,000 Baker units of the enzyme catalase to each 100 grams of starting soybean material.

The usual starting material in the process for making any of the soy concentrate, soy isolate and soy albumen material is defatted, dehulled soybean flakes or meal. The soybean materials are dehulled and defatted by conventional methods and result in a material normally having between 45 and 50% protein, 30 and 35% carbohydrates, 0.5 and 4% fats, 5 and 15% ash and fiber and the remainder moisture. If the dehulled soybean flakes or meal are not defatted, there will be present an additional 16 to 20% fat content in the starting soy material. The defatted and dehulled soybean material for making soy concentrates is immersed in boiling or near boiling water (175° to 212°F) for 20 to 35 minutes. The pH of the water is held close to 7. Under these conditions, the trypsin inhibitor in the soybean material is destroyed while very little denaturation of the soy protein occurs. Further, the off-flavor-potentiating materials are to a great extent destroyed. The boiled material may then be filtered or centrifuged and any insolubles containing small amounts of denatured protein, ash, fiber and other insolubles are discarded or sent to other processing steps.

The filtrate, containing all of the soluble protein, has added to it from 0.1 to 2 ml of a solution of the enzyme catalase (each ml of solution containing 1,000 Baker units of the enzyme catalase) per 100 grams of starting defatted dehulled soybean material. It is believed that some hydrogen peroxide is produced during the boiling phase of the process. To minimize the deleterious effect of the peroxide, catalase is introduced into the filtrate as the filtrate is reduced in temperature to 150°F or less. If the catalase is introduced when the filtrate temperature is above 150°F, the catalase will be ineffective.

The optimum range of filtrate temperature, for introduction of the catalase, lies between 120° and 140°F. The reaction is allowed to proceed for 1 to 15 minutes. It is preferred to use catalase derived from a plant source. Fermcolase, a concentrated red-amber aqueous preparation of the enzyme catalase, standardized at 1,000 Baker units of catalase per ml is an example of catalase employed.

To produce soy concentrates, the filtrate from the catalase reaction need then only be spray dried. The resulting concentrate is characterized by having nearly 100% dispersibility in water (when 5% of the concentrate by weight is added to water), 100% NSI solubility and very little off-flavor.

To prepare soy isolates, the same processing steps as described for making the soy concentrates are followed up to and including the catalase reaction step. After the catalase reaction step, the filtrate is acidified to a pH of 4.5 (the isoelectric point of the soy protein), preferably with malic acid, and the soy protein precipitated. The precipitate is then recovered and purified, usually by washing with water and then neutralizing with sodium hydroxide. The pH is adjusted to near neutral. The washed and neutralized soy protein is then spray dried and because the resulting product contains over 90% protein, it is by definition a soy isolate.

Example 1: One hundred grams of defatted untoasted soybean flakes were extracted with 1,000 ml of water at a temperature of 194°F for 25 minutes. The pH was 7. The mixture was centrifuged and the residue extracted with an additional 1,000 ml of water. The extracts were combined and when the temperature dropped to 125°F, 1.0 ml of Fermcolase (1,000 Baker units per ml) was added. After 10 minutes, the extract was spray dried. The resulting soy concentrate had 100% solubility and 100% dispersibility with good flavor.

Example 2: In this example, a soy isolate and soy albumen were prepared from soybean flakes. The initial steps of the process are the same as those set forth in Example 1 except that after the catalase reaction (Fermcolase) for the requisite amount of time and/or filtration, the pH is adjusted to 4.5 by the addition of a 10 N solution of malic acid. This solutions is then centrifuged. The supernatant liquid is replaced with an equal volume of water and centrifuged again.

The filtrate is removed and the residual precipitate material is washed. The pH of the precipitate is then adjusted to about 7 to 8 by the addition of a base such as sodium hydroxide and is then spray dried. The properties of the resulting isolate material include 50% solubility and 100% dispersibility. The filtrate is then spray dried and the solids left are termed soy albumen. The properties of the soy albumen so prepared include 100% dispersibility and an NSI solubility of 100%.

Soluble Denatured Soy Proteins

H. Sakai, H. Hara and A. Akioka; U.S. Patent 3,303,182; February 7, 1967; assigned to The Nisshin Oil Mills, Ltd, Japan have developed a process for the production of denatured soybean protein suitable for food uses which is sterilized, enzyme inactivated and readily soluble in water at a neutral pH.

Usually, in isolating soybean protein from soybean, soluble protein component has been extracted by water or dilute alkaline solution from defatted soybean meal. After removing insoluble matter by centrifuge or filter, protein is precipitated from solution by adjusting its pH. Then protein precipitate is washed and dried as it is or after adjusting its pH to neutral. The protein product thus recovered is not denatured and the digestibility of the protein is rather lower; also it is contaminated with bacteria and enzymes.

According to this process, the protein product is so denatured that it is changed in the configuration of the protein molecule, and it has different properties from untreated protein in colloidal behavior such as viscosity of dispersion, solubility for pH change and digestibility. Also, the protein product is satisfactorily sterilized.

When the protein dispersion extracted from defatted soybean meal which has a maximum solids content of 10% is heated at a temperature over 80°C for a predetermined time, the protein is fully denatured, increased in digestibility and yet maintains its solubility. Moreover, deleterious bacteria and enzymes in defatted meal are satisfactorily sterilized or inactivated. In this denaturing process, the time required to heat it is determined by the treating temperature, for example, at treating temperatures from 80° to 90°C, it takes from 30 minutes to 1 hour but at 135°C, it takes only from a few seconds to a few minutes.

Under these conditions discoloration and insolubilizing of protein are slight. When protein dispersion having more than 10% solids content is heated, the protein is insolubilized or gelled, and the resultant denatured protein has a reduced solubility. Therefore, the solids content of the protein dispersion has to be less than 10%, and it needs water more than 8 times by weight of the soybean meal.

The heated dispersion is then cooled to under 5°C as rapidly as possible and is precipitated at its isoelectric point, at a pH of about 4.5. However, if the denatured protein is precipitated at over 5°C, the resultant precipitate cannot fully recover its neutral dispersibility or solubility. According to this process, by isoelectric precipitation at under 5°C, redispersibility of protein at neutral pH remains nearly 100%.

The relation between precipitating temperature and protein resolubility is shown in the table. In these tests, the heat denatured protein dispersion of the process is isoelectrically precipitated at various temperatures and the untreated protein dispersion at 30°C and 80°C. Each protein precipitate is recovered and washed and adjusted to a pH of 6.5 by adding alkali in order to redisperse the protein.

Treatment	Precipitating temp., °C.	Protein resolubility, percent
Heat denatured	3	100
Do	10	87
Do	30	73
Untreated	80	18
Do	30	100

Example: Five kilograms of soybean high soluble protein meal (moisture, 7.5%; crude protein, 51% and NSI, 90) which had been defatted with n-hexane and desolventized by superheated solvent vapor were charged into a tank and 10 grams of sodium hydrogen sulfite and 50 liters of water were added. The mixture was stirred until in dispersion and heated at 90°C for 15 minutes. Filtrate was separated by basket type centrifuge from insoluble residue, cooled rapidly to 5°C by chilling brine under stirring and 450 ml of 10% hydrochloric acid was added to the filtrate until the pH was about 4.5. The protein was precipitated and recovered by basket type centrifuge. Protein precipitate, after washing with fresh water, was completely redissolved in 5 liters of water with addition of 20 grams of sodium hydroxide to the pH of 6.4 and spray dried to yield 1.3 kg of dry powder (moisture, 7%; crude protein, 86%; soluble protein, 95% and enzyme activity, negative). The growth efficiency of the denatured protein prepared as herein described is found to be the same as that of milk casein, when semipurified diets containing 15% protein were fed to male white rats of Wistar strain.

Dispersible Isolates by Jet Cooking

R.L. Hawley, C.W. Frederiksen and R.A. Hoer; U.S. Patent 3,642,490; February 15, 1972; assigned to Ralston Purina Company have developed a process of preparing bland flavored soy protein food of high dispersibility and food products made therefrom. The process preferably treats the soy material as an aqueous slurry with brief, controlled, rapid, dynamic heating to an elevated temperature range and with momentary physical working under dynamic conditions and under controlled positive pressure and elevated temperature conditions to expose and loosen the tenacious hold of the obnoxious substances by the complex protein molecules.

The released obnoxious substances are removed by causing subsequent instant pressure release to flash off, with vaporizing of some moisture that is with the entrained odiferous obnoxious substances, followed by separation of the vaporized materials from the slurry. The slurry is then dried to an attractive redispersible white powder. The product does not form a gel unless the range of solids, treating temperatures and holding times are regulated to form a product of controlled degree of gelatin.

The preliminary steps of the process form a soy protein isolate in a known manner. The isolate is made into an aqueous slurry and then treated as follows. The slurry has its pH adjusted which is important to obtain a product with high water dispersibility. Specifically, the pH is adjusted to a range of 5.7 to 7.5, preferably between 6.5 and 7.1. Below 5.7, the water dispersibility of the product is very low and not useful for many purposes. However, such a product, when specially heated as described later, can be used where low dispersibility and debittered flavor is useful, as for example, in baked goods, cereals, etc. At a pH above 7.5 and approaching 8.0, the product tends to assume an undesirable soapy taste. The degree of dispersibility in the product can be regulated by varying the pH within the pH range of 5.7 to 7.5, to suit the product to the final food being prepared. The pH may be easily adjusted by adding a food grade alkaline reagent such as sodium bicarbonate or even by washing with water for a period of time.

The slurry to be further processed should have a solids content of 3 to 30% by weight, and preferably 5 to 17% by weight. If it falls below 3%, subsequent processing steps are not economically advisable when a continuous process is employed. Drying is particularly costly. Above 17% solids content, the resulting product does not lend itself to the flash drying techniques so that other drying techniques must be employed which result in a product that is not as desirable in its functional characteristics.

Each tiny portion of this slurry is then subjected to dynamic instantaneous heating to an elevated temperature range while also being dynamically physically worked, preferably almost simultaneously. Currently, the most satisfactory way of achieving this is to pass the slurry through a device commonly known as a Jet Cooker. It includes adjacent jet nozzle orifices, normally concentric, through which the slurry and the pressurized steam used as a heating agent are ejected at high velocities in intersecting flow patterns. The temperature range to which the slurry is heated for the desired results is 220° to 400°F, although the temperature should not be in the lower region of this range unless the product is subsequently vented into a vacuum chamber after being held under pressure in a special holding chamber. Normally the temperature should be about 285° to 320°F for best results.

The product is introduced to the jet cooker nozzle at a positive pressure. This pressure should be near the pressure of the steam injected into the slurry, should be sufficient to cause high velocity discharge of the slurry through the jet nozzle, and must be greater than the pressure in the special retention chamber immediately downstream of the nozzle. Normally the steam pressure is 80 to 85 psig, the slurry line pressure is slightly above the steam pressure, usually 85 to 100 psig, and the discharge pressure in the chamber downstream of the nozzle is 75 to 80 psig. The pressure drop of the slurry across the nozzle is 5 to 15 psi, depending upon these other pressures, with 6 to 10 psi being common.

The time interval of the slurry in the nozzle is estimated to be about 1 second or less. The nozzle orifice for the slurry is small, being only a fraction of an inch, e.g., about one-eighth inch, so that the slurry solids are subjected to severe dynamic, physical working during passage. The steam intermixes intimately with the solids in the ejected slurry. The amount of steam required is not great, normally being an amount to lower the solids content of the slurry 1 to 2% by weight.

The steam and slurry are ejected into a special retention chamber. This may comprise an elongated tube through which the intermixed slurry and steam moves from the jet nozzle on one end of the tube to a pressure controlled discharge on the other end. The discharge can be controlled by a conventional preset pressure release valve to enable continuous process flow from the nozzle to and out of the discharge valve. This valve regulates the pressure in the holding chamber.

This chamber pressure must be great enough to prevent any significant vaporization of the moisture in the chamber, even though the temperature is well above the boiling point of water. A pressure of 75 to 80 psig readily achieves this. Since slurry and steam must continuously flow into this pressurized chamber, the pressure behind the slurry and the steam must be greater than the chamber pressure to cause this continuous flow. The heated slurry is retained in the holding chamber for a definite but relatively short time period of a few seconds up to a few minutes, normally 7 to 100 seconds. It is only necessary to retain the product in this heated condition for a few seconds for optimum product.

The pressure on the slurry is then instantly released by discharging the slurry to a reduced pressure zone, into a suitable receiving means. This causes flash-off of a portion of the moisture in the form of water vapor which is laden with the entrained odiferous, obnoxious pungent chemical components or substances of unknown composition from the soy product. The flash-off causes substantial cooling of the remaining slurry because of the heat of vaporization absorbed from the slurry so that the total time which the product is subjected to elevated temperatures is very short and controlled.

The reduced pressure zone into which the slurry is discharged is normally at atmospheric pressure, but it is sometimes subatmospheric, i.e., at a partial vacuum. In either case, the vapors should be instantly conducted away from the slurry, preferably by a moving current of air across the slurry or by drawing a continuous vacuum on the discharge zone to draw the vapors away.

The resulting slurried product can then be used directly for food products. It is an attractive white product. If the pH, prior to treatment, was within the stated range, the slurry contains most of the material in a partially dissolved state and a partially dispersed colloidal state that does not tend to settle out. Alternatively, it can be dried, with the dried product having excellent redispersibility in an aqueous medium.

Example: Soybeans are ground and the oil extracted with hexane to give flakes commonly called soybean meal. The flakes are added to an aqueous bath and a food grade alkaline reagent, sodium hydroxide, is added until a pH of 10 is reached. The material is cooked for 30 minutes and then centrifuged. The soy protein material is precipitated from the liquor by adding acetic acid until the isoelectric point is reached at about a pH of 4.7. The precipitate is washed with water and then added to water to make an aqueous slurry of 15% solids by weight. The pH is then adjusted to 6.7 by adding sodium bicarbonate. The slurry is then passed through a jet cooker under a pressure of 85 psig, simultaneously with steam ejection from the jet cooker under a pressure of 95 psig, into a pressure retention chamber at a pressure of 75 psig. The steam heats the slurry through the jet cooker to a temperature of 290°F. After 7 seconds progressive portions of the heated slurry are suddenly discharged into a receiver at atmospheric pressure or below causing flash-off of vapors laden with obnoxious smelling and tasting substances. The slurry is cooled by the flash-off vaporization. The substance laden vapors are removed from the purified slurry. The slurry is flash dried in a spray drier to a moisture content of 3%.

Pressurized Heating for Preparing Soluble Transparent Isolates

Another heating process for producing soluble protein isolates is described by *P.J. Magnino, Jr. and C.W. Frederiksen; U.S. Patent 3,645,745; February 29, 1972; assigned to Ralston Purina Company.* This process producing a soluble protein product from oleaginous seed involves altering the physical structure of the protein molecules by forming a solution having a pH from 1 to 2 or from 10 to 12 and heating the liquid portion to 85° to 350°F and later precipitating the protein. An aqueous suspension of this precipitated protein is subjected to instantaneous heat and pressure, then passed through a zone of lower pressure, thus converting the protein to a dry, particulate protein product. This soluble protein may be used to produce protein enriched food products and is particularly desirable for use in clear or transparent food products or foods where high solubility is desired, since this protein is very soluble and does not distract from the physical properties of a clear food product. The process is shown in detail in the following example.

Example: Defatted soybean flakes were slurried in water at a temperature of 85°F with a water to flake ratio of 9:1. To the slurry was added 1.6% of calcium hydroxide and 1.0% of sodium sulfite, these percentages being based on the total weight of the flakes. The slurry was agitated for 30 minutes and subsequently the liquid was separated from the solid materials in a centrifuge. The solid materials were reslurried in water at a ratio of 6 parts of water to 1 part of flakes. This slurry was agitated for a period of 10 minutes, after which time the solid and liquid materials were separated as before in a centrifuge. The two liquid portions or extracts were combined.

To the combined liquid portions which contain soluble proteins and carbohydrates was added 7.5% sodium hydroxide, this percentage being based on the estimated dry weight of protein present. The liquid portion has a pH of about 11.5 after the addition of sodium hydroxide. The liquid was then heated to 160°F and that temperature maintained for 40 minutes. After heating, phosphoric acid was added to the liquid to adjust the pH from about 11.5 to the isoelectric point of the protein, i.e., pH of about 4.5. The dissolved protein was precipitated out of solution and the precipitated curd was concentrated and washed in a centrifuge to remove the carbohydrate materials as completely as possible. The washed protein curd was diluted in a ratio of 1:1 by weight with fresh water and concentrated to about 17% by weight solids.

The concentrated curd was diluted to 15% by weight solids and the pH adjusted to 6.8 with sodium hydroxide. The neutralized protein curd was then passed to a Jet Cooker where the temperature of the curd was raised instantaneously to a temperature of 310°F and held at that temperature for about 5 seconds. The curd was then injected into a vacuum chamber under 27 inches of mercury vacuum. The heated curd was passed from the vacuum chamber to a spray drier under a pressure of 4,000 psi and dried at an exhaust temperature of 180°F. The resulting product was a finely divided particulate neutral protein product.

Several tests were conducted to compare the properties of the product with other similar products which have been produced by different methods. The products compared are designated A, B, C and D. Products C and D are made in accordance with this process. A is an isolated soy protein which has not been hydrolyzed but has been Jet Cooked and is commercially available, B is an isolated soy protein which has been hydrolyzed but not Jet Cooked, C is a product of the Example, i.e., an isolated soy protein which has been hydrolyzed and Jet Cooked and D is a product of C above which has been Jet Cooked a second time before spray drying. These tests are described below. The results obtained are set forth in the table which follows.

Solubility Index — Four grams of the protein product in 100 ml of water was dispersed with agitation in a Waring Blender for about 90 seconds. Fifty milliliters of the mixture was centrifuged for about 5 minutes at about 1,000 rpm. All but about 5 ml of the supernatant liquid was removed, water was added to make the volume up to 50 ml, the mixture was shaken gently and recentrifuged for 5 minutes at 1,000 rpm. The solubility index is expressed in terms of the amount of residue, expressed in ml, remaining undissolved. That is to say, the more the residue, the less soluble and less desirable the product.

Viscosity I — A 12% protein solution was prepared by mixing 30 grams of protein with 220 ml of water. The mixture was placed in an open beaker and heated in a hot water bath with agitation for about 30 minutes at a temperature of about 165°F. The mixture was cooled to about 75° and 40°F and the viscosity was measured with an LVT Brookfield viscometer using a No. 1 spindle at 60 rpm.

Viscosity II — A 12% protein solution was prepared by mixing 30 grams of protein with 220 ml of water. The mixture was sealed in a metal can and retorted without agitation at 250°F for 15 minutes. The product was cooled to about 75°F, the container opened and the mixture tested for viscosity with an LVT Brookfield viscometer using a No. 1 spindle at 60 rpm.

Product	Solubility index, ml.	Viscosity I (cps.)[a] 40° F.	75° F.	Viscosity II (cps.),[b] 75° F.
A	10.0	Gel	[c] 730	[c] 160
B	1.0	87.0	39.5	13.0
C	0.1	22.0	17.00	12.50
D	0.1	31.5	14.50	11.50

[a] 12% solution heated for 30 minutes at 165° F.
[b] 12% solution heated for 15 minutes at 250° F.
[c] Extrapolated to No. 1 spindle at 60 r.p.m.

GELABLE SOY ISOLATES

Thermoreversible Gels from Isolates

The alcohol extraction treatment of hexane-extracted soybean meal to remove an antiwhipping or antigelling factor is known, but the foams or whips made with the prior products lack stability and hydrogels prepared with these products are not heat reversible. Soy isolates which form thermoreversible gels have been produced by *A.C. Eldridge and A.M. Nash; U.S. Patent 3,218,307; November 16, 1965; assigned to the U.S. Secretary of Agriculture.*

It was found that by treating only the acid (isoelectrically) precipitated fraction of an aqueous whole extract of hexane-extracted soybean meal with certain critical ranges of lower alcohols, not only is an antigelling factor removed therefrom as expected but the alcohol-washed fraction possesses greatly improved properties over the similarly treated whole extract mixture of water-soluble soybean proteins. The most conspicuously different property is evident in the repeatedly thermoreversible character of hydrogels that are prepared by heating certain essential concentrations of the alkali-dispersed alcohol-treated product (alkali proteinate) of this process. It is believed that the heat reversibility of this improved soybean product depends on the presence of protein that is more resistant to heat coagulation and upon the absence of a heat-sensitive component present in the prior products. The flavor and color of these products are also superior to those of the prior products.

The following alcohol concentrations are highly effective: methanol, 80 to 100% by volume (95% v/v preferred); ethanol, 60 to 95% by volume (86% v/v preferred); isopropanol, 40 to 95% by volume (82% v/v preferred). To prepare this product, it is a matter of choice whether one starts with a highly pure commercial acid-precipitated soybean protein or whether one prepares such acid precipitated protein and treats it as part of a drying operation in the manner of Example 1.

Example 1: Two hundred grams of hexane-extracted soybean meal were mechanically stirred in 2,000 ml of water and the suspension was repeatedly adjusted to pH 7.4 to 7.6 with alkali during the one hour of stirring. The suspension was then centrifuged and the separated solids were stirred in another 1,000 ml of water for 30 minutes before centrifuging.

The two centrifugation supernatants were combined and HCl was added to lower the pH to 4.2 to 4.7. The precipitated protein was isolated by centrifugation and the moist curd (56 grams dry basis) was dispersed 5 separate times in 300 ml portions of 85% (v/v) aqueous ethanol in a blender. Then the filtered ethanol-moist cake was vacuum dried at 30°C yielding 53.2 grams of material which, for convenience, was then dispersed in alkali (pH 7.5), the residue from centrifuging was discarded, and the solubles were freeze dried to yield an alcohol-treated sodium proteinate product.

Example 2: The sodium proteinate product of Example 1 was treated in the following manner. 1.0 gram thereof was placed in each of 8 test tubes and 4 to 20 ml of water was added as shown in the following table to provide solids concentrations ranging from 4.8 to 20.0%. The proteinate was dissolved by stirring and the tubes were placed in a boiling water bath. As shown in the table, sodium proteinate concentrations of 12.2% gelled even when hot, (i.e., became irreversible) while the 11.5% concentration was gelled when cooled to or below 68°C after 5 minutes of heating at 92°C. The last described gel again became fluid (poured freely from its test tube) upon warming to 75°C, regelled upon cooling to 68°C, and liquified upon rewarming to 75°C. Similar results were obtained when various food dyes or flavors were added to the proteinate solution.

Tube	Wt. material added, g.	H$_2$O added, ml.	Solids, percent concentration	Comments
1	1.0	4.0	20.0	Boiling H$_2$O, 3 minutes, gave gel while hot.
2	1.0	5.0	16.7	Do.
3	1.0	6.0	14.3	Do.
4	1.0	7.1	12.2	Boiling H$_2$O, 5 minutes, gelled while hot.
5	1.0	7.7	11.5	Boiling H$_2$O, 5 minutes, was gelled at 28°C.
6	1.0	9.0	10.0	Boiling H$_2$O, 5 minutes, no gel hot, gelled in an ice bath.
7	1.0	10.0	9.1	Boiling H$_2$O, 10 minutes, no gel hot, slight gel in an ice bath.
8	1.0	20.0	4.8	Boiling H$_2$O, 10 minutes, no gel hot, no gel in an ice bath.

Heat Gelable Soy Protein

A soybean protein which is bland, heat gelable with excellent gelling characteristics has been developed by *F.M. Robbins, A.G. Bonagura and R.S. Yare; U.S. Patent 3,261,822; July 19, 1966; assigned to General Foods Corporation.* The protein products of this process due to their heat irreversible characteristics when heated at 90°C or higher and their ability to substantially duplicate gelatin gels in physical appearance may serve as substitutes in many cases for gelatin where it has been desired to have a gelatin-like product which is heat irreversible. Such uses may include the jelly coating on canned meats, films and the like.

This protein may be prepared by mixing soybean meal with water at a pH of 3.5 to 5.5, most preferably at 4.5. The amount of water employed may be 2 to 100 parts by weight, most preferably 7 parts by weight per part by weight of soybean meal. While it is possible to adjust the pH of the soybean slurry after the soybean meal has been added to the water, it is preferred to adjust the pH of the water prior to addition of the soybean meal and maintain the pH of the slurry at the desired pH as the soybean meal is being added. While the temperatures used may be below those at which denaturation will take place, it is preferred to use a temperature below 40°C and preferably in the order of 25°C. The slurry is maintained at the desired pH and temperature for a time sufficient to solubilize all of the soluble material which primarily consists of soluble proteins (albumins and albuminoids), soluble carbohydrates and soluble salts. Ordinarily less than 30 minutes will suffice to solubilize all of those constituents which are soluble.

The soluble fraction is separated from the insoluble fraction in as short a time as possible by ordinary separation means, typically filtration or centrifugation. The undesirable soluble fraction is discarded and the insoluble fraction containing the insoluble protein, cellulosic constituents and insoluble salts is suspended in water at a ratio of 2 to 100 parts by weight per part by weight of soybean meal, most preferably 7 parts by weight per part by weight of soybean meal to form a slurry. The pH of the slurry is adjusted to 6 to 8, and preferably to 6.5 to solubilize the desired soybean protein. The time required to solubilize the soybean protein will generally be less than 30 minutes.

The soluble protein fraction is removed from such insoluble constituents by ordinary separation means. When it is preferred to obtain a clear protein gel, all suspended matter should be removed from the soluble protein fraction. The pH of the soluble protein fraction is reduced to the isoelectric point of the protein which is in the order of 4.5. The pH is reduced by slow addition of acid with agitation to avoid localized concentration of acid which would possibly denature the protein in such localized areas. Adjustment of the pH to 4.5 causes a precipitation of the protein and the precipitated protein may be separated from the supernatant by any ordinary separation means. If desired, the protein obtained may be washed to remove any excess acid, although it is not essential to do so.

The soybean cake, if desired, can be dried under temperature conditions which do not denature the protein. Typical of the drying techniques which may be employed would be freeze drying, vacuum drum drying or low temperature spray drying.

The wet product is a bland, edible, white cake and in the dry form is a bland, edible, dry white powder which is free of the undesirable bitter and beany flavor usually found in prior art soybean proteins. When suspended in water at room temperature (25°C) and in a concentration of 14% protein, the product, as the pH is adjusted upwardly to 6.5, becomes more viscous and at 6.5 has the physical appearance of prior art heat gelable soybean proteins which have been heated to about 90°C.

The protein of the process at pH 6.5 and in a concentration of 14% when heated at 100°C forms a gel which is far stronger than any prior art soybean protein gel heretofore known and is a continuous, elastic, optically clear, bland gel which cleaves with a smooth face. The gel has a Bloom strength of about 202 as measured by standard gelatin Bloom techniques when the gel is prepared at pH 6.5 and the protein, prior to heat gelation, has a viscosity of about 34,153 centipoises as measured by a Brookfield viscometer with a heliopath spindle of 2.8 cm diameter rotated at 12 rpm.

Example: 350 grams of low temperature, solvent-extracted, defatted soybean meal was added to 2.5 liters of water which had been adjusted to pH 4.5 (preextraction pH) by the addition of normal hydrochloric acid. As the soybean meal was added with agitation, the pH was maintained at 4.5. The resulting slurry was held at a pH of 4.5 for 35 minutes under constant agitation. The slurry was centrifuged for 10 minutes at 700 RCF x G (relative centrifugal force times gravity). The supernatant at the end of the centrifugation was discarded and 1,725 ml of water was added to the centrifuged cake.

The material was stirred to form a slurry and the pH was then adjusted to pH 6.4 (solubilization pH) by the addition of sufficient normal sodium hydroxide. The slurry was maintained at 25°C for 30 minutes under constant agitation. The pH throughout was maintained at 6.4. At the end of 30 minutes, the slurry was centrifuged for 45 minutes at 10,000 RCF x G. The cake was discarded and the supernatant was adjusted to pH 4.5 to precipitate the protein by the addition of normal hydrochloric acid which was slowly added with agitation in order to avoid localized action. The precipitated protein and supernatant was centrifuged for 10 minutes at 700 RCF x G and the supernatant was then discarded. The protein cake obtained was a white, pasty mass which contained 25% protein (as measured by standard Biuret technique as described on page 545 of the 33rd edition of *Analytical Chemistry* (1961) and 75% water. The cake was diluted to provide a protein concentration of about 14%. The cake formed an optically clear, amber colored, elastic, continuous gel which cleaved cleanly when the pH was adjusted to 6.5 and the material heated at 100°C. The yield of protein obtained by the above technique was 45% of the protein in the meal.

Heat Coagulable Isolates and Concentrates

Soy products containing 75% or more protein (isolates or concentrates) are made heat coagulable in a process described by *J.D. Mullen, D.E. Smith and A. Ogrins; U.S. Patent 3,594,192; July 20, 1971; assigned to General Mills, Inc.* It was found that soy protein can be made more heat coagulable when used in combination with egg white by raising the pH of an aqueous dispersion to above 9.0 with an alkaline material and then reducing the pH to between 5.5 and 8.0. The resulting modified soy protein finds use as a replacement for up to about three-quarters of the egg white needed to provide the structures or matrixes of a variety of foods wherein the egg white functions primarily through its property of being heat coagulable.

The starting material can be a soy isolate or concentrate containing at least 70% by weight protein. Such materials can be obtained by removing at least a part of the nonprotein constituents of defatted soy flour, meal or flakes by various means. Isolates normally contain over 90% by weight protein and are preferred materials in this process.

It is to be noted that the improvement achieved by the use of this process varies somewhat depending on the precise method used in preparing the starting soy isolate or concentrate and especially as to the acid used in the precipitation step. Thus, for example, in this process, sulfurous acid precipitated soy protein can be treated with less alkali than is required to treat hydrochloric acid precipitated soy protein to obtain optimum heat coagulation properties or gel strengths.

Any of a variety of water-soluble alkaline materials can be used to raise the pH of the soy protein material to the indicated degree. Preferred materials are the inorganic bases and salts such as trisodium phosphate (Na_3PO_4) and the alkali metal hydroxides, especially sodium hydroxide and ammonia or ammonium hydroxide.

The temperature and the length of time of holding the soy protein in the alkaline state are not critical except that optimum results are obtained at certain temperatures and treating times for any particular soy isolate or concentrate at designated concentrations in the aqueous dispersions and at various alkaline material levels of usage and particular agent employed. Temperatures of from 25° to 85°C are preferred when treating an aqueous dispersion containing 5 to 20% by weight soy protein with sodium hydroxide at a soy protein solids to sodium hydroxide weight ratio of between 8:1 and 20:1. Under these conditions, it is preferred to carry out the alkaline treatment for from 2 minutes to 1 hour.

It is especially preferred to treat soy isolate containing 90% by weight or above protein at the designated concentrations in the aqueous dispersion at temperatures of 25° to 40°C for 2 to 12 minutes with the ratio of soy protein solids to sodium hydroxide being in the range of between 8:1 and 14:1. When ammonia gas is used as the alkaline agent, it is preferred to carry out the alkaline treatment for 15 minutes to 2 hours at temperatures of 25° to 85°C on aqueous dispersions containing 5 to 25% by weight soy protein. The ammonia gas is bubbled through the dispersion for the indicated times and then, if desired, the resulting alkaline solution can be held for 1 to 48 hours after the completion of the sparging and prior to reducing the pH of the solution.

When ammonia gas or ammonium hydroxide is used as the alkaline agents, the pH of the resulting solution can be reduced to near neutral simply by drying the solution. However, the pH of such alkaline solutions and also those where other alkaline agents are used can also be reduced by the addition of an acid which is preferably an edible inorganic or organic acid. Representative of these acids are phosphoric acid, hydrochloric acid, citric acid, lactic acid and the like.

During such neutralization it is desirable that the soy protein solution is stirred to disperse any beads of precipitated protein that form as the acid is added. In any case the pH of the solution is reduced to 5.5 to 8.0 and preferably to 7.0. At the near neutral point, the modified soy protein having improved gelling or heat coagulation properties can be used in combination with egg white in a variety of foods without effecting the alkaline or acidic nature of such foods.

The modified soy protein in the above aqueous solution can be used as such or can be dried prior to use in combination with the egg white. The gel strength is sometimes reduced by drying but such reduction can be partially or completely alleviated by adding a small amount of a phosphate salt, preferably sodium tripolyphosphate ($Na_5P_3O_{10}$) to the solution prior to the drying step. Any method of drying can be used although with the ammonia or ammonium hydroxide treated soy protein, it is preferred to use lyophilization or spray drying.

Example: An 8.3% by weight aqueous dispersion of a commercially available soy protein isolate (95.9% protein on a dry basis, alkaline extracted, HCl precipitated, Promine R) was continuously pumped into a high speed mixer where a caustic soda solution (5.4% by weight NaOH) was continuously added to yield a solution having an isolate solids to NaOH weight ratio of 11:1. The alkaline solution was maintained at 28°C for 4 minutes followed by continuous neutralization with phosphoric acid (77% by weight in water) to a pH of 7.0 in a second mixer. The resulting neutral slurry was spray dried.

Solutions were made up by dissolving 10% by weight of the modified soy in water, 5% by weight of the modified soy and 5% of dehydrated egg white in water, and 10% by weight of dehydrated egg white in water. The solutions were coagulated in the manner set forth above. The resulting gels had gel strengths of 15 grams, 243 grams and 185 grams, respectively. In the water binding test, the gel or coagulum from the 10% egg white solution bound only 35% of the water whereas the coagulum from the 5% modified soy to 5% egg white solution bound 86% of the water. The data of this example thus show that the combination of the modified soy and egg white yields results which are better than egg white alone.

Gelable Isolates from Soy Milk

Use of soy milk as the starting material for isolating protein has been described by *Y. Yamato, H. Taniguchi, S. Nakayama and T. Tateishi; U.S. Patent 3,607,860; Sept. 21, 1971; assigned to Fuji Oil Company Limited, Japan.* The protein attained has excellent water-solubility, gel-forming ability, water-binding property and emulsifying property. Briefly, the protein is obtained at a temperature below 80°C in an acidic condition from soy milk which is obtained from soybean meal or soybeans by water extraction, alkalizing the protein with an alkalizing agent, returning the pH to about neutral with acid and thereafter heating the resulting protein solution at a temperature above 60°C.

To obtain a final product having as high a water-solubility as possible, it is preferable to use soybean meal having a high water-soluble protein content. It is necessary to carry out the extraction at a pH above 6 to increase the extraction effect. When an alkalizing agent or sulfite is added to increase the bleaching effect and the protein extraction effect, it is economical to use a high temperature (up to 100°C). An alkalizing agent, such as sodium hydroxide, potassium hydroxide, sodium carbonate or sodium bicarbonate; and a sulfite such as sodium sulfite or sodium bisulfite is generally used.

The soy milk is then acidified with an acid to precipitate and separate the protein. The pH of the protein separating step is preferably 4.2 to 4.5. The acid used is preferably an edible organic acid or an inorganic acid. This operation is carried out at a temperature preferably below 70°C to recover the solubility of the protein by the following operation.

The precipitated and separated protein is then alkalized by an alkalizing treatment. The alkalizing agent used is the same as that used in the above water extraction step. In the alkalizing step, the pH is suitably from 9 to 12, preferably from 9 to 11. If the pH is below 9, the visco-elasticity may become insufficient, and if the pH is above 12, a browning reaction of the protein may become prominent, the salt concentration of the final product may be increased, and then the protein may be hydrolyzed, thereby decreasing its viscosity. The protein regains its water-solubility and at the same time a fibrous structure forming of protein molecules may occur partially. As a result the protein obtains a water-binding property and a gel-forming ability. It is considered that the globular protein is converted to the fibrous protein by stirring and the time elapsed, since a part of the soy protein is cleaved and the globular protein is loosened in this case.

The protein is then neutralized. Alkalinity is unfavorable in food and may cause rapid putrefaction and bad color. The pH of the neutralizing step is suitable from 5 to 8, preferably between 6 and 7. The neutralization is carried out by the use of an acid.

If the protein, separated by acidification as described above, is immediately neutralized and dried and no alkalization step is carried out, the protein may have an insufficient water-binding property and, at most, a hygroscopicity rate of 150%. In case the 150% hydrated or hygroscopic product is heated to thermally denature and coagulate the protein, a product with weak gel strength and inferior visco-elasticity may be obtained. Consequently, as described above, the successive steps of acidification, alkalization (to pH 9 to 12) approximate neutralization including weak acidity and weak alkalinity, are extremely important characteristics of this process. Furthermore, it is necessary not to raise the temperature as much as possible during the steps, each having a different pH after the isolation of the protein.

It seems economical that the protein could be extracted by alkali from soybean meal or soybean meal leached in an aqueous acid solution. In case a strong alkali is used for extraction, the protein molecules may be hydrolyzed and a decrease of the viscosity of the aqueous protein solution occurs. Therefore, this extraction process has a defect of decreasing the visco-elasticity in the heat coagulation of the protein. On the other hand, in the present process the protein is not hydrolyzed but becomes linear so that it is provided with visco-elasticity, provided that the pH of the isolated protein is regulated between 9 and 12. In order to obtain the effects of the present process, it is necessary that the protein, which is obtained as an aqueous solution by the water extraction of soybean meal or soybeans, is precipitated at its isoelectric point in order to increase the concentration of the protein and then the pH is regulated between 9 and 12. If other steps are employed in regulating the pH of the isolated protein, the effect in combination with the subsequent heating step cannot be obtained.

Example: One hundred kilograms of water and 20 grams of sodium sulfite were added simultaneously to 10 kilograms of soybean meal having a nitrogen solubility index (NSI) of 85 and soybeans having a NSI of 92. Thereafter the extraction was conducted at ordinary temperature. 250 to 300 ml of concentrated hydrochloric acid was then added to the obtained soy milk, and the pH was adjusted to between 4.2 and 4.6. The precipitated protein was separated by centrifuging and then suspending in water. The suspension was then neutralized with sodium hydroxide, and the pH adjusted to between 10 and 11 while it was stirred vigorously to homogenize the liquor. After being alkalized, the alkaline liquor was again neutralized with hydrochloric acid to a pH between 6 and 7. The temperature of the neutralized liquor was raised to 100°C by either blowing with steam or by indirect heating and then the liquor was dried by spray drying.

The qualities of the protein powder thus obtained were as follows: water content of the product, 4.68%; crude protein, 86.3%; NSI, 84.8; pH (in 1% aqueous solution), 7.25; ash, 5.93%. The gel which was formed by molding a 2.5% common salt (NaCl) aqueous solution containing 12% by weight of the protein powder product prepared as above, based on the entire weight of the solution and heating the molded material at 80°C for 30 minutes, had a self-supporting ability. A self-supporting gel was not formed when a soy protein product produced by the conventional techniques was used in the same way. The gel-forming ability is generally decreased when the solution contains salt but, nevertheless, an excellent gel-forming ability can be obtained in accordance with this process.

ISOLATES FOR SPECIFIC PRODUCT USE

Soy Isolates for Baby Foods Using Enzyme Digestion

M.L. Anson and M. Pader; U.S. Patent 2,802,738; August 13, 1957; assigned to Lever Brothers Company have prepared a soy protein having a soft, smooth texture useful as a meat substitute in baby foods. They have found that vegetable proteins such as soy protein can be made into a protein food product by a process which involves partial digestion by a proteolytic enzyme of a relatively highly concentrated aqueous suspension of protein.

Enzyme digestion of proteins is known and while it was to be expected that a heated protein suspension would be less firm if the protein were first partially digested by an enzyme, it was not predictable that it would be possible to achieve a texture simulating, e.g., strained meat or that such texture could be obtained in a controlled and reproducible way. However, by only partially digesting the protein within certain fairly critical limits, a product of the desired texture practically free from off-flavor was obtained.

In accordance with the process, an aqueous suspension of vegetable protein is subjected to hydrolysis by a proteolytic enzyme until 10 to 40%, preferably 20 to 30%, of the protein is digested as measured by the percent of nitrogen converted to a form soluble in water at the isoelectric point of the protein.

In the case of soy protein, however, a preferred extraction procedure comprises suspending flakes of soybean meal in an aqueous solution of calcium hydroxide of 0.003 molarity. Steam is then sparged into the suspension of flakes with agitation until the temperature is about 60°C. The suspension is then pumped through a centrifuge to obtain a clarified extract which has a pH of 6.8 to 6.9. The protein in the extract is then coagulated by the addition of an acid such as hydrochloric acid to bring the pH to 5 and subsequent centrifugation yields a solid aqueous suspension of coagulated protein. Preferably, the coagulated protein is then resuspended in water for purposes of washing, and centrifuged again.

For the purposes of this process, it has been found that protein coagulated at a relatively high temperature, for example, from 80° up to 100°C, has less off-flavor and provides a product of better texture and lighter color. It is preferred that the final product contain at least 15% protein, including digestion products, by weight up to about 25% by weight in order to have approximately the same protein nutritive value as meat, or better. In order to obtain a product of this protein content, the suspension of protein should have a slightly higher protein content since the subsequent addition of additives results in a lowering of the protein content, and therefore it is preferable to subject to enzyme digestion a suspension of coagulated protein having a coagulated protein content somewhat higher than 15%.

In the enzyme digestion of the suspension there can be used any of several known proteolytic enzymes, from animal, plant, fungal, or microbial sources, such as papain and trypsin. In carrying out the digestion the pH and temperature of the protein suspension are usually adjusted to those for maximum activity of the particular enzyme used. For example, with papain the protein suspension is adjusted to pH 7 and heated to about 60° to 70°C. Usually less than 1% of enzyme based on dry protein weight is added. The suspension is then held at the desired temperature with continuous agitation until the desired amount of hydrolysis has occurred, i.e., from 10 to 40%, preferably 20 to 30%, as measured by the amount of nitrogen of the protein that is converted to a form soluble at the isoelectric point of the protein or by a measurement correlated with this value. The enzyme action can then be stopped, e.g., by strong heating before proceeding to subsequent operations. Results of the process are given in the following example.

Example: Twenty-five kilograms of soybean flakes were dispersed in 490 liters of a 0.003 molar solution of calcium hydroxide made with filtered tap water. Purified steam was then sparged into the stirred suspension until a temperature of 60°C was reached, this requiring about 12 minutes. The condensed steam raised the volume of water to 500 liters. The 60°C suspension was then pumped through a solid basket centrifuge and the resultant 428 liters of clarified extract were pumped to a second tank, the total clarification time being about 1 hour. The residue was rejected. The pH of the suspension after extraction was 7 to 7.1 and the pH of the clarified extract was 6.8 to 6.9.

The extract was heated to about 100°C and about 6.8 ml of 3 normal hydrochloric acid per liter of clarified extract was next added to bring the pH to 5 ± 0.1 thereby coagulating the protein and the suspension was then pumped through a centrifuge and a paste of coagulated protein was obtained. The protein was then suspended in a small volume of water, passed through a colloid mill, diluted with additional water to 455 liters and heated to 100°C with live steam while being stirred vigorously. The washed coagulated protein was then centrifuged to obtain an aqueous cake containing about 31% solids having a pH of 5. The washed cake of protein was suspended in sufficient water to lower the protein content to 16% and the pH adjusted to 7 by the addition of sodium hydroxide.

The resulting suspension was next heated to 60° to 70°C and 0.3% papain calculated on the dry weight of protein was added and thoroughly mixed. The suspension was then held at 60° to 70°C for 30 minutes while stirring continuously and then hydrochloric acid was added until the pH of the suspension was about 6.5. At this point the protein had been digested to the extent that about 22% of the nitrogen in the protein had been converted to a form soluble at the isoelectric point of the protein. Next about 5% by weight of the suspension of hydrogenated vegetable fat was added (lowering the protein concentration to about 15%) and flavoring, coloring and nutritional materials were also added. The final product was then milled and canned in 3½ ounce cans. It was a mushy, rather smooth paste which due to the flavoring additives had a flavor similar to that of canned strained meat. The nutritive value of the protein in the product was as high as the starting protein as measured by rat growth tests.

Soy Isolates for Gel-Like Meat Products

M.L. Anson and M. Pader; U.S. Patent 2,813,024; November 12, 1957; assigned to Lever Brothers Company have also modified soy isolates for use in preparing chewy gels used in meat substitutes. Soy or peanut protein isolates as obtained by known methods may be used in the process. The washed and dried isolate is then mixed to form an aqueous alkaline suspension.

In order to convert the precipitated protein into a gel precursor, it is usually necessary to adjust the pH and protein concentration which are the principal factors in influencing the subsequent formation of chewy protein gel. It is preferred that the gel precursor have a pH from 6 to 7.5 and a solids content of from 14 to 35% by weight. The adjustment of pH and protein concentration ordinarily requires the addition of both water and alkali such as NaOH. In cases where the protein is dried prior to the preparation of a gel precursor, the adjustment of pH may be made at the time of drying.

By proceeding according to the above directions, there is obtained a protein gel precursor which varies in consistency from a semisolid paste-like mass up to a firm, shapable solid according to the protein content of the gel precursor and its pH history and the method by which the gel precursor was prepared.

In most cases, the protein gel precursor is too fluid to permit subdivision into small discrete particles and for this reason it is usually preferable to convert the gel precursor partially or completely into a chewy protein gel by the application of heat before further steps are taken. For this purpose it is preferred to steam the gel precursor until it has sufficient structure to be subdivided into discrete particles without coalescence of the particles. For this step temperatures in the order of 70° to about 120°C can be employed and steam is the preferred heating medium since the use of steam avoids any substantial amount of dehydration.

Example: Dry isolated soybean meal was prepared as follows. A protein extract was prepared from edible grade soybean meal whose nitrogen was practically completely extractable. First a 5% suspension of soy meal in an aqueous 0.003 M CaO solution was prepared. While the suspension was stirred gently, steam was introduced by a sparger until the temperature was 60°C. The suspension was stirred for 5 minutes at this temperature, then pumped to an efficient centrifuge where practically all of the insoluble matter was removed. The resulting extract contained over 90% of the nitrogen present in the original soy meal.

The extract, which was at 45° to 60°C, was stirred vigorously and approximately 3 N HCl was gradually added until the pH of the resulting slurry was lowered to 5.0. This precipitated the protein. The suspension of protein was then pumped to a basket centrifuge where it was collected as a wet cake containing about 20 to 30% protein. About 97% of the protein present in the extract was precipitated.

Finally the protein was washed at 60°C. The protein was next suspended in a volume of water approximately equal to that of the extract from which it was precipitated and the slurry was passed between the rotor and the stator of a colloid mill in order to disperse the protein completely. The dispersion, as it was stirred, was then heated to 60°C by means of steam and after it was at 60°C for 5 minutes, it was pumped to a basket centrifuge. The protein was collected in the centrifuge as a white cake containing about 25 to 30% protein.

The protein cake was then dried. A quantity of the protein cake was placed in an efficient mixer and while the cake was being mixed, a dilute aqueous solution of NaOH was added to raise the pH of the protein slurry to 7 and sufficient water to lower the protein content to about 10%. The mixture was then mixed until the protein was completely dispersed, and the resulting solution was spray dried. A gel precursor was made from the dried protein preparation.

The dry soy protein preparation was placed in an efficient paste mixer. While the mixer was operating, there was added a volume of water sufficient to prepare a protein dispersion containing 18% protein. Then a suitable amount of a red food dye was added and mixed in until the color of the mass was uniformly pink. The resulting gel precursor was a uniform plastic mass. The gel precursor was next converted to gel. The precursor was stuffed into molds about 4" by 4" by 6" and autoclaved at 5 psig for 2 hours. After cooling, the resultant gel was removed. Protein binders can be prepared, mixed with cubed portions of the above gel, and cooked to form simulated meats.

Alkaline Glycinin for Meat Processing

An alkaline glycinin is produced by *L. Sair; U.S. Patent 3,001,875; September 26, 1961; assigned to The Griffith Laboratories, Inc.* for use in processing meat products. This glycinin-base protein has better emulsifying properties and water-binding properties than glycinin precipitated at its isoelectric point. Glycinin is the principal protein of soybeans and methods for concentrating it from soybeans are well known.

It is a known characteristic of the extracted glycinin at its isoelectric point, that it is rendered soluble by alkali to raise the pH to values upwardly from 6. This is usually accomplished with sodium-base alkalies, and the altered glycinin might be termed sodium glycininate. To what extent and how the native glycinin of the bean is altered by mild alkali and by heat is unknown and precise knowledge of it is not material to this process. Accordingly, the protein extracted from defatted soybean material, and generally referred to as glycinin, and its forms which result from action of heat and mild alkali up to pH of 10.5, or the combinations thereof, are here covered by the term glycinin-base protein.

In general, meats of the preferred edible grades have a pH of 5.7 to 6.3, for example, comminuted fresh or cured meats used in preparing loaves, patties, and encased products, and fresh hams to be pickled in a brine bath or by artery pumping to provide either fresh or cured hams. In treating such meats, water-binding agents are important to minimize loss of water on storage in refrigeration, or in processing in the smokehouse. In compounding comminuted meats, emulsification of fat is important and the better the emulsification, the more the fat is prevented from separating in the product. The glycinin-base protein composition of the process is comparatively more powerful than known emulsifiers, and although it may dilute the meat content, it does not dilute the protein content.

The process may begin with wet glycinin concentrate having a pH near its isoelectric point. The glycinin concentrate, however it is prepared, is mixed with an edible agent to elevate the pH to 6 to 10.5, preferably from 6 to 7.5, for imparting solubility to the glycinin-base protein in the treatment of meat. Such solubility need not be complete solubility in water, but only solubility in the saline meat composition of the process. The wet insoluble glycinin mass may be mixed with the edible agent without adding water, and the resulting wet mixture may be used promptly in a comminuted meat mass. However, this requires a modification of the meat formulation to account for the water present in the resulting composition. So, it is preferred that such a wet mixture be dried and comminuted to provide a dry powder for use in a standard meat formulation without considering a change in water content.

Another method is to use enough water with the insoluble glycinin mass, either wet as prepared or previously dried, and also a pH raising agent in amount sufficient to effect complete solubility in water. Such a solution is then dried by spraying or other means to provide a dry composition.

The edible agents for elevating the pH may be alkaline materials or buffering compounds. Among them are sodium carbonate, sodium bicarbonate, sodium hydroxide, trisodium phosphate, disodium phosphate, sodium tripolyphosphate, sodium tetrapyrophosphate and ammonia. The amount of the agent to be used for elevating the pH of the resulting glycinin-base protein composition varies not only with the selection of the agent used, but also with the final pH which is to be achieved.

Example: Five pounds of defatted soybean flakes were mixed into a solution of 23 grams of soda ash and 3.4 grams of sodium sulfite in 6.75 gallons of water and the whole warmed to 140°F and agitated for one hour. It has a pH in the range from 7.4 to 7.6. Then, 230 grams of diatomaceous earth, as filter aid, are added and the mass filtered. The filtrate in the amount of 5 gallons is neutralized to a pH near the isoelectric point, as by adding 80 milliliters of concentrated hydrochloric acid to attain a pH of 4.4. A curd forms and rapidly settles. The supernatant liquid is decanted and the curd washed three times with water, each time to bring the volume to 5 gallons. The 5 gallon volume has a solids content of 10.8%.

Then the curd is treated to elevate the pH to 7 or above, for example, to 7.3 by adding 46 ml of 33% by weight caustic soda. On warming to 140°F, the curd all dissolves, and the solution is spray dried. A light colored powder with bland taste is an excellent emulsifier for ground meat products. A still lighter colored product may be obtained by bleaching the yellow-colored impurities with a mild bleaching agent without damage to the glycinin-base protein.

In the above process, it has been found that the temperature of extraction may be varied to affect the emulsifying property. Lowering the temperature from 140°F increases the solids content of the curd but lowers the emulsifying property per unit of weight. Raising the temperature above 140°F increases the emulsifying property per unit of weight but lowers the yield.

Copper Salts in Isolates for Baking

Isolates obtained from soybeans by ordinary methods produce heavy, low volume bakery products with poor crusts. *S. Kuramoto; U.S. Patent 3,252,807; May 24, 1966; assigned to General Mills, Inc and J.R. Short Milling Company* has found that by adding a copper-containing material to an aqueous alkali solution of the isolate, bakery products containing such isolates produce excellent breads and cakes.

The copper material can be metallic copper, alloys with other metals or any of a wide variety of copper salts such as cupric sulfate, cupric carbonate, cupric chloride, cupric glutamate and cupric nitrate. Cuprous salts can also be used but the cupric salts are preferred. The copper material is used in an amount sufficient to substantially eliminate the volume depressing action of the dried, isolated soy protein when used in yeast raised doughs. The amount used will ordinarily be about 30 to 150 ppm based on the solids content of the precipitate prior to the drying step.

The improvement in the isolated soy protein produced according to the process is believed to be due to oxidation, promoted by the copper material, of protein sulfhydryls to disulfides causing cross-linking of the protein molecules. Thus it is theorized that the following reaction takes place:

$$4R-S-H + O_2 \longrightarrow 2R-S-S-R + 2H_2O$$

where R–S–H is the original protein molecule. Regardless of whether or not the above reaction takes place, the isolated soy protein is vastly improved by this process. The isolated soy protein of this process is used to increase the protein content of yeast raised bakery products. The soy protein is added in any desired amount. However, it is preferred to add 5 to 20% by weight based on the weight of the flour and it is particularly preferred to add about 9% by weight.

Example: Hexane extracted soy flakes were extracted with 88% isopropanol at a solvent to flake ratio of 4:1. Extraction was effected at 35° to 45°C for 30 minutes with gentle agitation followed by decanting of alcohol and vacuum desolventizing to remove the last traces of alcohol. The flakes were slurried in water and the pH adjusted to 7 to 8. The temperature of the slurry was maintained at 35°C for 30 minutes and then the insolubles were removed by centrifugation and clarification was effected by another pass through the centrifuge. The protein was precipitated from the extract by adjusting the pH to 5.0 to 5.2 and separated from the soluble liquor by centrifugation. The glutenous cake was removed from the bowl and repulped in water.

At this stage, copper as $CuSO_4$ was added so that the slurried precipitate immediately prior to drying contained 100 ppm Cu based on the weight of the solids. The repulped precipitate was thoroughly washed by vigorous agitation and the protein again recovered by centrifugation. The solids content of the cake from the centrifuge was 50%. The cake was dried in a rotary drier using an inlet temperature of 60°C.

Partial Hydrolysis of Isolates for Cereals Using Mixed Enzymes

Soy protein for use in protein enriched cold cereals is made more palatable and tender by the process developed by *W.T. Bedenk and D.E. O'Connor; U.S. Patent 3,753,728; Aug. 21, 1973; assigned to The Procter & Gamble Company.* The soy protein is partially hydrolyzed using a mixed enzyme including papain and one additional proteolytic enzyme. The process is applicable to defatted soy proteins including the flour, concentrates and isolates.

The soy protein is made more palatable by forming a mixture of the soy protein, water, the proteolytic enzyme papain, and at least one other proteolytic enzyme. Quite unexpectedly, this specific mixture of proteolytic enzymes promotes a partial hydrolysis to the degree desired while an equal level of either papain alone, another proteolytic enzyme or a mixture of other proteolytic enzymes under the same reaction conditions does not result in the same degree of partial hydrolysis. Only at levels of proteolytic enzymes substantially greater than the total level of papain and at least one other proteolytic enzyme is there obtained the same degree of partial hydrolysis under the same conditions. In that proteolytic enzymes are relatively expensive, it is imperative that as low a level of proteolytic enzyme as possible commensurate with the proper degree of partial hydrolysis be used.

In this process, 15 to 2,500 ppm of papain and 5 to 2,500 ppm of at least one other proteolytic enzyme by weight of the soy protein is sufficient to cause the desired degree of partial hydrolysis when the reaction mixture is exposed to a temperature of 80° to 160°F for 1 to 120 minutes. The preferred levels of enzymes are 100 to 300 ppm of the papain and 100 to 300 ppm of at least one other proteolytic enzyme based on the weight of the protein. Temperatures of 120° to 130°F and times of 1 to 5 minutes are preferred.

The amount of water needed for the partial hydrolysis reaction is basically determined by apparatus limitations. That is, the lower limit of water is determined by the capability of the mixing equipment. The lower the level of water, the more viscous will be the resultant mixture. On the other hand, an excessive amount of water in the partial hydrolysis reaction would necessitate additional work in reducing the water level in subsequent processing steps. The preferred level of water is 50 to 80% based on the total weight of the mixture. The most preferred level is 55 to 60% based on the total weight of the mixture. Under the above conditions the soy protein is partially hydrolyzed to the extent that a cold cereal product containing the soy protein is acceptable with regard to taste, tenderness, crispness retention, and processability.

Example: 662 grams soy isolate, 830 grams water, 0.16 gram (242 ppm) papain, and 0.08 gram (121 ppm) alcalase are mixed and held at 100°F for 1 hour. To 1,350 grams of this mixture is added 1,250 grams of gelatinized corn grits. The resultant blend is then passed through an extruder at a temperature of 200°F. The extrudant is in the form of strands having a diameter of about 3/16 inch. These strands are sliced to form pellets 3/16 inch in length. The pellets are passed through a 2-roll mill to form flakes of 11 mil thickness. These flakes are now puffed in a salt puffer where the salt is maintained at 330°F.

After toasting the puffed flakes at 400°F for 0.8 minute, the resultant cold cereal is tested by an expert food panel. The cold cereal is tested for tenderness as measured on a 0-10 scale with a 10 rating being the most tender rating. The product is tested at 0 minute and at intervals of 2, 4, 6 and 8 minutes after being wetted with milk. The results are as follows.

Time	0	2	4	6	8
Tenderness	6.25	6.5	8.0	9.0	10.0

A cold cereal product is made by the same formulation and process steps and conditions as the above example with the exception that 0.24 gram papain (363 ppm) is substituted for the papain-alcalase mixture above. The same panel tested the resultant product in the same manner and rated it as follows.

Time	0	2	4	6	8
Tenderness	5	6	6.75	8.25	9

A comparison of the two tests shows that at all time intervals the first product, i.e., the formulation of the example containing the papain-alcalase, is more tender than the product of this comparative test containing papain at the same total proteolytic enzyme level.

SOY HYDROLYSATES

ENZYMATIC HYDROLYSIS

Protease Hydrolysis of Oilseeds

An overall process of nine steps is disclosed by *S.E. Sherba, R.B. Steigerwalt, W.T. Faith, Jr., and C.V. Smythe; U.S. Patent 3,640,725; February 8, 1972; assigned to Rohm and Haas Company* for separating nutritional components from soybeans or other oilseeds such as cottonseed, employing enzymatic hydrolysis of protein. The hydrolysate may be used as the basis for a nutritious beverage. These steps are carried out in the following manner.

Step 1: The beans are comminuted, i.e., ground, chopped, sliced, flaked, etc., to maximize the effective surface area exposed. For example, if soybeans are subjected to grinding in this step, the particles are desirably within the range of 10 to 100 mesh.

Step 2: The comminuted beans are heat treated at a temperature in the vicinity of boiling or above, advantageously from 90° to 140°C, and preferably at 120°C with steam. For example, the treatment is continued for 1 or more minutes, with the time depending on the temperature and the amount of extractable protein desired. At a temperature of 120°C, the heating is desirably continued for about 5 to 15 minutes. At lower temperatures, the heating is desirably carried out longer, and at higher temperatures, the heating step may be shorter.

Step 3: The heat-treated mass is cooled to incubation temperature, as by adding cooling water. The total amount of water added during the heating and/or cooling steps is desirably between about two and 20 times the weight of the ground beans.

Step 4: The proteolytic enzyme composition is then added to the slurry, which is maintained at incubation temperature and pH for a time sufficient to effect the desired degree of hydrolysis. Although substantially any proteolytic composition could be employed in the process of the present method, it is preferred to employ proteases of bacterial or fungal origin. Advantageously, the bacterial proteases may be derived from a member of the genus Bacillus, such as *B. subtilus, B. mycoides, B. amyloliquefaciens, B. cereus, B. macerans, B. megaterium, B. sphaericus, B. circulans,* etc. A particularly preferable Bacillus for the preparation of protease useful in the present process is *B. subtilis.* Alternatively, a protease of fungal origin may be employed. This may be derived from a crude culture of Aspergillus, and desirably from the *A. flavus-orzyae* and *A. niger* groups.

Step 5: Following incubation, the solids are separated from the slurry, e.g., by filtration through a screen having apertures appropriately selected relative to the size of the particles produced in step 1. If, for example, soybeans were ground to about 20 mesh particles in step 1, a 100 mesh screen might be employed to advantage in effecting the present separation. Optionally, filtration may be hastened by employing a basket centrifuge or other such conventional means.

Step 6: The fluid portion obtained from the preceding step is readily separated into an oil phase and an aqueous phase. The phase separation may be accomplished merely by permitting the fluid portion to settle, but in practical use is desirably facilitated by employing such apparatus as a centrifuge, or preferably a cream separator common in the dairy industry. Alternatively, countercurrent solvent extraction or other means known to the industry for separating the oil fraction may be used.

Step 7: The aqueous phase from the preceding step is acidified to about pH 4.5 with a food grade acid, e.g., lactic acid, to precipitate the isoelectric protein (ISP) remaining, which may then be separated by filtration, or preferably centrifugation. The moist solid product may be spray dried if desired. Although ISP is one of the products of the present process, it is not, however, intended to be the major product. Rather, the proteolysis of step 4 is intended to produce a maximum soy protein hydrolysate (hereinafter SPH), which is not precipitated at the isoelectric point.

Step 8: If it is desired to recover the SPH as a dry product or concentrated solution, the centrifugate from the preceding step is optionally adjusted to a pH of 6 to 7.5 and then is concentrated by conventional evaporation techniques, such as flash evaporation, triple-stage evaporation, reverse osmosis separation, etc. Optionally, the pH adjustment, if performed, may be carried out in step 9.

Step 9: The solution of SPH, preferably from step 8, but optionally from step 7 omitting step 8, may then be subjected to spray drying, or alternatively to lyophilization, air-toss drying, heat drying, etc., to recover the SPH in solid form, along with carbohydrates and other soluble materials.

The various enzyme compositions employed in the example are indicated in Table 1.

TABLE 1: ENZYME COMPOSITIONS

Composition	Major enzymatic activity	Measured proteolytic activity	Measured other activity
B—Derived from *B. subtilis*.	Neutral bacterial protease, as described by McConn et al., J. Biol. Chem., 239 3706 (1964).	62,500 HU; 510,000 GVU (pH 7.0).	750 PGU; 298,000 GVU (pH 4.7).	Pectinolytic: O APU; below 30 CU. Diastatic: 2900 BAU.
P—Derived from *A. oryzae*.	Neutral and alkaline *A. oryzae* proteases. Little of the acid protease. See Bergkvist, Acta. Chem. Scand., 17 1521 (1963). Produces extensive hydrolysis, but as measured by amino nitrogen liberated, hydrolysis to individual amino acids is far from complete.	194,000 HU	671 PGU; 309,000 GVU (pH 4.7).	Pectinolytic: 110 APU; 140 CU. Diastatic: 460 BAU; 613 SKB Lipolytic: 4.2 GSA (pH. 6.0).
H—Derived from *A. oryzae*.	High amount of the acid *A. oryzae* protease as well as neutral and alkaline protease. See Bergkvist, supra.	305,000 HU; 275,000 GVU (pH 7.0).	457 PGU; 167,000 GVU (pH 4.7).	Pectinolytic: O APU; 123 CU. Diastatic: 155 BAU; 183 SKB.
C—Derived from *A. niger*.	Carbohydrase. Hydrolyzes saccharide containing α 1,6 and/or α 1,2 linkage. See U.S. Patent Appln. Serial No. 725,497, filed April 30, 1968.	———————	Below 5 PGU.	530 Carbohydrase units.

Example: Ten grams of Kanrich number 1 soybeans containing 38.0% protein, 16.73% oil and 6.4% moisture was ground to 20 mesh particles and then added to 50 ml water, heated in an autoclave at 121°C for 5 minutes, and then cooled. An additional 50 ml water was then added, along with an enzyme composition containing proteolytic and carbohydrase activity. The slurry was incubated at 50°C at ambient pH 6.3 for 8 hours and then filtered through a 100 mesh screen. The fibrous residue (1) on the screen was washed with water until the solution collected amounted to 100 ml then the residue (1) was dried in an oven at 50°C. The solution was then centrifuged at 10,000 rpm for 1 minute, and the centrifugate was decanted. The residue was washed with 10 ml water and recentrifuged. The washed residue (2) was dried in a dessicator. The decanted centrifugate contained oil, which was removed by gravity filtering through Whatman number 1 paper. After drying, the paper was exhaustively extracted with petroleum ether. The protein remaining on the paper (3) was determined. The petroleum ether was then evaporated off, leaving oil to be determined by weight. It was found that 38% of the oil was extracted using a mixture of enzyme compositions B and C; 50.3% was extracted using corresponding weight amounts of compositions P and C. The filtrate (4) was then acidified (pH 4.5) to precipitate any ISP remaining, but none was obtained. The filtrate was then combined with an equal volume of 10% trichloroacetic acid at 4°C and no visible precipitate resulted.

TABLE 2: TOTAL SOLIDS AND PROTEIN CONTENT OF VARIOUS FRACTIONS

Enzyme composition (See Table I)	B (100 mg.) plus C (15 mg.)		P (100 mg.) plus C (15 mg.)	
	Weight	Protein Content	Weight	Protein content
Fibrous residue on screen (1), g..	2.40	0.500	2.60	0.536
Centrifugate (2), g.	1.50	0.563	0.922	0.236
Filter paper residue (3), g.	0.082	0.082	0.075	0.075
Filtrate [solids] (4), g.	4.93	2.400	5.27	2.590
Total, g.	8.91	3.545	8.87	3.437

Soluble Soy Protein Using Phytase

Phytase is used by *E.M. McCabe; U.S. Patent 3,733,207; May 15, 1973; assigned to Societe d'Assistance Technique pour Produits Nestle S.A., Switzerland* in his process for preparing soluble proteins. This protein fraction, particularly suitable for incorporation in carbonated beverages, is recovered from soy protein, by reaction of the soy with phytase, removal of matter insoluble at a pH of 4.6 and recovery of the fraction insoluble at pH of 5.0 to 5.4.

Most conveniently, the starting material is a soy protein isolate, containing at least about 90% protein and being soluble in alkali. However, both soy protein concentrates and even soy flour (desirably low-heat treated) may be used. For the enzymatic reaction, the soy protein starting material (isolate, soy flour or meal) may first be suspended in water or dilute alkali, for example at pH 7.0 to 8.5. The solids concentration should desirably not exceed 15% by weight as at higher levels the solutions become very viscous and less easy to handle. An acid is then added to adjust the pH of the suspension to 4.6 to 6.0. While the enzyme is most active at a pH of 5.0 to 5.5, it is preferred to work at pH values of 5.5 to 6.0 as in this range a greater proportion of the protein present is dissolved. Phytase is then added, preferably in an amount corresponding to 0.01% by weight of the protein present, and enzymic reaction is carried out at 50° to 55°C, this being the temperature of optimum activity of the enzyme.

The reaction time should be sufficient to secure breakdown of substantially all the phosphate linkages present in the protein. Depending on the activity and purity of the enzyme, as well as the reaction conditions, the total time may be up to 36 hours. The pH of the reaction mixture is then adjusted to a value of 4.6, and the resultant precipitate, containing unreacted material, is separated by filtration or centrifugation. Hydrochloric acid is preferred for pH adjustment, but any other nontoxic acid may be used. The pH of the recovered clear solution is then adjusted to a value of 5.0 to 5.4, preferably about 5.2. A precipitate is again formed, and it may be recovered as previously. This precipitate consists of the protein fraction soluble at low pH values (below 4.6), and it may be formulated in various edible compositions. It is particularly suitable for incorporation in beverages such as soft drinks which may be carbonated, for example at a level of about 2.0 to 2.5% by weight.

The desired protein fraction may be further purified by washing, molecular sieving or gel filtration to remove traces of salts formed during the process and to eliminate off flavors. The product has a bland flavor and is soluble in water as acidic solutions at pH values below 4.6. The process is further illustrated by the following example, in which the parts are by weight.

Example: 10 parts of soy protein isolate are dissolved in 100 parts of water to which sufficient sodium hydroxide has been added to raise the pH to about 8. Hydrochloric acid is then added to lower the pH to 6 and 0.1 part of a phytase-containing preparation [obtained from wheat meal by the method described by F.G. Peers, *Biochem. J.*, 53, 102 (1953)] are added. The temperature is maintained at 50° to 55°C for 24 hours, which is generally sufficient for the phytase to break down the phosphate linkages in the protein. Upon completion of the reaction the pH of the medium is adjusted to 4.6 by addition of hydrochloric acid. The resultant precipitate is removed by centrifugation and discarded, whereas sodium hydroxide is added to the supernatant to raise the pH to 5.2 where a precipitation occurs.

The precipitate thus formed, consisting of the protein fraction soluble at pH values of 4.6 and below, is recovered. It may be further purified by washing, reprecipitation or gel filtration to remove traces of salts formed during neutralization. The protein fraction may also be dried, giving a white powder with a bland taste, which is suitable for incorporation in different foods and beverages.

Dual Enzyme Solubilization and Hydrolyzation

A process which makes possible the solubilization and hydrolyzation of isolated protein has been developed by *F.F. Noe; U.S. Patent 3,761,353; September 25, 1973; assigned to Rohm and Haas Company.* The process provides an economical nondeleterious enzymatic method for solubilizing and hydrolyzing isolated protein. Still more specifically, the present process comprises: (1) contacting isolated protein with an enzyme preparation in an amount which is effective to solubilize such protein; and (2) contacting isolated or solubilized protein with an enzyme preparation in an amount which is effective to hydrolyze such protein.

The enzyme preparation employed in solubilizing isolated protein may be referred to as Solutase. It is derived from a bacterial organism and is further characterized by being able to render an isolated protein dispersible in water to the extent that a colloidal system will result. Furthermore, the protein so dispersed will be essentially recoverable therefrom by precipitation. This ability of a Solutase enzyme preparation to solubilize isolated protein shall be referred to as Solutase Activity. The degree thereof is expressed in Solutase Units (SU). One SU is that amount of enzyme preparation which will disperse 1 gram of isolated protein in 100 ml of water in 15 minutes at a pH of 7.8 when the mixture is agitated at 275 rpm at 40°C.

Hydrolyzation of the isolated protein and particularly the isolated protein which has previously been solubilized is achieved by contacting such with a fungal enzyme preparation. This fungal enzyme preparation is referred to as Hydrolase and is characterized by being capable of altering the protein to the extent that it will not precipitate from an aqueous solution.thereof. This ability of a Hydrolase enzyme preparation to hydrolyze isolated protein and solubilized isolated protein may, for the purposes hereof, be termed Hydrolase Activity. The degree thereof is expressed in Hydrolase Units (HU). One HU is that amount of enzyme preparation which will render 1.0 gram of solubilized protein unprecipitable when such is contacted for 16 hours at 40°C while being agitated at 275 rpm and maintained at a pH of 6.2. Thus, the Hydrolase Activity of an enzyme preparation can be readily determined.

The Solutase enzyme preparation will be a growing culture of Bacillus including, of course, the various strains and their mutants. Some typical members of the Bacillus genus include for example: *B. subtilis, B. mycoides, B. amyloliquefaciens, B. cereus, B. macerans, B. megaterium, B. sphaericus, B. circulans* and the like. In this respect, a particularly preferable organism for the production of Solutase is *B. subtilis.* Hydrolase is of fungal origin. Generally, it is derived from a growing culture of Aspergillus and particularly, from the *Aspergillus oryzae-niger* group. Typically, this will include microorganisms such as *Aspergillus oryzae* or *Aspergillus niger* with the latter organism being preferred. Both the Solutase and Hydrolase producing organisms may be grown in either a liquid submerged or surface culture on a medium comprising a utilizable source of energy, assimilable carbon and nitrogen. Of course, any of the regularly employed growth factors and mineral salts or combinations thereof may also be incorporated in such culture media. In this process the effective amount of Solutase and/or Hydrolase will vary not only with the activities of the enzyme preparations, but also with the particular protein to be treated in addition to related processing conditions. It has been found that for each gram of protein 0.1 to 100 and preferably 1.0 to 25 SU will completely solubilize them. To completely hydrolyze each gram of isolated or solubilized protein, about 0.1 to 50 HU will be utilized.

In general, the process is carried out at a pH of 6.0 to 7.8. Depending upon the protein being treated and the like, various substances can be incorporated therein for pH adjustment and maintenance. Typically, these would include ammonium sulfate, ammonium hydroxide, sodium hydroxide, sodium carbonate, hydrochloric acid, sulfuric acid, lactic acid, adipic acid and the like.

It is also highly desirable to maintain the protein-containing aqueous medium while being contacted with Solutase and/or Hydrolase at a temperature in the range of 20° to 60°C and preferably at 30° to 40°C. Usually, the solubilization and/or hydrolyzation process will proceed more rapidly at higher temperatures and under gentle agitation. For the most part, satisfactory protein solubilization will require at least five minutes and generally as much as 1 to 5 hours or longer. Hydrolyzation in accordance with this process usually necessitates about 6 hours and generally as much as 16 hours or longer. Under preferable conditions complete protein solubilization and hydrolyzation will be accomplished within 10 to 12 hours.

Example: In each of 5 flasks 3 grams of isolated soybean protein is suspended in 100 ml of tap water. This suspension is then adjusted to a pH of 6.2 with dilute lactic acid. In three of the flasks, the Solutase and the Hydrolase are added simultaneously in the proportions hereinafter set forth in the table. The percentage concentration of each enzyme preparation is based on the weight of the isolated protein suspended in the water. All of the flasks are then agitated at 275 rpm at 40°C for either 16 or 24 hours as indicated and then removed and filtered under reduced pressure through one layer of Whatman number 42 filter paper. The amount of insoluble protein on the filter paper is determined by dry weight and the amount of solubilized protein in the filtrate is calculated by weight difference.

To this filtrate there is added an equal volume of cold 10% trichloroacetic acid (TCA) to reprecipitate any isoelectric protein therein. A precipitate is apparent only in the nonenzyme treated samples. This is removed by centrifugation and all of the filtrates are then analyzed for protein nitrogen by the Kjeldahl Method of analysis so as to determine the percent of the protein solubilized which was hydrolyzed. The table below sets forth the results thereof.

Enzyme preparation—(percent by weight of isloated soybean protein)			Isolated protein Solubilized (gm.)	Isoelectric protein precipitated from filtrate (gm.)	Percent of protein solubilized which was hydrolyzed
Solutase	Hydrolase	Time (hr.)			
None	None	16.0	1.35	0.092	83.0
0.058	0.333	16.0	2.13	0.00	95.5
0.580	0.333	16.0	2.40	0.00	100.0
None	None	24.0	1.48	0.084	81.0
0.580	0.333	24.0	2.43	0.00	95.0

HYDROLYSIS UNDER ACID CONDITIONS

Controlled Incomplete Acid Hydrolysis

It is the object of the process disclosed by *L. Sair; U.S. Patent 3,391,001; July 2, 1968; assigned to The Griffith Laboratories, Inc.,* to subject protein to acid hydrolysis to a limited extent to provide a partial hydrolysate with superior flavor. An edible protein, such as soy protein and wheat gluten, is hydrolyzed in aqueous hydrochloric acid until the alpha amino nitrogen content is in the range from 35 to 58% of the total nitrogen content, and then a sodium alkali, such as sodium hydroxide, is added to a pH of 4.5 to 7.

In other words, protein is subjected to hydrolysis with hydrolyzing acid in any conventional procedure, but without completing the hydrolysis. The end point can be ascertained by calibrating a fixed procedure using predetermined amounts of water, selected acid, and selected protein. In one procedure a conventional amount of selected acid may be used which is suitable for completing a hydrolysis, and hydrolytic action then arrested at the time determined by the calibration, when a value in said predetermined range is attained. The second method is a modification using a predetermined smaller amount of the acid and likewise to stop the hydrolysis at the time determined by calibrating the procedure, when a value in the range is attained. This method is preferred when hydrochloric acid is used and when it is neutralized to sodium chloride and the hydrolytic products are neutralized to a pH in the range from 4.5 to 7. Sodium chloride in the neutralized product is thus minimized.

Commercial proteins vary in content of protein from about 60 to 100%. Upon complete hydrolysis the alpha amino nitrogen varies in the range from 63.5 to 67% of the total nitrogen of the 100% protein. The control is established by numerous experiments in which the range of N ratios is selected by judgment as to flavor. A fixed procedure has been used for such calibration in which the protein in a given volume of aqueous solution containing hydrolyzing acid is refluxed for 11 hours, neutralizing with caustic soda to pH of 5.5, and spray-drying the neutralized hydrolysate. By varying the acid content, the extent of hydrolysis is determined by determining the N ratio in the dried hydrolysate. Table 1 shows a mixture of two commercial proteins so processed with varying amounts of HCl (SG 1.1888, 37% HCl by weight) and varying amounts of water. Test 1 is designated as 100%, representing the amount of acid needed for the fixed procedure to produce the ultimate content of alpha amino acids where the N ratio is 67%.

TABLE 1

	Test			
	1	2	3	5
	Percent Acid Used			
	100	80	60	40
Soy protein (70% protein) in grams	99.6	99.6	99.6	99.6
Wheat gluten (85% protein) in grams	140.0	140.0	140.0	140.0
HCl (37%) in ml	250	200	150	100
Water in ml	180	230	280	330
Parts by weight of above acid (37%) per 100 parts of 100% protein	163	132	99	66
Percent total N in spray-dried hydrolysate	6.89	7.04	7.27	7.81
Percent alpha amino acid N	4.61	4.46	4.20	2.80
Percent amino N to total N	87.0	63.5	57.7	35.8

Table 1 shows that as the amount of acid is reduced the free amino acids are reduced. Table 2 shows the percent hydrolysis of Tests 1 to 5, repeating certain data of Table 1, and shows also the analytical amount of monosodium glutamate (MSG) in the hydrolysate. The flavor of the hydrolysates is tagged F. The table shows the character of F for the tests.

TABLE 2

Test	Percent Acid Used	Percent Amino N to Total N	Percent Hydrolysis	Percent MSG	Flavor F
1	100	67	100	14.1	Similar to a conventional complete hydrolysis
2	80	63.5	95	15.1	Similar to a conventional complete hydrolysis
3	60	57.7	86	14.9	Strong beefy flavor
4	50	44.1	66	9.0	Most desirable
5	40	35.8	53.5	1.5	Slight bitterness evident

Table 2 shows that in use of 50% of acid (Test 4), all of the potential MSG has not been liberated, the hydrolysate containing larger molecular bodies comprising glutamic acid components, as a mixture of peptides, and polypeptides and possibly some solubilized protein bodies. By lowering the acid to 40% only 1.5% of MSG is produced.

Acid Enzymatic Hydrolysis

A. Pour-El and T.C. Swenson; U.S. Patent 3,741,771; June 26, 1973; assigned to Archer-Daniels-Midland Company have disclosed the process wherein dispersed plant proteins are digested with an acid active enzyme at a pH below 4.6. Quiescent conditions are maintained in the reaction medium and digestion proceeds until the insoluble colloidal protein is substantially completely dissolved. The pH is raised to about 4.6 and the medium is allowed to stand causing additional insoluble protein to precipitate. The insoluble residue is removed and the soluble protein material is dried. A clear liquid, which can be carbonated, is formed containing the solubilized protein material at a pH corresponding to the isoelectric point of the protein.

Accordingly, it is the primary object of the present method to provide a protein which is adapted for forming sparkling clear solutions at acid pH's. These objects are accomplished in accordance with the present method by treating a dispersed plant globulin with an acid active enzyme at acid pH with controlled agitation wherein the enzyme digests the plant proteins in colloidal suspension and produces a protein which is approximately 100% soluble in the reaction medium and then raised to about 4.6 by adding a basic medium for 1 to 6 hours at 0°C to room temperatures. The method thus involves a proteolysis performed at a specific pH range followed by a chemical winterization at pH 4.6

The proteins which are intended to be treated by this process are all plant globulins, i.e., simple proteins which yield only alpha amino acids or their derivatives upon hydrolysis. Said simple proteins are more particularly those vegetable proteins, such as glycinin, which are derived from soybeans. However, the process is adapted to treat all globulins derived from plants. Plants, of course, include not only vegetables but cereals such as corn, wheat, rye, meal, and fruit kernels, etc. The enzymes which may be used to treat the plant globulins are proteinases, i.e., those which are active in breaking the peptide bond, particularly those which are active at acid pH of 2 to 5 such as fungal proteinases, pepsin, or any other acid active enzyme.

The digestion medium provided is generally an aqueous dispersion of protein containing 3 to 6% protein in colloidal dispersion. Any suitable acid may be added to provide the acidic pH required for digestion. An example is a combination of 6 normal hydrochloric acid and concentrated phosphoric acid. The addition of the acid active enzyme is performed by stirring approximately 0.1 to 1% based on the protein or 0.003 to 0.03% based on the total slurry of the enzyme and allowing the reaction medium to settle. During digestion, quiescent conditions are maintained. Since the pH gradually rises during digestion, it is sometimes necessary to add acid to maintain a low pH. This pH adjustment is performed by periodically adding concentrated phosphoric acid or other acids to adjust the pH; adjusting to pH 2.7 is generally adequate.

In U.S. Patent 3,741,771 the process is modified as follows. After sufficient time has elapsed for colloidal protein to be digested and for the larger solids to settle out of solution and either before or after the digested protein slurry is cooled and centrifuged and the solid residue is discarded, a base is added to raise the pH to 4.6 whereupon the mixture is then allowed to stand at temperatures varying from 0°C to room temperature, depending on the protein molecule size in solution, for a period varying from 1 to 6 hours. In both processes the solution is filtered or centrifuged to remove the insoluble residue. The base can be any metal oxide or hydroxide of food quality but is preferably calcium oxide, NaOH, or NH_4OH. Divalent bases are preferred to form insoluble protein complexes that are more readily precipitated rather than the plain protein aggregates formed by monovalent bases. The soluble material is dried and renders a powdery product which is adapted for forming sparkling clear solutions at all pH's, indicating substantially complete solubility even at the isoelectric point of the starting material.

In the following examples, which illustrate the process, Example 1 is covered by the process of U.S. Patent 3,713,843 and Example 2 by U.S. Patent 3,741,771.

Example 1: (a) Approximately 100 grams of Nutrisoy 7-B (a commercially available soy protein flour) is extracted by mixing with about 20 times its weight in water at 96° to 100°F for one hour with constant agitation. The resulting slurry is allowed to settle and is decanted. The solid residue is extracted by mixing with approximately 10 times its weight of water for one hour at 96° to 100°F with no agitation. The resulting slurry is centrifuged and filtered and the solid residue discarded. The supernatant liquid from the first extraction and the second extraction are combined and hydrochloric acid is added to bring the pH of the liquid to 4.75. This liquid is centrifuged (filtering may be substituted here) and the supernatant liquid is discarded. The solid residue is washed twice with 10 times its weight in water and the washings are discarded. The washed solid residue is used to make a 3% solids aqueous slurry to which is added a mixture of 6 normal hydrochloric acid and concentrated phosphoric acid to bring the pH to 3.5.

(b) To the acid slurry is added 0.03% by weight based on the weight of the slurry of acid fungal proteinase with stirring. After settling the medium is left for a period of 2 days wherein the pH is periodically adjusted to 2.7 to 4 by addition of concentrated phosphoric acid. The digestion medium is maintained in a generally quiescent state during the digestion period except during the acid additions and the originally present colloidal dispersion gradually is converted into a sparkling clear phase. After the 2 days, the digested protein slurry is cooled to about 15°C.

(c) Thereafter, the solution is centrifuged, and the solid residue is discarded.

(d) Then the supernatant liquid is converted by means of freeze drying into a powder product. Upon adding the powdered product to water and adjusting the pH to from 2.7 to 4.75, the proteinaceous material is substantially completely dissolved and forms a sparkling clear liquid when quantities up to 20% by weight of the soluble proteinaceous material are used.

Example 2: After step (b) of Example 1, solid calcium oxide is added to the cool medium at 6°C to bring the pH to 4.6 and the mixture is maintained for 6 hours. Then steps (c) and (d) of Example 1 were repeated yielding a product with the same solubility characteristics as found in Example 1.

NUTRIENT AMINO ACID COMPOSITIONS

Yeast-Hydrolyzed Protein Product

A process developed by *I.E. Witwicka, J.S. Pavuk and K.M. Gaver; U.S. Patent 2,999,753; September 12, 1961; assigned to The Ogilvie Flour Mills Company, Limited* provides a food product formed from a combination of yeast and hydrolyzed vegetable protein (HVP), such as the liquid product left over from the manufacture of glutamic acid from gluten, in which the characteristic flavor of the yeast is masked by the hydrolyzed vegetable protein and the salty taste of the hydrolyzed vegetable protein is eliminated by the yeast. The hydrolyzed vegetable protein (HVP) is obtained as described in U.S. Patent 2,828,336 and U.S. Patent 2,831,889, although other methods of obtaining such hydrolyzed vegetable protein are obviously usable.

In the separation of the glutamic acid the filtrate after the separation of leucine, isoleucine, tyrosine, etc., is acidified to the isoelectric point of glutamic acid. The glutamic acid is then crystallized out and separated. The end liquor has a pH of about 3.0 to 3.2. The end liquor is adjusted to a pH of 10.7 to 10.8 by the addition of sodium hydroxide, then warmed to temperatures ranging from 100° to 175°F under vacuum for between 8 and 20 hours to reduce the ammonia content to 0.01 to 0.05% or below, and neutralized to a pH of 5.5 to 5.6 with hydrochloric acid to obtain HVP. In lieu of alkalizing, heating, and neutralizing, the end liquor may be neutralized from its pH of 3.0 to 3.2 to a pH between 5.0 to 7.0 (e.g., 5.5) by the addition of sodium hydroxide and then warmed to a temperature of 100° to 175°F for from 8 to 20 hours under vacuum to eliminate the ammonia. This neutralized and pretreated glutamic acid end liquor is called hydrolyzed vegetable protein (HVP). Hydrated yeast is also obtained in any practical convenient manner. The preferred way of obtaining this hydrated yeast is by taking dry yeast and hydrating it. The hydrolyzed vegetable protein is then mixed with the hydrated yeast. The autolysis of the yeast is induced by the salt in the hydrolyzed vegetable protein. After autolysis, the autolysate is dried in any convenient manner. It is preferred to dry by vacuum evaporation or by spray drying.

Example: One thousand grams of compressed yeast (27% solids) was admixed with 1,000 ml of processed HVP (1,250 grams) and autolysis allowed to proceed with agitation at room temperature for four hours. Half of the autolysate was vacuum evaporated to a paste containing approximately 75% solids, and the other half was spray dried to a tan powder.

Analysis of the powder indicated:

	Percent
Solids	95
Protein	51.25
Total nitrogen	9.18
Amino nitrogen	8.35
Salt	31.25

Palatable Balanced Amino Acid Compositions

A process has been developed by *M. Winitz; U.S. Patent 3,698,912; October 17, 1972; assigned to Morton-Norwich Products, Inc.,* for making palatable nutrient compositions for human consumption which contain all of the essential amino acids plus nonessential amino acids in nutritionally balanced relationship. Aqueous solutions of acid or enzyme hydrolyzed proteins provide a starting point for the process. Amino acid based diets prepared with former protein hydrolysates possess highly objectionable taste properties, which have been erroneously attributed to the amino acids themselves, rather than to certain degradation products which arise during the process of hydrolyzing proteins.

It has been found that a formulation containing amino acids is unpalatable if it contains lower alkyl mercaptans, such as methyl mercaptan, in an amount that an aqueous solution of the formulation would have a concentration greater than 15 milligrams of mercaptans per liter, when the pH of the solution is 3.7. The tolerance level decreases slightly with increasing pH, with the tolerable concentration being 7.5 mg/l at a pH of 5.7. It has also been found that the sulfhydryl group of the amino acid cysteine likewise detracts from palatability, and the sulfhydryl group concentration should be below 0.05 gram per liter of solution at a pH of 5.7. The tolerable sulfhydryl concentration decreases at lower pH's, and at a pH of 3.7 it is 0.038 g/l. It has further been found that glutamic acid, or its salts imparts a strong characteristic flavor to an amino acid formulation, which can render a formulation unpalatable. The flavor renders the entire diet formulation unpalatable if it is present in an amount of more than about 1.43 grams per liter.

In accordance with this process, the protein hydrolysates which are used may be prepared from fairly abundant high quality proteins, such as casein, lactalbumin, soybean, fish, and bacterially produced protein from petroleum and the like. However, other suitable proteins may also be used, such as gelatin, egg albumin, whole blood protein, wheat gluten and zein, for example. Moreover, if desired, mixtures of more than one protein may be used. After an aqueous solution of the protein is formed, substantially all of the following amino acids, if present, are removed from the solution: tyrosine, cystine, aspartic acid and glutamic acid. These four amino acids may be removed in any order. The cystine and tyrosine may be removed separately but it is preferred that they be removed together. This pair of acids are preferably precipitated from the main hydrolysate solution while it is maintained at a temperature below 50°C and a pH between 5 and 7. It is also preferred to remove the glutamic acid and aspartic acid together, and these two amino acids are preferably removed by adsorption onto anion exchange resins. Removal of these four amino acids as pairs provides a preferred, simple and expedient method of removing these components from the main hydrolysate solution. Following removal of the aspartic acid, tyrosine, glutamic acid and cystine from the protein hydrolysate solution, water soluble forms of these components, or of nutritional equivalents are added to the hydrolysate solution, or to the dry mixture of amino acids remaining after evaporation, to alleviate the deficiency that now exists.

Tyrosine and aspartic acid are preferably supplied in forms having a water solubility at least equal to that of leucine. It is preferred that the tyrosine is converted to a water soluble tyrosine derivative (preferably tyrosine ethyl ester hydrochloride), that the aspartic acid is converted to a water soluble salt, such as a sodium salt, and that these water soluble components are then returned to the main protein hydrolysate. The two amino acids need not be returned to the same protein hydrolysate from which they were removed or in precisely the same amounts because water soluble L-isomers of these two amino acids are commercially available today. However, there are economic advantages to converting and returning the removed components, and this is the preferred method of operation.

Example: Sixty grams of purified soybean protein hydrolysate (enzymatically hydrolyzed) is dissolved in one liter of hot water. The solution is cooled to room temperature, and the pH adjusted to 2.8 by addition of concentrated hydrochloric acid. To this solution is added Amberlite IR4 resin (an anion exchange resin of the weak base type), while stirring, until the pH reaches 6.5. The ion exchange resin is pretreated by successive treatments of hydrochloric acid, water, sodium carbonate, and water. The resin is filtered from the solution, and the filtrate is acidified with hydrochloric acid to a pH of 2.8. The acidified filtrate is treated again with fresh ion exchange resin until the solution reaches a pH of 6.5. The resin is again filtered from the solution, and the two resin portions are combined. The filtrate is chilled to 4°C and kept at that temperature for twenty four hours.

Under these conditions, cystine and tyrosine precipitate from the solution. The solution is filtered using suction, the precipitate washed with a small amount of warm water, and the wash water is combined with the filtrate. The filtrate plus wash water is heated to boiling, Norit is added, and the hot solution is filtered. The volume of the solution is adjusted to 400 ml by either the addition of an appropriate amount of water or by evaporation. The cystine and tyrosine are separated from one another, and the tyrosine converted to tyrosine ethyl ester hydrochloride. Pure L-tyrosine ethyl ester hydrochloride, in an amount of 2.3 grams, is recovered from the cystine tyrosine precipitate. Approximately 6.5 grams of L-aspartic acid is recovered from the combined ion exchange resin portions which is converted to its sodium salt by dissolving in 60 ml H_2O in which 1.9 grams of NaOH is dissolved. The sodium L-aspartate solution and the L-tyrosine ethyl ester hydrochloride are added to the main solution. About 1.1 grams of purified L-methionine is added while the solution is at about 90°C. The heat is removed and the solution is rapidly cooled to 40°C or below by the addition of 400 grams of glucose, in increments of about 50 grams each, using continuous stirring to facilitate dissolution. Under these conditions, the temperature reaches 40°C about ten minutes subsequent to the initial addition of glucose. After the temperature of the solution is below 40°C, 0.34 gram of L-tryptophan is dissolved, followed by 5.0 grams of L-glutamine.

Analysis for amino acids showed: 0.84 gram L-tryptophan, 2.3 grams L-threonine, 3.1 grams L-isoleucine, 4.5 grams L-leucine, 3.7 grams L-lysine, 2.6 grams L-methionine, 2.9 grams L-phenylalanine, 2.3 grams L-tyrosine ethyl ester hydrochloride, 3.1 grams L-valine, 4.2 grams L-arginine, 1.4 grams L-histidine, 2.4 grams L-alanine, 6.5 grams L-aspartic acid, 5.0 grams L-glutamine, 2.5 grams glycine, 3.9 grams L-proline, 3.8 grams L-serine. The analysis shows that a substantially quantitative recovery of most of the amino acids from the 60 grams of soybean protein hydrolysate is obtained, with the exception of cystine, glutamic acid, and tryptophan. The solution is relatively clear in color, and tasting shows that it has a fairly sweet taste characteristic of high glucose content, but is clearly palatable to human taste. The amino acid glucose formulation is considered well suited for combination together with additional carbohydrate, vitamins, minerals and fat to provide a diet composition containing all the necessary nutritional requirements for the human species.

COTTONSEED PROTEIN

Cottonseed kernels, which contain up to 50% protein, have been used mainly in the production of oil with the residue, the meal, mainly used in animal feeds. The presence of gossypol in the pigment glands of the seeds is toxic to humans and single stomach animals. Its use has been therefore limited to certain animal feeds unless it has been detoxified. Recent agricultural research has resulted in the development of glandless (gossypol-free) cottonseed which could produce protein for human diet. Also, the liquid cyclone process as described in U.S. Patent 3,615,657 (page 147) has received recent notices as a method of producing protein concentrates for human use.

COTTONSEED PROCESSING

In addition to the following disclosures, the process of U.S. Patent 3,579,496 (page 6), is applicable to isolation of cottonseed protein.

Hexane-Acetone Extraction

Oil and gossypol are extracted from cottonseed meats in a mixed solvent extraction process developed by *W.H. King and V.L. Frampton; U.S. Patent 2,950,198; August 23, 1960; assigned to U.S. Secretary of Agriculture* Delinted cottonseed are first hulled in conventional equipment, and the resultant hulled, that is, raw decorticated cottonseed meats, are segregated in the usual manner. The raw decorticated cottonseed meats are then adjusted to a moisture content of from 10 to 15% by weight at ambient temperature to put them in suitable condition for flaking. They are then flaked in conventional flaking rolls to a thickness of from 0.003" to 0.010". The moist flakes may be used without further adjustment of their moisture content, or they may be partially or completely air dried at a temperature not to exceed 130°F.

Higher drying temperatures are not used because an undesired reduction in nutritive value of the cottonseed meal may result. In general, it is preferred to adjust the moisture content of the flakes to 7 to 15%, at a temperature not to exceed 130°F, for the subsequent solvent-extraction step. When the flakes are to be extracted using the direct solvent extraction of oilseeds, the flakes should be adjusted to a moisture content of 12 to 15%. In those instances where the conventional filtration-extraction of oilseeds is used, the preferred moisture content of the flakes to be extracted is 7 to 10% for optimum filtration characteristics. Under these conditions employed for preparing the cottonseed flakes for extraction, essentially none of the pigment glands of the meats are ruptured and consequently the gossypol does not intermix with and combine with other constituents of the meats to any great extent. However, the preparative steps produce raw cottonseed flakes having physical properties which allow the extraction solvent of the process to make intimate contact with the oil cells and pigment glands of the meats, thus permitting efficient extraction of both oil and pigments. Following the above preparative steps, the raw cottonseed flakes are extracted, at ambient temperature, with a homogeneous, constant boiling solvent mixture consisting of 53 parts by volume of acetone, 44 parts by volume of hexane, and 3 parts by volume of water. This solvent mixture (AHW mixture) can be used in any of the extraction equipment and operational procedures commonly used in the solvent extraction of oilseed materials. Once the oil and pigments are dissolved in the AHW mixture, the meats are washed free of oil and pigment-bearing miscella with further portions of the same solvent mixture by one of the usual oilseed extraction procedures.

The commercial solvent normally used in processing cottonseed is a petroleum ether which contains a large proportion of hexane, and which boils in the range of 62° to 78°C and with a median boiling point of about 70°C. For recovery of petroleum ether from, and in removing the solvent from the solvent-damp marc (the extracted cottonseed flakes damp with solvent), temperatures considerably above this boiling range are generally used. These temperatures are sufficiently high to cause heat damage to the protein in the resultant meal. On the other hand, the AHW mixture boils within the narrow range of 48° to 52°C and it can be removed from the marc at a considerably lower temperature. Consequently there is less heat damage to the protein during removal of solvent when the AHW mixture is used.

Another advantage of the AHW mixture is that it is not a dehydrating solvent. In other words it does not remove water from moist cottonseed flakes since the solvent mixture is already saturated with moisture. Furthermore, due to the fact that the flakes maintain their integral nature in the presence of the AHW solvent, the flakes settle rapidly in the solvent mixture, and thus may be washed by decantation or countercurrent extraction in a completely closed system. This property makes possible an extraction procedure, exemplified by the following example, employing countercurrent extraction of flakes by the solvent.

Example: Freshly decorticated cottonseed meats were adjusted to a moisture content of 15% and flaked on spring-loaded smooth rolls to a thickness of 0.003". The flakes were then air-dried to a moisture content of 10.0% by weight. An equal weight of the AHW mixture (consisting of 53 parts by volume of acetone, 44 parts by volume of commercial hexane and 3 parts by volume of water) was added to a portion of the flaked meats. While the resulting slurry was stirred gently, fresh solvent mixture was added slowly and continuously from the bottom of the container and oil-pigment-solvent miscella was allowed to overflow from the top. The solvent was recovered continuously from the miscella in a laboratory-size Struthers-Wells type evaporator and fed back to the bottom of the extraction chamber. This was continued until tests made on portions of the effluent showed that essentially all of the gossypol had been extracted from the flakes. The extracted flakes were then drained, air-dried, and ground into a meal. Results of analysis of the meal are shown in the table below.

Free gossypol, %	0.00
Total gossypol, %	0.40
Lysine, % of protein	4.3
Soluble nitrogen, % of total	92.0
Oil content, %	0.1

Samples of solvent recovered from the miscella in the Struthers-Wells type evaporator were analyzed and found to consist of 53 parts by volume of acetone, 44 parts by volume of hexane and 3 parts by volume of water, the same composition as the original AHW mixture used for the extraction of the flakes.

Dynamic Gaseous Separation of Reduced Pigment Flour

A nonsolvent process has been disclosed by *W.W. Meinke and R. Reiser; U.S. Patent 3,124,461;*

March 10, 1964; assigned to The Texas A & M Research Foundation for processing a low gossypol content cottonseed flour. The process comprises subjecting comminuted cottonseed containing unruptured pigment glands and having a total volatile and moisture content not substantially in excess of 10% by weight and a cottonseed oil content not in excess of 10% by weight, to a substantially dry, moving, nondeleterious gaseous medium, e.g., air, whereby a fraction rich in cottonseed meal or flour is preferentially separated from a fraction rich in unruptured pigment glands.

To prepare the cottonseed for this gaseous classification step, cottonseed meal or flakes are subjected, under controlled conditions to considerable mechanical working or comminution without excessive rupture of the pigment glands. Thus, contrary to the prior art that comminution of deoiled cottonseed meal or flakes must be carried out in a liquid medium, e.g., slurry grinding in hexane, to avoid excessive gland rupture, cottonseed meal or flakes may be comminuted for the gaseous classification step without substantial rupture of the pigment glands by conventional mechanical crushing or grinding techniques, provided that the moisture and volatile content is maintained in the range of 4 to 10% by weight, preferably 5 to 8% by weight.

Fortunately, this comminution without gland rupture may be carried out on conventional equipment, such as nonshearing, seed-crushing rolls, attrition grinders, and/or the like. When employing conventional nonshearing rolls, the cottonseed is preferably, but not necessarily, comminuted in a single pass. In attrition grinders, particle size is reduced by impingement of the particles upon each other and/or upon the sides of the grinder itself.

The resulting comminuted particles are then deoiled by conventional solvent extraction techniques using a nonpolar solvent which will not rupture the pigment glands, e.g., hexane, heptane, pentane, octane, iso-octane and similar aliphatic hydrocarbons; carbon tetrachloride, ethylene dichloride and similar chlorinated hydrocarbons; benzene, toluene and similar aromatic hydrocarbons; and the like, preferably at extraction temperatures of 30° to 70°C. The deoiled material is then desolventized to less than 10% solvent, preferably to a solvent level of less than 1.0%, optimally approaching 0% by weight, by steam stripping, air drying, or the like. The deoiled cottonseed particles are then further comminuted without excessive gland rupture by passing them through the same or similar conventional seed-crushing rolls, attrition grinders or the like as were previously used for the initial comminuting step so as to reduce the particle size of at least a substantial proportion of the material, preferably substantially all of the material (except the unruptured pigment glands) to about 50 microns or less, e.g., 5 to 50 microns.

The resulting deoiled cottonseed meal containing unruptured pigment glands is then subjected to a moving gaseous stream of sufficient velocity whereby the meal or flour particles are preferentially and dynamically separated from the unruptured pigment glands. The particular gas velocity used depends, in part, upon the characteristics of the gas, e.g., density, viscosity, and/or the like, and also on the size of the pigment glands (which normally range from 100 to 400 microns) and flour particles, as well as their configuration. For any particular case, the gas velocity can be determined by routine experimentation.

Any gaseous medium which is substantially inert to, and otherwise not deleterious to, the pigment glands and other constituents of the cottonseed may be used for the gaseous classification step. The gaseous medium should of course, be substantially dry so that the meal will not become sticky, tacky, or the like. Examples of operable gases are air; oxygen; nitrogen; carbon dioxide; carbon monoxide; ozone; the rare gases such as argon, krypton, and xenon; helium; normally-gaseous, nonreactive (to cottonseed constituents) hydrocarbon gases such as methane, ethane, and propane; and the like. As a practical matter, air is used almost exclusively in the process. It was found that effective separation of a gland-rich fraction from a cottonseed meal or flour fraction was achieved in a 4" diameter vertical, cylindrical classifier when air was moved upward through the classifier at a rate of about 4.0 to 5.0 ft^3/min at 14.7 psia and 70°F, preferably 4.2 to 4.4 ft^3/min. In practice, air rates are readily determinable in such equipment by simply commencing and increasing the air rate until a velocity is reached at which satisfactory separations are being effected.

Cottonseed Protein

Example: Cottonseed kernels, essentially free of hulls and having a moisture content in the range of about 5 to 8%, were disintegrated in an attrition-type mill, wherein disintegration of the kernels was accomplished, without rupture of a significant portion of the pigment glands, by impacts between the cottonseed particles themselves and by impact of the particles with the walls of the mill. The resulting kernel fines, a substantial portion of which had a particle size in the range of about 5 to 50 microns, were subjected to extraction with hexane so as to reduce the cottonseed oil content to about 4 to 5% by weight. After solvent removal, the resultant cottonseed meal was subjected to classification, employing air at ambient conditions as the classifying medium.

Classification was carried out in a flask which was connected by an adaptor to seven 1 ft sections of 2" plastic tubing, which were connected in series and arranged horizontally. The end of the tube was covered with a porous fabric bag to recover fines from the exiting air. Air from a compressed-air line was introduced into the flask. Particles entrained in the air thus passed into the tube and settled out in the different sections thereof. After classification, free gossypol and protein values for samples collected from the component parts of the classifier system (flask, adaptor, sections of the plastic tube and fabric bag) were obtained and are as follows:

	Free Gossypol, Wt. percent	Protein, Wt. Percent
Cottonseed Meal Charge	1.080	50.6
First Section of Tube	0.124	53.6
Second Section of Tube	0.120	56.6
Third Section of Tube	0.080	52.9
Fourth Section of Tube	0.092	61.5
Final 3 Sections Plus Bag	0.088	62.4

The above data demonstrate that free-gossypol values decrease as distance from the flask increases. Microscopic examination of the fines in the tube also showed that the number of intact pigment glands or gland fragments decreased as distance from the flask increased. Intact pigment glands, however, were readily visible in large numbers in both the original meal and the flask residue.

Liquid Cyclone Process for Protein Concentrates

High protein concentrates suitable for human consumption are produced by *E.A. Gastrock, E.L. D'Aquin and P.H. Eaves; U.S. Patent 3,615,657; October 26, 1971; assigned to U.S. Secretary of Agriculture* from dried cottonseed meats in a multistep process. Briefly, this process is characterized by an integrated sequence of drying, flaking, disintegrating, screen separating and gravity separating steps. The process removes intact cottonseed pigment glands and as a consequence a gland-free material, which may have a protein content as high as 73% by weight on an oil and moisture free basis may be isolated.

The salient features of the process, which is applicable to either defatted or undefatted cottonseed as a starting material, comprise: rigorous control of moisture in the starting material (meats) at moisture levels well below those levels previously considered feasible in conventional oilseed milling practice; precisely controlled, practically instantaneous disintegration of the material being processed while maintaining the integrity of the gland structure to avoid dispersal of gland contents in the processed material; disintegration in a nonaqueous, nonpolar, fluid medium by use of a high-speed stone mill; and the use of vibrating screens and liquid cyclones in series.

There are four embodiments of the overall process. The first is one where the cottonseed meats are dried, flaked, and extracted for removal of the oil prior to the steps of disintegrating the extracted material in a solvent, separating and concentrating the protein fraction. This embodiment yields a concentrate that exhibits a protein content of about 65% by weight. A second embodiment parallels the several steps of the first embodiment but incorporates the additional step of passing the process stream of the protein concentrate through a series of liquid cyclones to increase the protein content of the finished concentrate product to about 70% by weight. A third embodiment involves the steps of drying and flaking the cottonseed meats but immediately thereafter the stream enters the phases of fluidization, disintegration in a solvent, protein separation, and concentration, with oil enriched solvent (miscella) being withdrawn from the process stream at several appropriate steps. This embodiment like that of the first embodiment is designed to yield a product concentrate of about 65% protein. A fourth embodiment parallels the several steps of the third embodiment but like the second incorporates the additional step of passing the process stream of the protein concentrate through a series of liquid cyclones to increase the protein content of the finished concentrate product to at least 70% by weight. The steps of the various embodiments of the process are given below.

Drying Meats — Meats are dried to 2 to 4% moisture at a temperature not exceeding 180°F. Drying meats prior to extraction prevents the increase in moisture of meats tissue resulting from removal of oil, i.e., meats at 8% initial moisture and 33.3% oil, when extracted as is, would yield oil free marc having a moisture content of 12%. At this high level of moisture, pigment glands are weakend and ruptured simply by transfer of moisture to the gland walls.

Flaking — Meats are flaked preferably to a thickness of 0.008" to 0.012" while still warm from the drying operation. Flaking the meats while they are still warm mitigates pickup of moisture following drying.

Extracting Oil — The oil is extracted from the flakes with hexane to a residual lipids content of 2% or less. The solvent damp extracted marc is routed to a feeder which feeds the wet marc to the liquid cyclone system through a fluidizer.

Fluidizing — Fluidizing the wet marc is necessary to obtain a material of the proper consistency to feed smoothly to the stone mill and to provide a material of the proper viscosity for maximum disintegration in the mill without rupture of the pigment glands. Fluidization is accomplished by passage of the wet marc through a pug type, baffled mixer which provides vigorous nonimpact agitation. Best results have been obtained with wet marc containing 45% solids and 55% hexane.

Disintegration — Disintegration of the meats is accomplished by passing the fludized marc through a high-speed stone mill. This mill consists of two horizontally mounted coarse grit carborundum stones 4¾" in diameter. The upper stone is stationary and has a center hole about 2⅓" in diameter through which the fluidized marc is fed. The lower stone is mounted on an adjustable spindle which permits adjustment of the clearance between the stones from contact of the horizontal plane surfaces to 0.25 of an inch. The lower stone revolves at 3,600 rpm.

The stones are set for a clearance of from 0.002" to 0.015" (preferably 0.006" to 0.008") so that there is no contact between the stones and no grinding action as such. The force exerted on the material is a torsional, rolling, fluid shearing action which has been found to effectively disrupt the meats into micron size particles and to separate the glands cleanly with essentially no breakage or permanent deformation of the glands. The milled marc is discharged directly into a tank provided with an agitator. Initially hexane is pumped to this tank at a rate such as will provide a slurry containing about 15% of total solids.

Screening — The diluted milled marc is pumped from the feed tank to a vibrating screener fitted with 24 mesh and 80 mesh screens. The vibrating screener discharges three streams of slurry as follows:

(a) The on-24 mesh material contains about 1% of the solids in the feed to the screener and contains 60 to 70% solids. For embodiments one and two this cake is routed to dryers. For embodiments three and four it is washed free of oil on the filter and is then routed to dryers.

(b) The on-80 mesh solids is 15% or less of the total solids of the slurry fed to the screens and contains about 50% solids and 50% hexane. This material is returned in toto to the system via the fluidizer for reworking as described above under Fluidizing.

(c) Through 80 mesh. This stream contains 85 to 90% of the total solids of the input wet marc. Total solids content amounts to about 11 to 14%, with the solids being made up of the ultrafine meats particles which are free of pigment glands and constitute the desired end product, coarser meats particles containing some embedded pigment glands, pigment glands free of adhering meat particles, and some fine hull particles. This through 80 mesh slurry discharges from the screen directly into the feed tank for the first liquid cyclone.

First Liquid Cyclone — The through 80 mesh slurry is initially diluted with hexane to a solids content of 7.5%. When on stream conditions are attained the hexane for dilution may be replaced in whole or in part by overflow feedback from the second liquid cyclone (a 10 mm Doxie), which contains about 1% solids. The diluted slurry is maintained under vigorous agitation in the tank to keep all solids in suspension and is fed to a 50 mm diameter liquid cyclone (P50) at 15 to 40 psi pressure (preferably 20 to 30 psi) by a pump. Classification and separation of the suspended particles in the slurry takes place in the liquid cyclone to deliver an underflow and an overflow stream.

The underflow preferably amounts to between 5 and 14% of the total slurry entering the feed aperture of the P50 liquid cyclone and contains from 25 to 45% of solids. The overflow discharges from the upper, or the vortex finder outlet, of the P50 liquid cyclone. This overflow stream preferably amounts to 86 to 95% of the total slurry entering the feed aperture of the liquid cyclone and contains from about 3.5 to 7.0% of solids. The weight ratio of overflow to underflow is defined as the split and preferably ranges between from 6 parts of overflow to 1 part of underflow to 20 parts of overflow to 1 part of underflow.

The underflow contains essentially all of the intact pigment glands of the feed slurry, relatively coarse (but smaller than 80 mesh) particles of meats many of which contain embedded pigment glands, and hull particles. These solids range from 3% to as much as 8% in gossypol content and from 45 to 60% in protein. The underflow stream is removed from the system and filtered. For embodiments 1 and 2 the cake is routed to dryers. For embodiments 3 and 4 it is washed free of oil on the filter and then routed to dryers. The overflow stream discharges from the upper, the vortex finder outlet, of the P50 liquid cyclone into an agitated feed tank. This overflow stream contains the extremely fine solids comprising the desired high protein, low gossypol portion of the feed stream. The overflow from the liquid cyclone may also be further processed through one or more additional liquid cyclones (Doxie, 10 mm diameter) if desired.

Filtering — The high solids stream containing the desired protein concentrate product is fed to a rotary vacuum filter which yields a cake containing about 50% solids.

Drying — The cake is heated in a suitable dryer to about 225°F in 1 hour to coincidentally remove solvent and destroy microorganisms.

Grinding — After heat treatment as above, the cake is ground through a sanitary stud mill to a fine flour and packaged. The final product flour has a protein content on the order of 65% or higher for embodiments 1 and 3, and 70% or higher for embodiments 2 and 4, and for all embodiments, a total gossypol content of 0.30% or less.

GOSSYPOL REMOVAL OR DETOXIFICATION

Alkali-Protein Treated Meal

E. Eagle and J.W. Bremer, Jr.; U.S. Patent 2,797,997; July 2, 1957; assigned to Swift & Company found that small amounts of protein containing materials added in an aqueous alkaline solution, followed by a heating operation, give a detoxified meal far superior in feeding properties to meal detoxified by any of the prior methods. In this method the alkaline protein solution is added at a relatively small level to the toxic cottonseed material either before or after the extraction of the oil. Following this in plant operation, the cottonseed material is subsequently toasted or desolventized and toasted in conventional equipment to complete the detoxification process. To demonstrate the effectiveness of the process in detoxifying a toxic seed meal, a quantity of air-dried hexane-extracted cottonseed flakes which had not been heated during processing was subjected to a series of tests.

68 grams of the protein (on dry basis) was blended with 1,000 grams of distilled water, with sufficient alkali (in this instance caustic soda) being added to bring the pH to 9.0. The weight was then adjusted to 1,800 grams with distilled water. The aqueous solution in each instance was added to 6,810 grams of the cottonseed flakes. Each of the proteins in several diets was added at a level of 1% based on the weight of the cottonseed meal. The meal was then heated in an atmospheric toaster under agitation with a jacket steam pressure of 15 lb/in^2 (gauge). Cooking or toasting time was for a period of 90 minutes, with the temperature of the toaster after the first 20 minutes being held within 215° to 250°C. The meal was then ground in a Mikro-Samplmill through a screen with $1/16$" round perforations. Each of the several alkali protein-treated cottonseed meals was substituted in a test diet for the untreated meal at the same 67% level. None of these protein-treated meals had a toxic effect on the animals, and all of the rats fed these test diets gained weight as shown in the following table:

Diet	No. of Rats	Dietary Variable	Average Starting Weight	1 wk	2 wk	3 wk	4 wk	5 wk	6 wk	7 wk	8 wk
1	15	Untreated control	55	–	–*	–	–	–	–	–	–
2	10	Dry heated	60	46**	–***	–	–	–	–	–	–
3	15	Moist heated	58	77	92	111	130	148	159	161	166
4	5	Stick hydrolysate	55	75	103	133	151	168	198	207	229
5	5	Soy protein	55	91	139	175	201	232	269	280	303
6	10	Condensed beet solubles	55	92	133	172	213	247	271	292	314
7	10	Casein	55	102	151	188	228	262	286	315	328
8	10	Corn steep liquor	73	121	161	203	240	274	303	320	342

*All rats died within 8 days
**Average of 7 rats; 3 rats dead on seventh day
***All dead within 9 days

Soda Ash-Soda Soapstock Treatment

Cottonseed meal is detoxified and improved as to feed value in the process disclosed by G.C. Cavanagh; U.S. Patent 2,934,431; April 26, 1960; assigned to Ranchers Cotton Oil. It has been known that the pretreatment of moist cottonseed meals with soda ash, subsequent to cooking and comminuting but prior to oil extraction, results in a cottonseed residue of improved nutritional value for livestock. It was also known that it is beneficial to treat cottonseed meal by the addition of soapstock. Such treatment has not only resulted in improved nutrient value and lowered toxic principle content but in improved physical properties avoiding excessive dust and permitting ready pelleting.

The broad essence of the process resides in the discovery that soda ash pretreatment followed by this addition of caustic soda soapstock achieves synergistic effect in converting the toxic principles of cottonseed into nontoxic materials to an extent far greater than the aggregate individual effects of the soda ash and soapstock treatments when performed on different meals. To illustrate this, an homogeneous quantity of comminuted cooked cottonseed meats was divided into two equal portions identified as A and B. 1% by weight of solid soda ash was added to A and intimately mixed therewith. Subsequently, the oil was extracted from A with normal hexane. Portion B received no treatment whatsoever prior to the extraction of its oil with normal hexane in precisely the same manner. Portion A was then divided into equal quantities designated as samples A-1 and A-2. Similarly, B was divided into equal quantities designated B-1 and B-2. B-1 was tested for free gossypol using the American Oil Chemists official method Ba 7-50 and was found to contain 0.1125% free gossypol. Sample A-1 was similarly tested and found to contain 0.1063% free gossypol. This demonstrated a reduction in free gossypol incident to the use of the solid soda ash of 0.0062%. Sample B-2 had 5% caustic soda soapstock of the free fatty acids of cottonseed added in the presence of sufficient hexane to achieve uniform dispersal and was dried by stirring on a steam bath to volatilize the hexane. The resultant material was tested for free gossypol and found to contain 0.10625%. This obviously demonstrated that the employment of the caustic miscella refining soapstock reduced the free gossypol by 0.00625%.

It is well known that the gossypol, being water-soluble after treatment, usually concentrates in the soap. Thus, it would be expected that the addition of the soap would increase the free gossypol content of the meal rather than decreasing it in the manner achieved. Sample A-2 was treated with 5% caustic miscella refined soapstock in precisely the same manner as that described for sample B-2. The resultant material was found to contain 0.0887% free gossypol. From the foregoing it is evident that the described soda ash treatment reduced the free gossypol 0.0062% of the total weight of the material employed. The treatment with caustic miscella refined soapstock reduced the free gossypol by 0.00625% of the material. From this, in the absence of any synergistic effect, the total reduction in free gossypol by the employment of both soda ash and the soapstock would be expected to be 0.01245%. Actually, the employment of both the soda ash and soapstock achieves a reduction of 0.0238% of the gossypol in the total weight of the material.

Alkali-Solvent Treatment of Meal

A process was developed by *W.H. King; U.S. Patent 2,873,190; February 10, 1959; assigned to U.S. Secretary of Agriculture* for reducing the free gossypol of a desolventized, oil-free, organic solvent-extracted, preferably hexane-extracted, cottonseed meal, containing in excess of 0.04% free gossypol, to a free gossypol content not exceeding 0.01%. The process involves forming a mixture containing the meal, aqueous alkali, preferably sodium hydroxide, and a hot water-miscible, volatile organic solvent, preferably 2-propanol. The quantity of the aqueous alkali used in the mixture is that needed to adjust the pH of the wet meal to at least 7, a pH of 7 to 10 preferred, and the quantity of the water-miscible solvent used in the mixture is that amount sufficient to adjust the moisture content of the meal to about 5 to 14%. Thereafter, the meal is separated from the mixture, as by filtration, and then extracted, preferably a plurality of times, with a hot water-miscible volatile organic solvent until the free gossypol content of the meal does not exceed about 0.01%. The resulting meal is then freed of liquid and desolventized. As a result of the above treatment the pigment glands are ruptured and gossypol and related pigments are released. These pigments rapidly combine with cottonseed meal constituents to form nontoxic bound derivatives.

Example: The meal used was a solvent-damp cottonseed meal produced by the hexane extraction of raw cottonseed flakes. When desolventized in the usual way, the meal contained 0.800% free gossypol. A 400 gram sample, dampened with about 300 ml of hexane, was mixed while stirring slowly in a Model C-10 Hobart food mixer with a solution consisting of 82 ml of water, 400 ml of constant boiling 2-propanol (9% by volume of water and 91% 2-propanol) and 4.0 gram NaOH. This provided 28% moisture on the basis of the wet meal. Stirring was continued at surrounding temperature (80°F) for 15 minutes.

While the stirring was continued, the vessel containing the wet meal was heated and a light current of air at 80°F was blown over the top of the mixing bowl, so that the temperature of the mixture rose to 170°F in 5 minutes. This temperature was held for 10 minutes and then raised to 212°F in the next 15 minutes. The mixture was then cooled to 100°F in 3 minutes. The resulting solvent-free meal contained 11% moisture. The finished meal consisted of sandy granules of tan colored meal which were free from the dustiness usually associated with direct-solvent-extracted cottonseed meal. The free gossypol content was 0.041% and the pH was 8.3.

Amine Extraction

According to the process disclosed by *W.H. King, V.L. Frampton and A.M. Altschul; U.S. Patent 2,934,432; April 26, 1960; assigned to U.S. Secretary of Agriculture*, oil extracted cottonseed meal is tested with an organic primary amine. In general, the meal is contacted with an excess of an organic primary amine for at least 5 minutes, to convert gossypol and gossypol-like pigments in the meal to derivatives which are soluble in the relatively nonpolar organic solvents such as liquid hydrocarbons like benzene, hexane, heptane, etc., and the liquid chlorinated hydrocarbons like chloroform, phenyl chloride, etc.

The contacted meal is then extracted with the indicated solvent which removes the gossypol derivatives as well as any of the unused organic primary amine. No other treatment of the meal is required. The process thus eliminates the need for severe mechanical or similar treatments designed to rupture the gossypol cells as a requisite in removal of the gossypol. The amines which may be used in this process are any primary organic amines capable of reacting with the residual free gossypol under mild conditions of temperature, at normal moisture content (3 to 12%), to form derivatives which are soluble in the relatively nonpolar organic solvents.

The time of contact between the treating amines and the meal can vary over wide limits (from 5 to 60 minutes or even longer) depending upon the temperature employed and upon the amount of pigment in the meal undergoing treatment. The reaction can be facilitated by mild agitation during the treatment with amines. The amine treatment can be carried out at room temperature or at higher temperatures. Preferably temperatures ranging from ambient (room) temperature to temperatures of about 150°F are used. The lower temperatures are preferred in order to protect the protein in the cottonseed from heat denaturation and consequent destruction of the nutritive value of the meal treated.

Example 1: 67 parts of solvent-damp cottonseed meal (marc from hexane extraction at a commercial prepress-solvent extraction cottonoil mill) containing 50 parts of meal was mixed with 67 parts of octylamine and allowed to stand for 15 minutes at 125°F. The mixture was extracted with chloroform and then desolventized by spreading the chloroform-damp marc in contact with the air at 80°F. The free gossypol content of the meal in the marc before treatment was 0.060%. The free gossypol content of the resultant meal, after the above treatment, when analyzed by the AOCS Tentative Method Ba7-55 for free gossypol was found to be 0.007%.

Example 2: 67 parts of prepress-solvent extracted marc as described in Example 1 was slurried with 67 parts of a 50-50 mixture (by volume) of aniline and chloroform and allowed to stand at ambient temperature (80°F) for 24 hours. The mixture was then extracted with warm chloroform (135°F) and desolventized at ambient temperature by aeration. The free gossypol content of the resultant meal was found to be 0.006%.

Amide Extraction

B.H. Thurman; U.S. Patent 2,958,600; November 1, 1960; assigned to Refining, Unincorporated has detoxified cottonseed meal for use in animal feeds by the use of an amide such as urea or biuret. The use of a small amount of urea or other amide-radical compound in detoxifying cottonseed meal is of importance not only because the detoxification can be effected at lower temperatures but also because the resulting meal has a higher protein rating as a result of the amide-radical compound it contains. Pure urea contains 46.67% nitrogen, the usually-used feed grades containing 41.9% thereof. Adding 1.0% of commercial urea will increase the protein rating of the meal by 2.62%. As an example of a high-protein meal, a 50% protein cottonseed meal can be produced by adding 4 parts of urea to 96 parts of 41% protein cottonseed meal. This yields a 50% protein meal that is detoxified when part or all of the urea is added to the meal in the usual desolventizer. In this process, the amide may be added at various times in the overall processing of cottonseeds and can be used in the seed meal as well as in the gums and foots obtained from the extracted oil. The meal and gums may also be combined to form the final feed. However, it is preferred to add the amide to the meal before the gums or soapstock are added, particularly when processing cottonseed meal to effect detoxification. In some instances, if the amide is first mixed with the foots and the mixture added to the meal followed by heating to the extent mentioned above, little or no detoxification of cottonseed meal or cottonseed foots will result. However, if the same amount of the amide-radical compound is first mixed with the meal and the foots added, the same heating will effectively detoxify the meal.

Detoxifying temperatures, will ordinarily be in the range of 85° to 120°C and will be applied for a period ranging from about several minutes or more at the higher temperatures of 30 to 120 minutes at the lower temperatures, if small amounts of urea or other amide-radical compounds are present. The presence of 0.25 to 2.5% of urea, sometimes as low as 0.125% will be found to be effective. Temperatures of 85° to 110°C, usually 100° to 110°C, applied for 20 to 40 minutes, typically 30 minutes, are commonly employed and, with the small amounts of urea mentioned above, will reduce gossypol contents of meal or meal-foots mixtures from 0.5% or more to values well below 0.1% usually 0.03 to 0.07%. It is often desirable to effect the detoxifying heating in the presence of added water. The addition to the meal, to the foots or to the meal-foots mixture of about 10 to 50% water, by weight, typically 20%, will benefit the process, render the product more fluid and facilitate mixing and uniform heating.

An example of detoxifying a cottonseed meal at quite low temperatures is as follows. A solvent extracted cottonseed meal was mixed with 20% of water containing 0.5% of urea, the mixture being then heated at a temperature of 87° to 90°C for 30 minutes. The amount of free gossypol was reduced from 0.37% before the treatment to 0.04% after the treatment. The presence of the urea made detoxification possible at temperatures much lower than would have been needed in the absence of the urea. As examples of detoxifying extracted cottonseed meals at somewhat higher temperatures, the following table gives results of using various amide-radical compounds dissolved in the designated amount of water, amounts being based on the weight of the meal, the mixture in each case being heated to 107° to 110°C for 30 minutes.

Additive	Free Gossypol Content	
	Untreated Meal, %	Treated Meal, %
2.5% biuret in 20% H_2O	0.43	0.054
2.5% urea in 20% H_2O	0.43	0.076
1.0% urea in 20% H_2O	0.43	0.070
0.5% urea in 20% H_2O	0.43	0.033
0.5% urea in 20% H_2O	0.50	0.032
0.25% urea in 20% H_2O	0.50	0.032
0.49% acetamide in 20% H_2O	0.37	0.080
0.61% propionamide in 20% H_2O	0.37	0.078
1.19% salicylamide in 20% H_2O	0.37	0.061

Fermentation by Rumen Microorganisms

R.S. Geister and F.A. Norris; U.S. Patent 2,960,408; November 15, 1960; assigned to Swift & Company have reduced the gossypol content of cottonseed meal to an amount harmless to monogastric animals. Generally, the process uses the discovery that rumen microorganisms when mixed with cottonseed meal in the proper manner will initiate fermentation which detoxifies gossypol and synthesizes vitamins, amino acids, carbohydrates and fatty acids. The cottonseed meal, preferably oil extracted, is moistened to the consistency of a slurry, rumen microorganisms numbered in the billions are inoculated into the slurry, a certain percentage of rumen fluid is added, a minute amount of a reducing agent such as cysteine hydrochloride or sodium-thioglycollate to induce anaerobic conditions may be added and the slurry is incubated for a number of hours. The incubated slurry may then be press dried in cake form or dried in some other manner in which the product is not subjected to a high temperature for prolonged periods of time.

Example: 1,000 parts by weight of oil extracted cottonseed meal is moistened to the consistency of a slurry. 10 parts by weight of dried rumen culture is added to the slurry and thoroughly intermixed. 100 parts by weight of rumen fluid and $\frac{1}{10}$ part of cysteine hydrochloride are incorporated with further mixing. The slurry is incubated for 48 hours at 40°C and then press dried in cake form. This process not only detoxifies the meal but also increases its food value. Raw cottonseed meal also contains a large amount of cellulose or cellulosic material. The ordinary digestive juices of animals are incapable of converting cellulose into glucose sugar and therefore, the former is largely unavailable as foodstuff to most animals. The process by also breaking down cellulose into glucose, offers a valuable addition to the high protein content of cottonseed meal.

Alkali-Peroxide Process

A two step process has been developed by *R.A. Johnson and P.T. Anderson; U.S. Patent 3,084,046; April 2, 1963; assigned to Food Techniques, Inc.* for eliminating gossypol from vegetable seeds. Oleaginous seed meals or extracted or pressed meals from oleaginous seeds are treated first with sodium hydroxide followed by subsequent treatment with either hydrogen peroxide or hydrochloric acid. In the principal embodiment of this process bar hulled oleaginous seed meats such as cottonseed meats or partially or fully expelled or extracted seed cakes are fed into a continuous steam jacketed blender of the ribbon or thermoscrew type.

The meal or meats are then sprayed with 2.5% aqueous solution of sodium hydroxide to obtain a pH of approximately 10.5. It has been found that if the 2.5% aqueous solution of sodium hydroxide is applied at a rate of 0.92 lb of the solution per 1 lb of meal feed that a pH of approximately 10.5 will be obtained. The jacket temperature for the blender is maintained at approximately 160°F and the sodium hydroxide processed meal is retained within the blender for approximately 10 to 30 minutes.

The meal is then conveyed to a second blender of the same type where it is sprayed with 130 volume hydrogen peroxide to obtain a pH of approximately 7.0 to 8.5. A spray of 130 volume hydrogen peroxide at the rate of approximately 0.05 to 0.18 lb of hydrogen peroxide per pound of wet material feed from the previous operation will obtain the requisite pH of 7.0 to 8.5. The temperature within the hydrogen peroxide treating blender is maintained at between 160° to 190°F. The hydrogen peroxide treated meal is maintained within the blender at between 160° and 190°F for approximately 10 to 30 minutes.

The meal may then be dried by the use of a rotary, tray, or similar type of drier to approximately 7% moisture content or may be passed directly for treatment for other purposes without drying. In the resultant product the free and bound gossypol has been eliminated and has a lighter color and protein injectability than nontreated material. If after the sodium hydroxide treatment the meal or meats are treated with hydrochloric acid it has been found that a product will be obtained which is substantially absent of free gossypol and having a small reduction of bound gossypol. Such a product has a much more acceptable AGU (available gossypol units) because of the substantially complete absence of the free gossypol.

Example 1: 20 ml of water was added to 25 grams of the cottonseed meal to make a more workable cake. 3.075 ml of 19% sodium hydroxide was added to the press cake while continuously stirring until a pH of 10.9 was reached. The mixture was then heated to 160°F for ½ hour with stirring. Thereafter 9.35 ml of 130 volume hydrogen peroxide was added to the cake wherein the temperature was elevated to 190°F and held for 15 minutes with continuous stirring. The final pH of the mixture was 7.62. The mixture after being dried in an oven at 160°F was found to have the following analysis: no free or combined gossypol. Prior to the peroxide treatment but after the alkali treatment the material was found to have 0.012% free gossypol and a total gossypol content of 0.78%.

Example 2: 20 ml of water was added to 25 grams of the cottonseed meal. 3.075 ml of 19% sodium hydroxide was added to the meal and the material heated to 160°F where the temperature was held for 30 minutes with continuous stirring. Thereafter 8 ml of 3% hydrochloric acid was added to the material to provide a pH of 7.1. Thereafter the material was dried in an oven at 160°F wherein the final analysis of the dried cake showed 0.002%

free gossypol and a total gossypol content of 0.67%. Prior to the hydrochloric acid treatment but after the alkali treatment the material showed a free gossypol content of 0.012% and a total gossypol content of 0.78%.

REMOVAL OF SOLVENTS

Removal of Mixed Solvents from Marc

An improved process for removing aqueous-organic solvent mixtures from extracted marc is described by *W.H. King; U.S. Patent 3,459,555; August 5, 1969; and U.S. Patent 3,481,743; December 2, 1969; both assigned to U.S. Secretary of Agriculture.* It was found that by adding a small amount of food acid or alkali, during or prior to the drying operation, to the moist proteinaceous material to give a pH of about 7.0 to 9.0 when alkali is used and a pH of about from 4.0 to 5.5 when acid is used, either before or after removal of organic solvents by distillation, the tough, and extremely hard, brittle stage is avoided and comminution of the continuous phase of homogeneous plastic material occurs at moderate and practical expenditure of mechanical power, at the same time converting the mixture into a granular meal which is easy to handle and which can be further dried to any desired moisture content.

In carrying out the details of the process the solvent-damp marc is subjected to agitation with application of heat to supply heat of vaporization to the volatile components of the mixture, either at ambient or reduced pressure, depending on the maximum temperature desired, to remove the volatile organic solvent components with part of the water. The amount of water removed in this preliminary stage depends on the particular solvent combination used and whether or not an azeotropic mixture is formed. A suitable proportion of food acid, such as orthophosphoric acid (H_3PO_4), or alkali such as sodium hydroxide (NaOH) is added during thorough mechanical agitation of the mixture either in the beginning of the drying operation before the organic solvent has distilled or at a later stage in the desolventization procedure. Other acidic or alkaline edible substances suitable for use in foods such as citric, tartaric, and lactic acids and calcium, potassium and lithium hydroxides, or their alkaline salts such as carbonate and phosphate may also be used.

The addition of only small proportions of the agent (e.g., 0.5% to approximately 10% by weight of the proteinaceous material in the marc) is sufficient to have a pronounced effect on the physical properties and behavior of the mixture during the agitative drying process. Agglutination of the protein is inhibited and the tough, plastic stage, which would otherwise develop and which would require excessive power to comminute by physical action, is avoided.

Example 1: A control was run using 8 lb of marc (from extraction of 4 lb of raw cottonseed meats with a mixture of acetone, hexane, and water in the volume proportion of 53:44:3) which contained 70% total volatile matter (TVM) by weight were added to the mixing chamber of a Baker-Perkins size 6-AN-2 Laboratory Mixer. This equipment is a steam-jacketed steel mixer equipped with a vapor-tight, removable cover. It consists of a cubical mixing chamber with two rotating sigma blades which knead or mix the contents, depending on the physical state of the charge. The bottom of the chamber contains two close-fitting rounded channels to provide minimum clearance for the rotating Z-shaped blades so as to insure complete mixing of the contents. These rotating blades are vacuum-sealed and are rotated by a 220 volt electric motor through a speed-reducing link-belt drive.

The capacity of the mixing chamber is ½ ft^3, part of which is occupied by the mixing blades. The steam jacket was preheated to a temperature of 350°F and maintained at this temperature throughout the drying operation. Rotation of the mixing blades was begun immediately and was continued until the dried material was discharged. The apparatus is equipped with a vacuum gauge and a thermometer well which extends into the material being dried and agitated. Vacuum was applied to the system from a steam aspirator line. An ammeter in series with the power line to the motor was used to measure current required to operate the stirring motor during the drying experiment. The charge of marc was found by analysis to consist of 71% by weight of volatile matter and 29% of dry cottonseed meal with an oil content of 1.3%. The volatile matter consisted of 9% water, and 91% of acetone-hexane mixture. The cover was placed on the apparatus and the vapor discharge valve was opened to allow the constant-boiling mixture of acetone, hexane, and water (53:44:3 volume percent) to discharge at atmospheric pressure. This phase of the desolventizing procedure was allowed to continue for 8 minutes. Temperature of the marc, which was being continually agitated and heated by the steam-jacketed walls of the chamber, remained at approximately 122°F.

The charge of partially desolventized marc remaining had a composition of 18% moisture, 5% acetone, and traces of hexane. This mixture had a plastic, doughy consistency which was continuously kneaded by the rotating mixer blades. Vacuum was then applied and the following temperature, time, pressure, and power requirement (in amperes at 220 volts AC) relationships were recorded in the following table.

Time (minutes)	Pressure (inches of Hg)	Temperature (degrees F.)	Power required [1] (amperes at 220 v.)
0	2	90	0–4
1	7	93	4–4
2	7.5	95	4–6
3	7	97	4–8
4	7	97	4–9
5	6	98	1–11
6	7	99	1–12
7	8	100	0
8	9	100	0
10	10	120	0
12	10	123	0
16	11	130	0

Discharged, lumpy to fine, dry powdery meal, 7% moisture

[1] In excess of current of 1 ampere required by motor when idling.

Example 2: Another 8 lb batch of the marc was charged to the drying apparatus in the same manner as Example 1. This time, however, immediately after putting the charge in the drying chamber 30 grams of NaOH (sodium hydroxide) dissolved in 100 ml of water were added. The added chemical agent was immediately and thoroughly mixed with the solvent-damp marc. The pH of this mixture was 9.0. After 3 or 4 minutes during which the bulk of the volatile organic solvent mixture distilled off the solid marc crumbled into small discrete particles (meal). Stirring was continued at atmospheric pressure for 5 more minutes. No measurable power in addition to the 1 ampere required to rotate the motor and mixing blades under idling conditions was required during this period. Vacuum was then applied to the mixing chamber and the following time, pressure, temperature and power relationships were recorded in the table below.

Time (minutes)	Pressure (inches of Hg)	Temperature (degrees F.)	Power required [1] (amperes at 220 v.)
1	8	98	0
2	10	115	0–1
3	11	126	0–2
4	11	130	0–3
5	11	135	0–2
6	12	136	0–1
8	14	135	0
10	16	137	0

Discharged, granular meal, no odor of solvent—moisture content 10%

[1] In excess of current of 1 ampere required by motor when idling (no charge).

Example 3: Another 8 lb charge of the marc was charged to the drying chamber in the same manner as in Example 1. This time, however, 25 ml or 43 grams of syrupy phosphoric acid (85% orthophosphoric acid, SG 1.72) made up to 100 ml with water were added immediately. The added chemical immediately became thoroughly mixed with the charge. The pH of this mixture was 5.0. After 6 minutes of stirring and heating, the vapor discharge valve was closed and vacuum was applied to the chamber. The granular mass was subjected to the pressure and temperature conditions for the times shown with the power requirements listed in the table on the following page.

Cottonseed Protein

Time (minutes)	Pressure (inches of Hg)	Temperature (degrees F.)	Power required [1] (amperes at 220 v.)
1	6		0
2	6	90	0
9	7	99	0
10	8	103	1-0
11	12	110	0-1
12	12	112	0-1
13	13	117	0-1
14	13	121	0-1
15	14	125	0-1
25	14-17	125-137	0-1
Discharged, granular meal, no odor of solvent-moisture content 10%			

[1] In excess of current of 1 ampere required by motor when idling (no charge).

Removal of Water

W.H. King; U.S. Patent 3,604,123; September 14, 1971; assigned to U.S. Secretary of Agriculture have adapted the process of U.S. Patent 3,459,555 and U.S. Patent 3,481,743 for removal of water alone from vegetable protein. It was also found that by adding a small amount of nonvolatile acid or alkali, during or prior to the drying operation, to the moist proteinaceous material to give a pH of about from 7.0 to 9.0 when alkali is used and a pH of about 4.0 to 5.5 when acid is used, either before or after removal of part of the water by distillation, the tough, and extremely hard, brittle stage is avoided and comminution of the continuous phase of homogeneous plastic material occurs at moderate and practical expenditure of mechanical power, at the same time converting the mixture into a granular meal which is easy to handle and which can be further dried to any desired moisture content by conventional methods.

Example 1: Using the equipment of the previous method, 5 lb of defatted cottonseed flaked meats containing 30% water were placed in the mixer. The steam jacket was preheated to a temperature of 350°F, and maintained at this temperature throughout the drying operation. Rotation of the mixing blades was started immediately upon addition of the charge and was continued until the dried material was discharged. The material being stirred became plastic, viscous and dough-like, and the electric current required to keep the motor running, as measured by an ammeter in series with the 220 volt power line to the motor, varied as shown in Figure 7.1a.

FIGURE 7.1: REMOVAL OF WATER FROM WET VEGETABLE PROTEIN

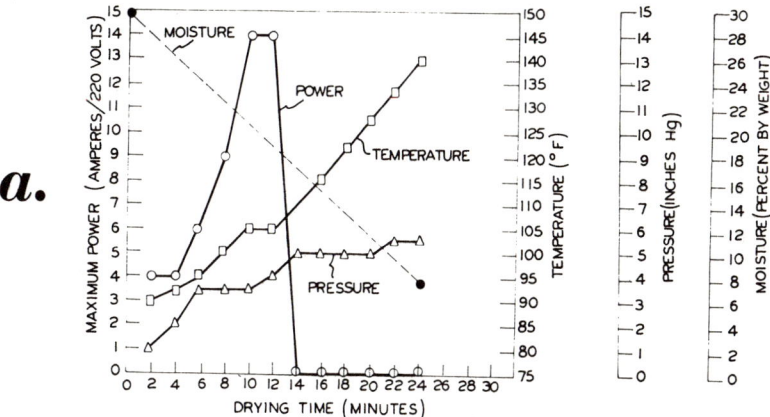

(continued)

FIGURE 7.1: (continued)

b.

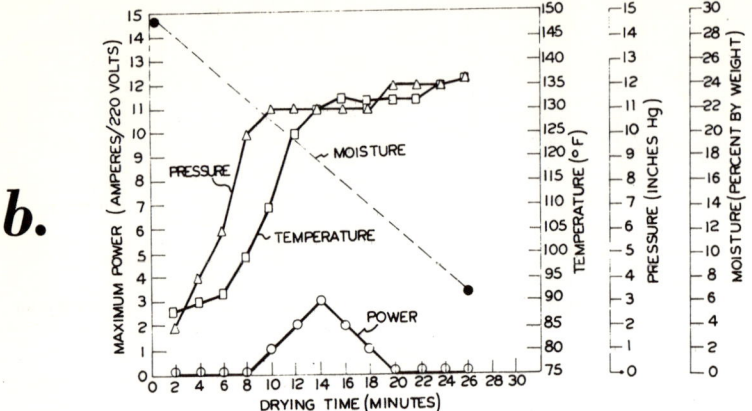

Source: W.H. King; U.S. Patent 3,604,123; September 14, 1971

The time of stirring and the pressure within the chamber, as well as the temperature of the mass of material being dried, are also shown in this figure. It will be noted that as the drying progresses the power, measured in amperes of electric current at 220 volts, rises sharply as the moisture is reduced. Then, as the tough material breaks up the power drops to that required to operate the electric motor while it is simply idling.

Example 2: To show the effect of raising the pH of the moist protein, a 5 lb charge of the same material was placed in the drying apparatus as in the foregoing example. Then 30 grams of NaOH dissolved in 150 ml of water were added when stirring began. Due to the efficient mixing action of the apparatus, this reagent was immediately distributed throughout the plastic mass. The cover was replaced, vacuum was applied, and the electric current data of Figure 7.1b were obtained. The comminuted, soft, granular, free-flowing meal which was discharged from the apparatus was found by analysis to contain 7% moisture. Similar results were also obtained when 25 ml of 85% H_3PO_4 was added to 5 lb of marc and processed as in Example 2.

GRAIN PROTEINS

HIGH PROTEIN FLOURS

Wheat Flour of Bran-High Protein Endosperm Blends

A nutritional high protein, low calorie wheat flour is produced by *T. Zacharia; U.S. Patent 2,895,831; July 21, 1959; assigned to Proto, International Hygienic Food Co., Lebanon.* The process involves treating endosperm with water to form (1) a moist endosperm residue containing a high amount of protein, a low amount of starch and minerals; and (2) a starch solution containing starch removed from the endosperm; separating the starch solution from the moist endosperm residue; and combining the moist endosperm residue with a flour component containing fine bran and germ.

This may be readily combined with conventional food components, including patent flour, milk, etc. in the manufacture of various food products such as cereals, biscuits and the like.

The wheat, conditioned by hot air, is milled by a conventional sieving operation to produce three fractions, namely, a coarse broad bran fraction containing husks; a finer flour containing bran and germ and a still finer flour of extraction consisting essentially of endosperm. More particularly, the wheat, after milling is sieved using a series of sieves in the range of 50 to 100 mesh or higher in size where the number for a particular sieve indicates the number of meshes per square inch. Employing a series of sieves of No. 50, 70, 90, 100, 110, 120 and 130 in size, the following fractions are obtained for 100 parts of whole wheat:

- (a) About 10 to 15 parts are retained by sieve No. 50; this material being fibrous husks (broad bran), which is not used in this process.

- (b) Of the material passing through sieve No. 50, about 15 to 20 parts are retained by sieves 50 to 120 and this retained flour component is a mixture of shorts, red dogs and first and second clears separated from straight flour middlings. This fraction in addition to containing the bran and germ, contains a small amount of starch which permits uniform mixing with other starch-containing flour components.

- (c) The material which passes through sieves of 120 or higher in number is the endosperm.

In producing the moist endosperm optimum nutritive qualities are obtained if the water employed is acidic in nature and of certain hardness characteristics. Ordinary tap water may be used, but the nutritive quality of the resulting moist endosperm is not as high as when the preferred conditions of the pH and hardness of the water are used. The pH of the water should be sufficiently high that significant amounts of starch will go into solution and also so as to not deleteriously affect the proteins present in the endosperm. On the other hand, the pH of the water should be sufficiently low so that the removal of starch may be controlled. In general, it has been found that satisfactory results are obtained when the pH is in the range of about 5.2 to about 6.8.

Another factor in obtaining the controlled extraction of starch from endosperm is the hardness of the water used. If the water is too hard, it causes a separation of valuable proteins. On the other hand, if the water is too soft, there is excessive removal of nutritive minerals. Optimum results have been obtained when the water has a hardness of 10 to 30 parts (in terms of calcium carbonate) per 100,000 and, preferably 15 to 25 parts per 100,000.

It has been found necessary for optimum starch removal and minimum protein depletion to soak the endosperm in the water, having the above pH and hardness characteristics for sufficient time to permit any protein tending to separate from the endosperm to be present in suspended form rather than in solution. In general, a soaking time in the range of from 1 to 2 hours has been found to be adequate. In general, the kneading operation is carried out at a temperature of 25° to 30°C. At temperatures well below the indicated minimum the starch becomes less soluble in the water while at temperatures well above the indicated maximum there is danger of deleteriously affecting the components of the endosperm.

Example: One kilo of endosperm of the following composition:

	Grams
Moisture	130
Protein	120
Sugar and pentoses	45
Starches	700
(Minerals, crude fibers and miscellaneous components)	5

is mixed with water having a pH of 6.0 and a hardness of 20 parts per 100,000 and subsequently kneaded at a temperature of 32° to 35°C. The kneading is done by an electrical kneading machine capable of operating at three speeds in which the speed is progressively increased during the kneading operation. Water is continuously added while kneading. After a 15 minute soaking period, the water is removed by any conventional manner, such as by draining, to a centrifuge and a fresh quantity of water is added and the above procedure repeated until about 80% of starches are removed, to produce an endosperm residue having a protein content of 29.5 to 30% and a moisture content of 65%.

When the starchy liquid is separated, the moist endosperm residue, when calculated on the same basis of moisture content of the original endosperm would have the following composition per 1,000 grams:

	Grams
Moisture	130
Protein	643
Sugar and pentoses	15
Starches	207.8
(Minerals, crude fibers and miscellaneous components)	4.2

To control the protein and mineral content of the above residue and to impart vitamin qualities thereto, 200 to 250 grams of flour containing fine bran and germ are added per kilo of moist endosperm.

High Protein Wheat Endosperm Fraction Blended with Soy Flour

T. Zacharia; U.S. Patent 3,091,538; May 28, 1963; assigned to Proto, International Hygienic Food Co., Lebanon has used the moist endosperm as produced in the example of U.S. Patent 2,895,831 and mixed it with soy flour in place of the fine wheat bran to produce high protein baked goods.

It was found that the combination of moist endosperm and defatted soy flour results in a product of high protein content and good baking properties. The introduction of defatted soy flour to the resulting moist endosperm serves, moreover, to help correct the amino acid deficiency of wheat flour notably in lysine, which is in low proportion in wheat, and high proportion in soy flour. Thus, the protein content of baked goods made with this product will have more nearly the constitution of a whole protein, with a consequent higher nutritive value.

Alternatively, the process of producing the moist endosperm, which involves the use of water, permits an effective method of enhancing the lysine content to any desired degree, by dissolving lysine hydrochloride in the desired proportion in water and adding it to the moist endosperm at the conclusion of the extracting process, thus insuring an effective dispersion superior to that afforded by dry mixing. Another advantage claimed for this wet procedure is that it allows for the treatment of the endosperm in the wet state, with proteolytic enzymes, trypsin, pepsin, papain and the like, for the purpose of further tenderizing the enhanced protein content and promoting better baking qualities.

Example 1: The moist endosperm residue formed in accordance with the Example of U.S. Patent 2,895,831 (i.e., one kilo of endosperm is kneaded with water to produce 1,000 grams of moist endosperm) is mixed with 150 grams of defatted soy flour to form a dough. When this dough is baked there results a biscuit containing about 40% protein.

Example 2: One kilo of endosperm is kneaded with water in accordance with the procedure of U.S. Patent 2,895,831 to produce 500 grams moist endosperm. This moist endosperm is mixed with 150 grams defatted soy flour to form a dough. When this dough is baked, there results a biscuit containing about 62% protein.

Air Fractionation at Critical Cuts of Milled Cereal Flour

The fluid dynamic properties of milled cereal flours are used by *T.A. Rozsa, C.G. Harrel, W.T. Manning, A.B. Ward and R. Gracza; U.S. Patent 3,077,408; February 12, 1963; assigned to The Pillsbury Company and T.A. Rozsa, C.G. Harrel and W.T. Manning; U.S. Patent 3,077,407; February 12, 1963; assigned to The Pillsbury Company* to separate the protein particles from the high starch content particles.

Prior to this process, it was not known that the most concentrated protein-matter particles of cereal flour are contained within the fines or "throughs" of the "sub-sieve" size (passable through test sieve having 400 meshes to the linear inch and of what is termed 38 micron size) and that, secondly, such minute protein-matter particles may substantially all be separated from the parent flour stock by air separation with the help of fluid-dynamic measurements and principles. In fact, it was believed that the more concentrated proteins are found in wheat particles over 38 microns in size and which will not pass through the 400 mesh experimental sieve.

In this process efficient air separation is used at unexpected and previously unused "critical-cuts." The critical-cut of commercial air separation as used here is the graphically derived particle size, expressed in F-D units, at which the total percentage of the oversize particles in the fine fraction and the percentage of the undersize particles in the coarse fraction are at a minimum.

The F-D or flow dynamic units used for determining the critical-cuts for the separation process is measured as a function of three factors: (1) shape, (2) density, and (3) size.

The numerical results cannot be unequivocally expressed in known units of measurement such as definite units of length (while the physical dimension of this characteristic *is* length) and, therefore, the result is expressed in terms of units which are arbitrarily referred to as flow-dynamic units. These units correspond only in a general way with what is regarded as the effective diameter of the particle expressed in physical units of length such as microns.

Wheat or other cereal flour particles have a wide diversity of shapes ranging from substantially spherical to particles having most irregular surfaces. The resistance of a particle to fluid-dynamic flow will be the result of shape and size. The third particle characteristic, i.e., density, influences the magnitude of the propelling force. The purpose of the method herein described is the differentiation and comparison of the fluid-dynamic property of particles moving in a liquid medium and the numerical expression of this property.

The theoretical explanation of this measurement of particle size is given in some detail in the patents. The following figure illustrates the particle distribution obtained from a milled flour using conventional air separation means.

The cumulative particle size curve of the coarse fraction (representing 85% of the original material) is plotted and the second or upper curve is plotted representing particle size distribution of the smaller and finer fraction constituting 15% of the sample or parent stock material air classified. By such plotting of actual air separator performance to determine the critical particle size of separation that particle size is selected from the curves at which the total of the oversize percentage in the fine fraction and the undersize percentage in the coarse fraction are at their minimum.

That is what a critical separation should accomplish, self evidently at such a critical particle size (31 F-D units in this instance) the vertical distance is greatest between the two cumulative curves. This vertical distance is the "sharpness of the separation" – 81% in this instance. The oversize in the fine fraction may be read on the graph as 6% and the undersize particles in the coarse fraction are shown by the graph to constitute 13%. It is very easy and rapid to find, with a straight edge, the place of the greatest vertical distance between the cumulative curves of the coarse and fine fraction. This is the method for determining at what critical particle size expressed in fluid-dynamic units (F-D units) the separation took place and, furthermore, what the efficiency or sharpness of the separation amounted to.

FIGURE 9.1: FRACTIONATION AT CRITICAL CUTS OF MILLED CEREAL FLOUR

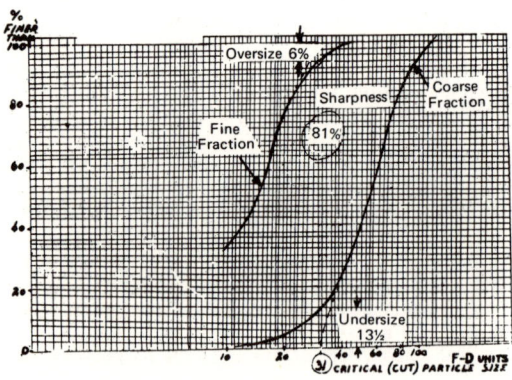

Source: T.A. Rozsa, C.G. Harrel, W.T. Manning, A.B. Ward and R. Gracza; U.S. Patent 3,077,407; February 12, 1963

Using these F-D units, one can define the ranges of critical-cut for optimum results, first in the concentration of maximum protein-matter ingredients and secondly in the depletion, in the coarser and larger fraction, of protein and high ash-containing ingredients. These ranges are as follows:

	F-D Units
For protein concentration	
Hard wheat flour	18 to 30
Soft wheat flour	15 to 25
White rye flour	15 to 25
Dark rye flour	18 to 25
Corn flour	20 to 35
For protein and ash depletion in coarse fraction	
Hard wheat flour	25 to 40
Soft wheat flour	20 to 35
White rye flour	20 to 35
Dark rye flour	20 to 35
Corn flour	25 to 40

Example: The production of two valuable flour fractions by two-stage air separations from a commercially milled hard wheat patent flour out of straight Nebraska winter wheat is described. The protein content was 10.08%, ash 0.371%, with a Fisher value of 18.1. The first-stage air separation was made at approximately 25 F-D with 15% by weight, fine-particle fraction less than 25 F-D (A) and an 85% coarse fraction (B). (A) contained 18% protein, 0.745% ash, with a Fisher value of 4.4. (B) contained 8.5% protein, 0.322% ash, with a Fisher value of 19.5.

A second-stage air separation was then made on (B) at a critical-cut of approximately 64 F-D, and (B) divided into a 33% second-stage fine fraction of from 25 to 64 F-D particles, (C) and into a 52% second-stage coarse fraction (D) containing the particles above 64 F-D (the last percentages being related to the total weight of the original or parent flour stock). (C) contained 6.41% protein, and 0.344% ash, with a Fisher value of 13.75. It should be noted there that the protein was far below the level of the protein of the original parent stock. (D) had a protein of 9.72% and an ash of 0.307% with a Fisher value of 25.1.

The first-stage fine fraction (A) was then blended with the second-stage coarse fraction (D) in the natural proportions enumerated (15% + 52% = 67%) for the production of an excellent bread flour having a higher protein content than the parent flour stock, to wit: 12.4% protein, 0.420% ash, with a Fisher value of 14.6. This blend, by test, baked a better bread than the original parent flour. The second-stage fine fraction (C) (33%) which was no part of the aforementioned blend is usable, for instance, as a blended part of a southern soft wheat family or all-purpose flour, mainly utilized in biscuits and cakes.

Agglomerates of High Protein Flour

The apparatus and process of *E.L. Galle; U.S. Patent 3,397,067; August 13, 1968; assigned to The Pillsbury Company* involves agglomerating high protein fine cereal flour by increasing the moisture content, briefly mixing and agitating the moistened flour in a first zone, immediately transferring the flour to at least one other mixing zone, continuing the mixing and agitation until sufficient bonding has occurred, then drying the agglomerates.

Flour in general and high protein fine fraction flour in particular has always been exceedingly difficult to handle in conventional flour processing and handling systems. The fine fraction flour particles exhibit very poor flow characteristics, being even more difficult to handle than most conventional types of flour, and incapable of being sifted by conventional sifting means. For convenience, the high protein fine flour fraction under consideration is referred to as HPFF. High protein fine fraction flour (HPFF) contains 18 to 23% protein and has a maximum Fisher particle size value of 5.0. The problems can be overcome and HPFF placed in a form capable of being readily handled in conventional handling systems by assembling the fine flour particles into porous agglomerates. These agglomerates consist of a number of fine particles of HPFF randomly clustered together and bonded at their interfaces by protein. These particles define a multiplicity of interstices and voids which provide the agglomerates with the porous structure and provide a liquid access to the agglomerate interior so as to facilitate the wetting of the particles forming the agglomerate and the dispersion thereof in a liquid.

HPFF in this agglomerated form is free-flowing, dust-free and easily handled in conventional flour processing and handling systems. The agglomerated HPFF has a higher bulk density than the nonagglomerated HPFF and occupies less space, thereby simplifying and reducing the cost of storing and packaging. The agglomerates also exhibit improved baking characteristics as compared with the nonagglomerated material.

To obtain the most desirable type of agglomerate and to achieve maximum efficiency in the agglomeration, the amount of moisture added to the material to be agglomerated is carefully controlled so as to provide enough moisture to achieve the degree of adhesiveness to form a strong bond between the particles comprising the agglomerates, and also to prevent over-moisturization and the resultant excessive glutenization of the protein.

Experimentation has indicated that agglomeration can be accomplished by increasing the moisture content of the material to a total moisture content of 20 to 35% during agglomeration. When the total moisture is 22% or lower, the agglomerates are quite soft and fragile and are generally considered unsuitable for bulk handling because of their tendency to break down. Total moisture contents of 35% and higher represent excessive moisture not needed for successful agglomeration which may cause some excessive glutenization of the protein and put a greater (and normally unnecessary) load on the drying equipment used to remove the added moisture from the agglomerates used after the formation thereof. The optimum moisture content for agglomeration is dependent on the form of the process. A two-stage mixer that can be used in one form of the process is shown in Figure 9.2. The operation of this mixer is illustrated in the following example.

Example: A high protein flour (20% protein) having a Fisher value of about 4 was agglomerated in an apparatus similar to that described in Figure 9.2. The mixer **72** had a diameter of 8" and a length of 30". The mixer **76** had a diameter of 10" and a length of 14". The speed of the periphery of blades **72b** was 6,250 ft/min (2,990 rpm). The mixer **76** was operated to provide a peripheral speed of 1,730 ft/min at the periphery of blades **76b** (660 rpm). High protein flour was fed to the duct **90** at the rate of 10 lb/min. Water was introduced through line **72c** at the rate of 3.15 lb/min. Air was introduced in duct **90** at the rate of 2,775 cfm. The temperature of the air in duct **90** was 292°F before the introduction of the product through duct **85**. The moisture content of the finished product was 16.8% by weight. The flowability of the product was 5.75 seconds. The finished agglomerates, in size ranged between about 110 and 947 microns.

FIGURE 9.2: AGGLOMERATES OF HIGH PROTEIN FLOUR

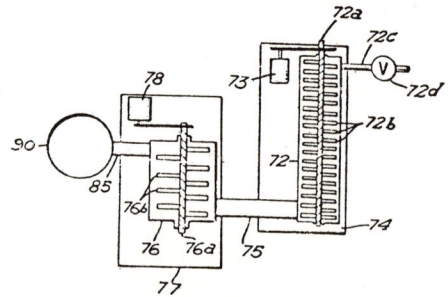

Source: E.L. Galle; U.S. Patent 3,397,067; August 13, 1968

After formation of the agglomerates, they are preferably dried, preferably immediately and preferably to a total moisture content of not more than 14%. The temperature of the agglomerating system normally should be controlled and maintained at a sufficiently low temperature so that no undesirable changes occur in the physical and chemical makeup of the material, such as gelatinization of the starch or degradation of the protein.

After the agglomerates have been formed and dried, they may be used either in the dried form or may be reduced in size if desired such as by grinding, the ground agglomerates retaining their agglomerated form and free-flowing characteristics, the reduction in size as by grinding reducing only the size of the agglomerates and not the character thereof.

Protein Addition to Wheat Flour

Salts of acyl lactylates are used by *C.C. Tsen and W.J. Hoover; U.S. Patent 3,780,188; December 18, 1973; assigned to Kansas State University Research Foundation* to incorporate protein supplements into wheat flour used in bread and other baked goods.

Supplementation of the protein content of wheat flour based bread and baked or fried product doughs to a level to significantly improve nutrition has not heretofore been successful because of the adverse affect on the quality of the food product attributable to the supplement added, not only from the organoleptic standpoint but also the appearance, physical quality and shelf life of the baked goods.

It was found that by incorporation of from 0.1 to 3% of either the sodium or calcium salts of an acyl lactylate of C_{14}-C_{22} fatty acids (preferably stearoyl-2-lactylate or to a slightly lesser degree calcium stearoyl-2-lactylate), or the condensation product of from 10 to 95 parts by weight of ethylene oxide and correspondingly from 90 to 5 parts by weight of a partial glycerol ester of a C_{10}-C_{24} fatty acid containing at least 10% monoglyceride content with diglycerides, triglycerides and glycerine constituting the balance (preferably polyoxyethylene (20) mono- and diglycerides of C_{14}-C_{18} fatty acids having an acid number within the range of 0 to 2, an hydroxyl number of 65 to 80, a saponification number of 65 to 75, and an oxyethylene content of 60.5 to 65.0%) in the bread or other baked or fried goods dough, a sufficient amount of nonwheat or wheat protein supplement may be added to the composition to improve the nutritive quality of the bread without a loss in loaf volume and organoleptic characteristics.

Incorporation of an additive such as sodium stearoyl lactylate, calcium stearoyl lactylate, or the specified ethoxylated glyceride composition permits supplementation of the wheat flour based dough with any one or more of a number of protein additives which may be selected from the group consisting of soy flour, soy isolates, nonfat milk solids, whey products, fish protein concentrate, cottonseed flour, chick pea flour, sesame seed flour, corn-soy-milk blend flour, wheat protein concentrate, wheat gluten, defatted wheat germ, Torula yeast, wheat soy blend flour, edible single cell proteins compatible with wheat flour for baking purposes, and mixtures of these.

Generally speaking, the quantity of protein supplement added will not need to significantly exceed that required to furnish a quantity of protein essentially equal to the quantity of protein provided by the wheat flour content of the dough.

Sodium stearoyl-2-lactylate is generally prepared by admixing lactic acid in an aqueous medium to commercial grade stearic acid (an admixture of myristic, palmitic and stearic fatty acids) at a sufficiently elevated temperature to remain in a molten condition. In general, 1.0 equivalent of fatty acid is used for each 1.2 equivalents of lactic acid as monomer for each lactyl group desired. In this case therefore 2.4 equivalents of lactic acid are provided for each 1.0 equivalent of fatty acids. The mixture is stirred with heating whereupon about 1 equivalent of sodium hydroxide is added. The mixture is then heated to bring the temperature thereof up to about 200°C to complete the reaction. The reaction is carried out under an atmosphere of an inert gas to remove water vapor and prevent oxidation of the stearic acid. A solid, slightly cream colored material is produced upon cooling of the reaction products and it is then ground to a fine powder for use. The powder is a mixture of sodium salts of a homologous series of stearoyl lactylic acids, in which the number of lactyl groups in the molecule is a function of the relative ratios of the constituents brought into admixture. Calcium stearoyl-2-lactylate is prepared in a similar manner except that calcium carbonate is used as the neutralization agent.

For example, when bread is prepared by the standard sponge dough method as prescribed by the American Association of Cereal Chemists using a formula comprising in parts by weight, 100 parts of wheat flour, 12 parts of soy flour, 2 parts of yeast, 5 parts of sugar, 2 parts of salt, 0.5 part of a yeast food containing an oxidizing agent such as potassium bromate (i.e., Arkady yeast food), and 0.5 part of sodium stearoyl-2-lactylate, the resulting product compares very favorably with bread made from the same formula minus the soy flour and the lactylate additive.

Baking tests have shown that when the standard bread formula specified above is supplemented with 12% soy flour (baker's weight) the addition of 0.5% sodium stearoyl-2-lactylate (baker's weight) cause the specific volume of a loaf of bread made from wheat flour supplemented with the soy flour to increase from 5.38 where no lactylate additive is used, to 6.25 cc/g when the lactylate is added to the formulation. The latter figure represents a volume that is even larger than bread made from the same amount of wheat flour without soy supplementation. In this case, It can readily be appreciated that a significant amount of protein has been added to the bread composition since the soy contains 52.7% protein.

It is common practice in the baking industry to regard "specific loaf volume" as an important parameter for gauging the marketability of bread. A marketable bread should have a specific volume over 6.00 cc/g, provided it also has acceptable appearance, crumb texture, grain and organoleptic properties. Although incorporation of 12% soy flour (baker's weight) into bread dough is adequate for most commercial purposes, it has been found that the level of soy flour can be increased to 16% if desired using only 0.5% sodium stearoyl-2-lactylate as an additive for use therewith, and levels as high as 20% soy flour can readily be adopted although in this instance it may be desirable to slightly increase the amount of lactylate used to assure retention of the necessary physical and taste properties of the bread.

Similar results were obtained using 0.5% of polyoxyethylene (20) mono- and diglycerides of C_{14}–C_{18} fatty acids. Usually, the amount of calcium stearoyl-2-lactylate added should exceed the level of sodium stearoyl-2-lactylate in the same dough formulation to obtain generally equivalent results.

ISOLATION OF GRAIN PROTEINS

Aqueous Ammonia Extraction of Gluten

A process developed by *P.H. Johnston and D.A. Fellers; U.S. Patent 3,574,180; April 6, 1971; assigned to U.S. Secretary of Agriculture* separates the starch and protein components of wheat flour wherein the flour is agitated with water and ammonium hydroxide and centrifugation is applied to the resulting slurry.

A particular advantage of this process is that coherence and stickiness of the wheat flour are obviated by the use of ammonium hydroxide, which is volatile and readily removable from the products. In contrast, various agents which have been advocated for the purpose, e.g., lime, sodium hydroxide, salt, malt extract, etc., are nonvolatiles which remain in the final products. A particular disadvantage in using a fixed alkali such as potassium or sodium hydroxide is that this base causes a modification or denaturation of the protein (gluten) component so it cannot be used in bakery products.

In practice of the process a mixture of wheat flour, water, and ammonium hydroxide is formed and subjected to agitation or shear forces to form a smooth slurry. In preparing the slurry, one generally uses 1 to 2 parts of water per part of flour. For flours derived from hard wheats (high protein) a preferred ratio is 1.5 parts of water per part of flour. For flours derived from soft wheats (low protein), a preferred ratio is 1.25 parts of water per part of flour. Enough ammonium hydroxide (or ammonia gas) is added to the mixture to provide a pH of 6.5 to 9.5, preferably 7.5 to 9.0. It may be noted that without addition of NH_4OH, the slurry will generally have a pH of 5.8 to 6.0. The slurry is ordinarily prepared at room temperature (25°C) as being convenient and giving excellent results; however, temperatures somewhat lower or higher, e.g., 15° to 45°C, may be used if desired.

The slurry is subjected to centrifugation whereby there is formed a dense bottom phase and a supernatant liquid phase. The bottom phase contains essentially all the prime starch from the flour in almost a pure state. This starch being granular, completely free from stickiness, can be readily separated from the remainder of the system, and can be readily processed as by washing with water and drying to prepare a high-grade starch for any desired use. Any ammonium hydroxide remaining in the starch fraction will, of course, be vaporized when the starch is dried. Essentially all the protein from the flour is contained in the liquid supernatant. Since this phase is free from stickiness and gumminess, it pours readily and thus can be easily removed from the starch phase. The supernatant liquid is dried to produce a protein product.

Example: The starting material was a hard red spring, straight-run unbleached flour having a protein content of 14.25% (measured on the undried basis, moisture content was 13.3%). A mixture was made of 1,250 g of the flour, 2,031 g of water, and enough ammonium hydroxide to provide a pH of 8.0. The mixture was blended for 10 minutes in a 1-gallon capacity Waring Blendor. The resulting smooth, easily-pourable slurry was then centrifuged for 5 minutes at 1,500 rpm (about 500 g). It was observed that there was formed a compact starch phase and a supernatant liquid phase. No problem with stickiness or coherence was encountered; the supernatant could be easily poured off the lower starch phase.

The supernatant liquid was freeze-dried to a moisture content of 2.15%. A yield of 498 g of protein concentrate was obtained. Analysis indicated a protein content of 38.5% on a dry basis. The starch fraction was freeze-dried without washing. A yield of 720 g of starch was obtained. Analysis indicated it contained only 0.96% protein, on a dry basis.

Glyceride Oils for Extracting Gluten

D.A. Fellers; U.S. Patents 3,463,770; August 26, 1969; 3,493,384; February 3, 1970; 3,498,965; March 3, 1970; 3,501,451; March 17, 1970; and 3,542,754; November 24, 1970; all assigned to U.S. Secretary of Agriculture has disclosed a process in which protein concentrates are prepared from wheat flour by the following technique. The flour is slurried with water and an edible gluten-modified agent. The slurry is then centrifuged, yielding a supernatant liquid which contains essentially all the protein from the flour. This liquid may be dried and used as a protein supplement in bread and other foods.

Various substances may be used as the gluten-modifying agent, for example: Water-soluble proteins such as soybean protein, gelatin, sodium caseinate, and substances containing casein, e.g., dried milk; edible (glyceride) oils, such as soybean oil, peanut oil, corn oil, cottonseed oil, safflower oil, olive oil, and the like; water-soluble cellulosic gums such as methylcellulose and carboxymethylcellulose; phospholipids such as lecithin and cephalin. These substances may be used individually or as admixtures of two or more.

The gluten-modifying agent is operative over a wide range of proportions. Usually, it is preferred to use the minimum effective amount of the agent and generally this will be in the range from about ½ to about 5 parts thereof per 100 parts of flour. Larger proportions may be used, for example, up to 30 parts per 100 parts of flour, but they exert little, if any, extra benefit.

In practice the following procedure is used. A slurry of wheat flour, water, and an edible gluten-modifying agent is prepared. In preparing this slurry, one uses about 1 to 3 parts of water per part of flour. A preferred proportion which provides a slurry thin enough for easy handling, yet employs a mimimum of water (to reduce the expense of drying the protein concentrate), is 1.5 parts of water per part of flour. The critical component in the slurry is the gluten-modifying agent. This substance exerts the important effect of modifying the wheat protein (gluten) so that its tendency to form a dough or other sticky material is markedly decreased.

In preparing the slurry, the three components, flour, water, and gluten-modifying agent, may be combined all at once. Alternatively, the other components may be added sequentially to the water. Preferably, where the gluten-modifying agent is water-soluble, the following technique is used: The gluten-modifying agent is dissolved in the water. Then, the flour is incorporated in the solution to form the slurry. Where the gluten-modifying agent is a water-insoluble material, for example, an edible oil, the following technique is preferred: A part of the flour (for example, ¼ of the total amount) is incorporated with the water, the oil is then stirred in, and finally the remainder of the flour is incorporated into the system.

Having prepared the slurry it is then centrifuged. This causes formation of two or more phases: A dense bottom phase; a supernatant liquid phase; and, in some cases, a sludge phase between the bottom phase and the supernatant liquid phase. The bottom phase contains essentially all the prime starch from the flour in almost a pure state. This prime starch being in a granular condition, completely free from stickiness, can be readily separated from the remainder of the system, and can be readily processed as by washing with water and drying to prepare a high grade starch for any desired use.

Essentially all the protein from the flour is contained in the supernatant liquid, or divided between the supernatant and sludge phases where both are present. Both the supernatant and sludge phases are free from stickiness and gumminess so that they can be readily handled, removed from the starch phase, etc.

Essentially all of the added gluten-modifying agent appears in the protein concentrate; that is, it is in the supernatant liquid or divided between the supernatant and sludge layers where both of these phases are present. The occurrence of the additive in the final product is not a detriment because it is an edible material and may even add to the nutritive value of the protein concentrate. This is the case, for example, where the additive is a protein or edible oil. The presence of the edible oil in the protein concentrate is especially desirable where the latter is to be converted into a milk-like beverage of full-fat character in that no extra fat need be added to achieve this end.

Ordinarily, for convenience, the slurry of flour, water, and gluten-modifying agent is prepared at room temperature. However, special benefits are attained at lower temperatures, i.e., the intermediate sludge layer has a more liquid character and is easier to remove from the centrifuge. Accordingly, in a preferred modification of the process, the slurry is prepared in the cold, that is, below room temperature but not so low as to cause the slurry to become frozen. For best results, the temperature is lowered to a point just above freezing, for example, about a degree above 32°F so that the slurry remains a liquid. The benefit of the lowered temperature is primarily achieved when the slurry is prepared in the cold. However, it is also beneficial to maintain the low temperature when the slurry is centrifuged.

Example 1: Wheat flour (1 part) was slurried with 1.5 parts of water. To different lots of the slurry was added corn oil in various proportions (as given below) and the mixtures stirred in a blender for 1 to 5 minutes. Each of the slurries, at room temperature, was then centrifuged for 5 minutes at 1,000 rpm. In each case, it was observed that the slurry separated into a supernatant liquid layer and a lower layer of prime starch. The supernatant layer was removed to yield the desired protein concentrate. The results are shown below.

Run	Corn oil added, percent (on wt. of flour)	Starch removed, percent (dry basis, on wt. of flour)	Increase in concentration of protein, percent (dry basis)
1	4.2	55.9	208
2	8.4	52.0	178
3	12.5	56.3	178

Example 2: The procedure of Example 1 was repeated substituting various other agents for the corn oil. In all cases it was observed that the slurry on centrifugation separated properly into the supernatant liquid and prime starch phases. The results are summarized in the following table.

Run	Additive and proportion thereof, percent based on wt. of flour	Starch removed, percent dry basis on wt. of flour	Increase in concentration of wheat protein, percent dry basis
4	Soybean protein, 5	55.0	200
5	Soybean protein, 15	52	159
6	Methylcellulose, 1.3	46	180
7	Methylcellulose, 3.9	44	167
8	Soybean protein, 5; Corn oil, 4.5	60	201
9	Gelatin, 15	58	140
10	Lecithin, 5; Corn oil, 4.5	48	163

Example 3: A series of runs were made, using the following technique: Wheat flour (1 part) was slurried with water (1.5 parts). To different lots of the slurry was added, with stirring, nonfat dry milk solids in various proportions (given below). In one of the runs, nothing was added to provide comparative data. The slurries were then centrifuged for 5 minutes at 1,500 rpm. The results are tabulated below:

Run	Amount of non-fat dry milk solids added, percent based on wt. of flour	Starch removed, percent dry basis on wt. of flour	Increase in concentration of wheat protein, percent dry basis
11	None added	51	204
12	2.5	57	220
13	5.0	54	192

It was further observed that in run 11, the centrifuged material was sticky and gummy so that it was difficult to separate the starch from the supernatant. In runs 12 and 13 the centrifuged material was in two phases: Prime starch and supernatant liquid. Moreover, in these runs there was no stickiness or gumminess and the supernatant and starch layers could be separated easily and completely.

An important advantage of this process is that it yields protein concentrates which contain valuable components which are normally lost in conventional procedures for separating the starch and protein components of wheat flour. The centrifugation causes a segregation only of the prime starch; the desired components, including the gluten and the water-soluble proteins, vitamins, minerals, and sugars, are concentrated in the supernatant liquid phase (or in the supernatant and sludge phases, where both are present).

Protein from Ruptured Cereal Grain Germ

It is known that corn germ is potentially a good source of protein. The protein content in cereal germs is of much better nutritional quality than that protein which may be obtained from the endosperm portion of the cereal grain. However, to date, there is no simple, low cost yet efficient method of upgrading the protein content in germ to produce a protein rich fraction relatively low in fiber content such that it will be useful for human consumption.

Spent germ, such as corn or grain sorghum germ may be produced by either dry or wet milling schemes. In the usual wet milling system cleaned whole grain is softened by steeping for about two days in warm water that is acidified with a small amount of sulfur dioxide. Spent germ may be also derived from a dry milling technique. Here no preliminary steeping is involved. Usually the grain is tempered or conditioned to bring the moisture up to a predetermined level. After moisture adjustment, the grain may be passed through a series of cracking and reducing rolls, or break rolls, to break the kernels apart and give a mixture of bran, germ and endosperm particles.

J.E. Freeman and R.M. Olson; U.S. Patent 3,615,655; October 26, 1971; assigned to

CPC International Inc. have disclosed a method of preparing from a cereal grain germ a protein rich product useful for human or animal consumption. In its broadest aspects, the process includes the steps of rupturing the cells of a cereal germ, abrading the ruptured germ to free protein particles which adhere to the ruptured germ and separating the freed protein particles from the remainder of the ruptured germ cell fragments. A protein rich germ fraction is thus produced which has a relatively low content of fibrous material. Also, a fibrous germ residue fraction is produced which is relatively low in protein content. Both fractions are useful, with the high protein content portion being particularly suitable for human consumption in a number of food materials such as bread, and the fibrous fraction being useful as a source of cellulose and hemicellulose, each of which has many well-known useful applications.

Cereal germ cell rupture can be effected by subjecting the germ to either a hammermill or an impactmill, through means of pressure such as by a mechanical screw press, by means of an extruder, by subjecting the cereal grain to abrasion by corrugated rolls, by conditioning the germ through exposure to any sharp surface such as a knife or blade, etc. In essence, all that is necessary is to create a new surface, thereby exposing the protein and making it available for release in the subsequent step.

The next step involves freeing the protein particles from the remainder of the germ cell. This can be best accomplished by subjecting the particles to a type of frictional force. Conventional impact and attrition mills are most suitable. Thus, a spent cereal germ may be subjected to an impactmill to free the protein particles, that is, by rubbing or abrading the surface of the germ cell the tightly adherent protein fraction is released. For best results, this step should be carried out when the already ruptured germ is in a liquid slurry form. As an example, spent cereal germ such as spent corn germ may be slurried up in water and subjected to an impactmill to free the proteinaceous material.

The last step involves separation of the two germ fractions. The protein rich fraction of the germ is of materially smaller average particle diameter, compared to the fibrous or protein lean fraction. Specifically, the particles forming the protein rich fraction have an average particle diameter of less than 40 microns. The greater proportion of the proteinaceous material has a particle diameter less than 20 microns, with the greatest percentage of the product falling within 1 to 10 microns in particle size. The fibrous fraction, on the other hand, has a particle diameter substantially greater than 40 microns, and usually is of a diameter greater than 100 to 300 microns.

Thus, it is a simple matter to take advantage of this difference in physical size of protein rich and protein lean fractions to thereby separate them by some classification means. If, for example, the mixed fractions are in dry form they may be screened or classified, such as via air classification. Again, any type of centrifugal separation could be carried out here to effect a relatively clean separation of fractions. For example, the materials may be subjected to liquid hydroclones, to cyclone treatment or to centrifugation by means of horizontal bowl centrifuges.

Example: Corn germ was first softened by steeping for approximately 2 days in sulfur dioxide-containing warm water. The grain was then ground in a mill to liberate the germ. The germ was washed, dried, heated and subjected to a mechanical screw press to release the bulk of the oil contained therein. The residue was then flaked and extracted with hexane via a countercurrent flow method.

100 grams of the above extracted spent corn germ flake was then wetted with sufficient water to obtain a slurry. In order to release the proteinaceous fraction the slurry was then subjected to abrasion by means of an attrition mill. The attrition mill in this particular case was a Quaker City Laboratory Attrition Mill. To classify the two fractions described above, the germ slurry was then screened on a bolting cloth having a 53 micron mesh size. The tailings retained on the cloth were again passed through the attrition mill and also screened. The fibrous protein lean fraction which was retained on the cloth was washed with tap water and freeze-dried. The proteinaceous fraction which passed through

the screen was centrifuged in a Sharples Supercentrifuge. The solids of this fraction were also freeze dried. The two samples analyzed as follows:

	Extracted corn germ flake, percent dry basis	Fibrous fraction		Protein rich fraction	
		Percent composition, dry basis	Percent of starch, protein, fat, and pentosans of starting germ flake recovered in fibrous fraction	Percent composition, dry basis	Percent of starch, protein, fat, and pentosans of starting germ flake recovered in proteinaceous fraction
Starch	25.7	0.5	1.0	16.7	23.1
Protein	25.7	16.0	31.2	32.9	45.6
Fat	2.0	1.2	28.5	1.55	26.0
Pentosans	28.5	42.0	73.6	14.7	18.4

Protein from Cereal Endosperm Using Buffered Solutions

An improved process of preparing protein enriched products from cereal endosperm materials has been developed by *D.D. Christianson, A.C. Stringfellow and J.S. Wall; U.S. Patent 3,661,593; May 9, 1972; assigned to U.S. Secretary of Agriculture.* Generally, endosperm materials (grits, meal, and flour) are milled and classified (e.g., sieving or air classification) into products having varying protein concentrations.

In this process, the endosperm products are soaked in a buffer comprising 0.1 M potassium phosphate buffer (pH 7.5) containing 0.006 M magnesium chloride. This buffer is approximately isotonic relative to corn tissue. In this process, the cereal products are prepared from corn, sorghum, and wheat.

The process has essentially only three parameters: amount of buffer used in relation to the cereal endosperm product; time of soaking; and temperature of soaking. A quantity of buffer sufficient to cover each cereal endosperm particle at from 4°C to room temperature for a period of from 1 to 24 hours will slough the protein sheath from the starch granules as seen by the scanning electron microscope. Amounts of protein found in the various fractions obtained from classification procedures such as sieving or air classification are also used as criteria for showing this starch-protein agglomerate disruption. Since endosperm materials from normal milling give only a small separation when subjected to classification procedures, particle size weight distribution of the treated materials subjected to the same procedures provide a good qualitative measure of disruption.

Table 1 compares percent protein and weight distribution in sieved fractions obtained from high-lysine corn grits (12.4% protein) which have been hammermilled through a 0.027" mesh screen followed by either buffer treatment, water treatment (stirred 24 hours at 4°C), or no further treatment. Percent protein analysis and the weight distribution in the sieved fraction obtained from water treated hammermilled grits and those having no treatment show that little or no disruption of the starch-protein agglomerates has taken place, Table 1.

TABLE 1

	High-lysine corn grits [1]							
	Nontreated		Water		Buffer,[4] 24 hr.-4°C.		Buffer,[4] 1 hr.-22°C.	
Fraction	Percent P [2]	Percent wt. [3]	Percent P [2]	Percent wt. [3]	Percent P [2]	Percent wt. [3]	Percent P [2]	Percent wt. [3]
+90μ	12.8	46	10.4	30	18.0	8	17.9	16
+75μ	9.9	19	10.8	25	17.2	10	11.1	11
+60μ	9.1	19	15.4	20	15.4	10	9.9	13
+45μ	12.6	13	15.2	21	12.6	19	9.6	21
+30μ	12.0	4	15.3	4	10.4	29	7.8	28
+20μ		0		0	8.0	21	8.3	10
+10μ		0		0	5.6	4	0	2

[1] Hammer milled through 0.027-inch mesh screen before treatment.
[2] Percent P = percent protein in fraction.
[3] Percent wt. = percent of total sample weight.
[4] Buffer is 45% of the sample weight.

The similarity between these data indicates that the distribution of particle size is due to hammermilling and not the water treatment. The fact that there is little difference in percent protein in any of the fractions from these two samples means that the protein has not sloughed off the starch particles. The protein and starch are essentially evenly distributed among all fractions. Data from the buffer treated samples in Table 1 show considerable distribution in both percent protein and in particle size. There are fractions that vary from 5.6 to 18% protein. Since the original grits contain 12.4% protein, it is evident that the protein has been separated from the starch by the buffer treatment.

Example 1: Fifty grams of Ponca hard red winter wheat flour analyzing 10.3% protein was stirred with 0.5 liter of isotonic buffer for 24 hours at 4°C. The isotonic buffer is composed of 0.1 M KH_2PO_4 and 0.006 M $MgCl_2 \cdot 6H_2O$ at pH 7.5. At the end of the 24-hour soaking period, the resulting mixture of starch and protein was dried. A portion of the dried material was sieved through a series of BMC micromesh screens (30, 45, 60, 75, and 90μ) in a Ro-Tap sieve shaker for 10 minutes. A portion of the original flour was sieved in the same way and the results of weight distributed into the fractions compared in Table 2. The sieved fractions from the treated flour were subjected to protein analysis and these results also appear in Table 2.

Example 2: Fifty grams of experimentally milled high-lysine corn (opaque-2) grits having 12.4% protein were hammermilled through a 0.027" mesh screen. The milled grits were then treated, sieved, and analyzed for protein and compared to grits which were hammermilled and sieved without buffer treatment, as in Example 1, Table 2.

TABLE 2

Fraction	Example 1 Treated Percent P[1]	Example 1 Treated Percent wt.[2]	Example 1 Untr. Percent wt.[2]	Example 2 Treated Percent P[1]	Example 2 Treated Percent wt.[2]	Example 2 Untr. Percent wt.[2]
Orig.	9.5			12.4		
+90μ	24.4	11.5	36.2	18.0	8.2	56.0
+75μ	10.9	24.7	21.9	17.2	9.7	28.6
+60μ	7.5	24.9	15.7	15.4	9.6	14.0
+45μ	4.6	26.1	17.3	12.6	18.8	1.0
+30μ	2.8	16.0	8.9	9.1	54.2	0

[1] Percent P = Percent protein in fraction.
[2] Percent wt. = Percent of total sample weight.

Two-Stage Extraction Process for Wheat Gluten

This process disclosed by *G.V. Rao and O.B. Gerrish, Sr.; U.S. Patent 3,790,553; Feb. 5, 1974; assigned to Far-Mar-Co., Inc.* is directed primarily to the recovery of vital wheat gluten from the whole wheat kernel instead of from wheat flour, as is the practice. However, it will be appreciated that the process could also be practiced using wheat flour as the starting material by eliminating the tempering and crushing steps to be described in detail hereinafter.

Selective separation of gluten from starch, and of low molecular weight protein from high molecular weight protein, is accomplished according to this process by a two-step extraction procedure which utilizes the differences in solubility of the respective proteins in water and in mild base solutions. The vitality of both the high and low molecular weight portions of the gluten recovered is preserved. Inasmuch as high molecular weight protein yields greater loaf volume and greater hinge strength, it is of more value to the baking industry than is low molecular weight protein. The separation effected by the process, based upon molecular weights, permits subsequent blending of high and low molecular weight components in any desired ratio and custom tailoring of the gluten characteristics.

The first step of the two-step extraction process comprises tempering the whole wheat kernel in water to achieve a moisture content in the range, by weight, from 15% to saturation and preferably from 20 to 22%. Unprocessed whole wheat kernels have a moisture content generally from 11.2 to 14%. The saturation moisture level depends upon the characteristics of the kernel and, at room temperature, is generally in the range from about 45 to 55%. Tempering in accordance with this process involves immersing the kernels in water for a period of time sufficient to allow the kernels to take-up the necessary water to reach the desired moisture content. It is preferred, although not required, to temper the kernels in precisely the amount of water necessary to reach the desired water content.

Tempering the whole wheat kernel is necessary as a preliminary to the crushing and/or flaking step wherein the kernel is effectively reduced in particle size. Crushing or flaking may be accomplished in a conventional roller mill, having clearances from 0.001 to 0.05". While the ultimate particle size resulting from the crushing or flaking is immaterial, it has been found that particles in the 10 to 100 mesh range may be satisfactorily processed in accordance with this process.

The crushed wheat is slurried with water and the pH adjusted to 4.5 to 8.0, preferably 6.0 to 6.5. Depending upon the extent of tempering and the use of antimicrobial agents, such as ammonium hydroxide, during the tempering process, the pH of the slurry will vary and, therefore, the extent of, or need for, pH adjustment will likewise vary. The ratio by weight of crushed wheat to water in the slurry may vary over the range from 1:1 to 1:5. Preferably, the slurry is diluted with 2 parts water per part of crushed wheat.

At this point in the process it has been found to be convenient to screen the slurry, such as by using screens from 30 to 300 mesh, to remove the bran from the crushed wheat. During the crushing or flaking process, the bran maintains its physical integrity and is susceptible of ready removal by screening at this point in the process.

The wheat-water slurry is thoroughly agitated and exposed to high shear forces in any high speed electric mixer for 5 to 30 minutes, after which the slurry is centrifuged in batch or continuous centrifugation apparatus until a distinct phase separation is observed. The supernatant liquid resulting from the centrifugation is water containing dissolved water-soluble, low molecular weight protein. The protein may be recovered by separating the supernatant liquid from the solid residue and evaporating the liquid under vacuum at room temperature or by use of other suitable drying techniques.

The residue resulting from the centrifugation is next slurried with a mild base, such as ammonium hydroxide. While strongly basic substances might be suitable for effecting the dissolution of the high molecular weight protein from the wheat, they will adversely affect the vitality of the gluten protein. The slurry composition may have a residue to ammonium hydroxide proportion by weight in the range from 1:1 to 1:5, with the preferred dilution being 1 to 2 parts by weight residue to 2 parts by weight ammonium hydroxide. The pH of the slurry is adjusted to 9.0 to 12.0 and preferably from 9.5 to 10.2.

To extract the high molecular weight proteins from the wheat, the residue-ammonium hydroxide slurry is agitated vigorously in an electric mixer to expose the slurry to high shear forces, after which the slurry is centrifuged to effect a distinct phase separation. As in the water extraction step, agitation is continued from 5 to 30 minutes as necessary and centrifugation is continued until phase separation is accomplished. The supernatant ammonium hydroxide solubles are separated from the residual solid and dried at room temperature in a vacuum oven or other suitable apparatus to recover the high molecular weight protein. The residual solid is wheat starch, containing only a fractional percentage of protein. The starch may be air-dried at room temperature.

Example: One thousand grams of hard red winter wheat and one hundred-fifty grams of water were mixed uniformly in any suitable mixer over a period of twenty minutes and left overnight to temper. The tempered wheat (moisture 22.5%) was crushed or flaked in a roller mill. Two hundred and fifty grams of crushed or flaked wheat was made a slurry in five hundred grams of water and most of the bran was removed by passing through a 30 mesh screen. The screened slurry was mixed in a high speed electric mixer for twenty-five minutes and centrifuged until phase separation was observed. The supernatant aqueous phase was evaporated under vacuum and analyzed for protein content. The solid residue from centrifuging was mixed with five hundred grams of ammonium hydroxide solution at 10.2. The slurry was adjusted to pH 10.2 and subjected to high shear forces in an electric mixer for five minutes. The slurry was centrifuged until phase separation was observed. The supernatant ammonia solubles were vacuum dried and analyzed for protein content. A material balance showed a total protein recovery of 81% from the wheat kernel. Analysis: (dry basis) gluten: % protein (water-solubles), 26.8; % protein (ammonia-solubles), 73.6; starch: % protein, 0.18.

GLUTEN MODIFICATION

Wheat gluten is heterogeneous in character and is usually considered to be made up of two major protein groups, glutenin and gliadin. Commercially marketed undenatured or vital wheat gluten is removed from flour by one of several washing processes and usually contains 75 to 80% protein, 6 to 8% fatty-like phospholipids and related compounds, some fiber, residual starch, and a small amount of mineral matter in addition to phosphorus. Modern vacuum drying procedures are generally used in gluten production.

Powdered gluten has found application in the food industry. Particularly, such powder is used to fortify flour for bread or other yeast-raised products and to increase the protein content, and thereby the strength, of macaroni products. Wheat gluten is also processed into textured fibers in U.S. Patent 3,645,747 on page 242.

Carbon Dioxide for Drying Gluten

The drying or removal of the firmly bound and entrapped water of the raw gluten has heretofore been extremely difficult. The raw gluten is extremely sensitive to heat. Many processes of drying have been proposed, but no matter what process of drying was heretofore proposed, considerable difficulty was encountered. In drying by the use of vacuum, for example, it has been found difficult and expensive to keep a vacuum high enough to accomplish the purposes desired, difficulties are often encountered in maintaining satisfactory quality, and a great deal of hand labor is required which in turn creates the necessity for stringent sanitary safeguards.

W.M. Miley, L.M. Thomas and G.M. Bierly; U.S. Patent 2,797,212; June 25, 1957; assigned to The Keever Starch Company have discovered that when carbon dioxide under superatmospheric pressure is added to the raw wet gluten as it comes from the washing process the gluten becomes less cohesive, can be more easily separated into distinct or discrete particles, can be shaped more easily as desired and in general is more easily handled.

Seemingly, its chemical composition is permanently affected only slightly, if at all, because after drying, the end product has the essential characteristics of pure undenatured gluten. However, it is possible that the CO_2 reacts temporarily with certain end groups of the gluten in somewhat the same manner as it reacts with hemoglobin in the animal living process. If gluten so treated is introduced into water, or if wet gluten from the washing process is introduced into water in the presence of carbon dioxide under superatmospheric pressure, the gluten is easily dispersed or dissolved.

The dispersion is then readily dried by any of conventional drying methods such as vacuum oven, conveyor oven, spray drying, roll drying, flash drying, centrifugal extrusion drying, or rotary drying. The carbon dioxide treated gluten can also be treated in other ways. It will react with certain chemicals which do not react with untreated gluten. For example, it reacts with alkaline earth metals such as calcium hydroxide to obtain an alkaline earth metal salt of a protein complex (e.g., a calcium salt of a gluten complex).

$$2C_5H_9O_4N + Ca(OH)_2 \longrightarrow Ca(C_5H_8O_4N)_2 + 2H_2O$$

The treated gluten also undergoes different enzymatic reactions from corresponding reactions of untreated gluten.

Example 1: Wet wheat flour gluten, in masses as it came from the starch washing processes, in the mill, was introduced continuously into a tank of water maintaining an atmosphere of carbon dioxide above the water under a pressure of 30 pounds per square inch. This dispersion was pumped continuously through spray dryers wherein it was forced into the atmosphere in small streams in the presence of heat thus producing small beads and flakes of dry pure gluten, the water and the carbon dioxide evaporating.

Example 2: A dispersion of wheat gluten in water dispersed by means of carbon dioxide under pressure as described in Example 1 was dried by being passed through a Proctor and Schwartz conveyor belt dryer to obtain pure dry undenatured gum gluten.

Rehydrating Gum Gluten

Dehydrated vital gluten or gum gluten is extensively used in bread doughs as a protein supplement to obtain improved physical characteristics and added nutritional value for the resulting bread items. This vital gum gluten is generally available commercially in finely divided form like flour. Such dehydrated vital gum gluten has not been previously used to obtain a gluten mass for use in high protein vegetable base foods or synthetic foodstuffs resembling, in finished form, meat or meat products, because it has yielded a gluten inferior in all respects to freshly washed gluten made from a good grade of flour. Moreover, it has lacked uniformity of moisture content after rehydration, and the hydration has never been as complete as with freshly washed gluten.

The universally used mixers of dough made from flour and vital wheat gluten are too slow and, moreover, do not produce a uniform quality rehydrated gluten nor one where other ingredients may be uniformly incorporated in the gluten mass simultaneously with the rehydration of the vital gum gluten used in making such gluten mass with a very high percentage of gluten.

W.E. Hartman; U.S. Patent 3,290,152; December 6, 1966; assigned to Worthington Foods, Inc. has found that the apparatus needed for the process must keep the gluten thoroughly broken up and agitated so that the water used in rehydrating the gluten, is kept in contact with the fine gluten particles so that the rehydration can be uniform. If other materials are to be blended into the gluten mass, this can be done simultaneously with the rehydration during this rapid mixing, cutting, chopping, or beating of the mass.

A machine which makes this rehydration possible is one that is known as a "Silent Hopper." This machine has been designed for preparation of comminuted meat and has a bowl with a round bottom and a raised center portion of the bottom such that it has a bottom contour similar to the bottom half of a doughnut laid on its side. This bowl is rotatable about a central vertical axis, while mounted over the bowl at one side of the center of same is a rotary shaft in a plane perpendicular to the bowl shaft and with such shaft mounted over the bowl having mounted thereon closely spaced knives, the cutting edges of which closely fit the outer periphery of the inner surface of the bowl so that as this shaft and knives are rotated the knives will closely follow the inner periphery of the bowl along the outer edge portion of same and substantially full depth at the mid-portion of the knife-carrying length of the shaft.

In the use of this cutting or chopping apparatus in the process, the bowl was rotated at 8 rpm and the shaft carrying the cutter knives at 1,750 rpm, with the mixing time being approximately five minutes as given in the examples, although it is obvious that these speeds could be varied over a wide range so long as the cutting or chopping action is rapid enough to attain the intimate rapid mixing required in order that the fine particles of gluten being rehydrated are kept substantially separated so that the rehydrated water can get to those particles for thorough and uniform rehydration during the mixing procedure.

A preferred range of speeds for the rehydration of the gluten and mixing of the ingredients would be 1,400 to 3,500 rpm of the knives and a corresponding bowl speed of approximately 5 to 16 rpm. The time of mixing in this range of speeds would preferably be in the approximate range of 7 to 5 minutes. In carrying out the rehydration process, the water or moisture content of the rehydrated gluten may be varied over a much wider range than has been possible with the standard practices previously used; for instance, the working range for water in this process varies substantially between 50 and 80% of the finished mass. Still lower percentages of water, due to the more efficient absorption and adsorption of same by the vital gluten, results in a gluten mass which is so stiff and resistant to working that manipulation of the gluten mass becomes substantially impossible. On the other hand still higher percentages of water may be added to the gluten mass than usual but beyond that the water is not fully retained and, accordingly, some of the starch is washed out or lost from the gluten mass under those conditions.

Example:

>   50 pounds of dehydrated vital gluten (80%+ protein)
>   20 pounds of standard patent white flour (13% protein)
>   106 pounds of water at 70°F
>   280 grams caramel coloring
>   85 grams lecithin

All of the above ingredients are placed in the bowl of a high speed cutting and chopping machine such as a Silent Chopper and the machine run for five minutes with the bowl rotating at 8 rpm and the cutter knives at 1,750 rpm. This yields a uniform gluten mass of predictable physical characteristics, weight, and analysis which may be duplicated or modified as desired.

Lipid Coating of Undenatured Wheat Gluten

Particle cohesion of vital wheat gluten has been overcome in a process disclosed by *B.W. Landfried and J.R. Moneymaker; U.S. Patent 3,362,829; January 9, 1968; assigned to Top-Scorer Products, Inc.* The gluten is coated with a nonionic hydrophilic lipid selected from monoglycerides, salts of lactylic esters of fatty acids, polyoxyethylene stearate, and stearyl monoglyceridyl citrate. The coated gluten particles are characterized by stability against particle cohesion in neutral aqueous dispersion, i.e., better dispersibility. A minor portion of a highly dispersible surface active agent may be included in the coating material.

Materials found to be satisfactory for treating the gluten are of the nonionic type and consist of monoglycerides (both of the molecularly distilled type and commercial mixtures of mono, di and triglycerides), edible salts of lactylic esters of fatty acids, polyoxyethylene stearate, stearyl monoglyceridyl citrate, and mixtures of the above. When hydrophilic lipid esters of essentially fully saturated fatty acids containing from 16 to 22 carbon atoms are used, improved results are achieved by admixing in addition a small quantity of a highly water-dispersible, edible, surface active agent such as lecithin, hydroxylated lecithin or polyoxyethylene sorbitan monostearate.

The proportion of lipid material necessary to satisfactorily treat a dry vital wheat gluten powder appears to be dependent not only on the particular lipid used but also on the average particle size of the material undergoing treatment. This would support the hypothesis that the improved results are primarily concerned with a surface phenomenon.

The product produced by this process performs in a normal manner when introduced into a dough type system, for example, a typical bread or macaroni dough. The characteristics of the treated product make it ideally suited for use in continuous dough making systems where the powdered gluten may be added to the brew and maintained in a dispersed state until final development of the dough takes place.

Example: The gluten treated was from one lot. Analysis indicated that 87% of the powdered gluten would pass a standard U.S. 100 mesh screen. Protein content was 80% of the total. The lipid ingredients were melted in a jacketed laboratory sigma blade mixer, although coating of the gluten particles can be obtained with lipid heated so as to be partially in a soft plastic state and partially liquid, or totally in a soft plastic state. Temperature was carefully controlled at 68° to 70°C which, in absence of water, will not substantially denature the gluten during the time period involved. The powdered gluten was introduced over a period of 15 minutes with continuous stirring although much shorter periods may be used depending upon the type of mixing equipment and other factors. The resultant homogeneous product was cooled to room temperature while mixing and then passed through a hammermill.

In the following preparations parts are by weight. Preparation 1 contains 5.0 parts distilled lard monoglycerides (Myverol 18-40) and 95.0 parts of powdered gluten; preparation 2 contains 5.0 parts calcium stearyl lactylate (Verv CA, commercial grade calcium salt of lactylic esters of fatty acids) and 95.0 parts powdered gluten; preparation 3 contains 5.0 parts polyoxyethylene stearate (Myrj 45, commercial grade polyoxyethylene 400 monostearate) and 95.0 parts powdered gluten.

For comparison, powdered gluten was similarly treated with 5 parts/100 distilled propylene glycol monoester. Untreated powdered gluten was used as control. The effect of these treatments on the stability of aqueous gluten dispersions was tested by wetting five grams of material with 100 ml of distilled water at 24°C while stirring continuously. Following formation, the dispersions were allowed to stand for 120 minutes prior to an evaluation of cohesion control. Results of the various preparations follow:

Test Material	Dispersibility	Hydration	120 Minutes Cohesion Control
Preparation 1	Instant	Very good	Very good
Preparation 2	Fair	Good	Very good
Preparation 3	Instant	Very good	Very good
5% Propylene Glycol Monoester	Good	Good	None
Untreated Gluten Powder	Fair	Fair	None

Dispersible Agglomerates of Gluten

A process disclosed by *R.J. Hampton, J.R. Rolland and T. Gallo; U.S. Patent 3,704,131; November 28, 1972; assigned to The Ogilvie Flour Mills Company, Limited, Canada* improves the dispersibility of gluten without the use of lipids. The agglomerated gluten is ideally suited for the manufacture of yeast-leavened bakery products, especially by continuous dough-making processes.

The gluten is subjected to a treatment wherein the individual particles are exposed to a wetting agent to make the surfaces sticky. The superficially agglutinated particles are then brought into intimate contact to cause them to adhere in aggregates of random size and irregular shape, but always of a size substantially greater than the size of the individual particles. The agglomerates are then dried to adjust the final residual moisture content to the preferred level of less than 10%.

Thereafter, the material is classified, and agglomerated gluten of the desired size and structure is delivered as a finished product and suitably packaged. Nonagglomerated and oversized material is recycled. The agglomerated gluten structures obtained are sufficiently strong and firm to withstand normal handling without giving rise to any serious disintegration problem.

This preferred process will be described with reference to Figure 9.3 which schematically illustrates a suitable apparatus assembly wherein it may be performed. Referring to this figure, the apparatus assembly includes a supply hopper **11** for storing the powdered gluten and a reactor **12**. This reactor is a vertical, cylindrical structure divided by a perforated diffuser plate **13** into an upper zone wherein the fluidized bed **14** is formed, and a lower zone in the form of a plenum chamber **15**. A downwardly-extending duct **16** is provided in the roof of the reactor to receive the gluten from the supply hopper, while an upwardly-extending duct **17**, also in the roof, is provided for the discharge of the fluidizing gas,

178 Vegetable Protein Processing

FIGURE 9.3: PREPARATION OF DISPERSIBLE AGGLOMERATES OF GLUTEN

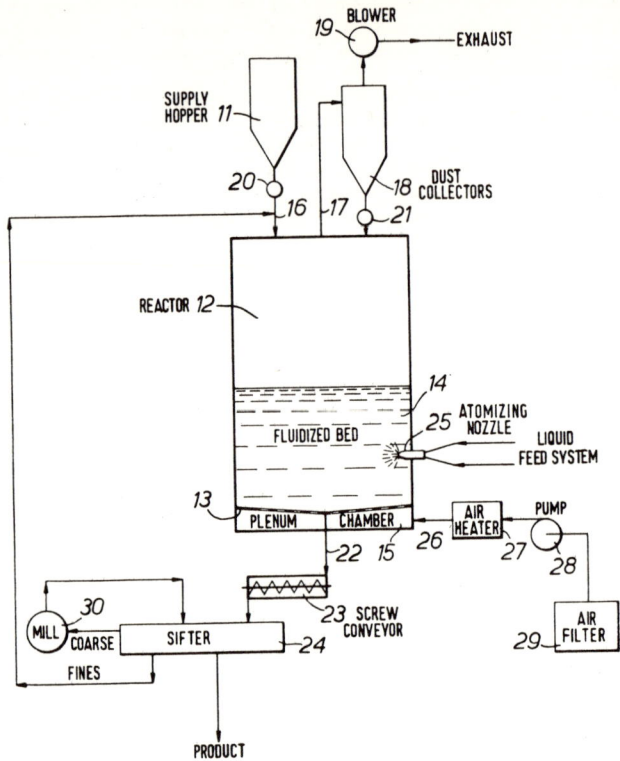

Source: R.J. Hampton, J.R. Rolland and T. Gallo; U.S. Patent 3,704,131; Nov. 28, 1972

compressed air, from the reactor. This duct **17** connects with a dust collector **18** and a blower **19**. Interposed in each of ducts **16** and **17** are rotary air locks **20** and **21**. The floor of the reactor includes a discharge duct **22** for removal of the gluten, a major proportion of which normally will be in an agglomerated state. The discharge duct leads to a screw conveyor **23**, fitted with a variable speed device to control the discharge rate, and a sifter **24**. Located in the wall of the reactor, in a median position, are a series of atomizing nozzles, one of which, **25**, is shown, for introducing droplets of the wetting agent directly into the fluidized bed **14** located in the upper part of the reactor.

The preferred system involves gas atomization. In the wall of the plenum chamber **15**, below the atomizing nozzles, a duct **26** is located for introducing the fluidizing gas into the chamber. This duct is connected to a heater **27**, a pump **28** and an air filter **29**.

Example: Preparation of Agglomerated Gluten — The apparatus assembly described in Figure 9.3 was used in a typical run to process a powdered gluten. The gluten was fed into the reactor at a rate of 600 pounds per hour and formed into a fluidized bed by admitting heated air to the chamber through the diffuser plate at a constant rate of 40 fpm. Water in the form of droplets of 10 to 100 microns was sprayed into the fluidized bed through ten, two-fluid nozzles, each delivering 1.5 gph of water, utilizing air at a pressure of 35 psig, as the atomizing agent. The temperature of the air entering the plenum chamber below the diffuser plate was about 115°C and the temperature of the gluten in the treatment chamber was about 46°C. The bulk of the gluten recovered was in the form of dry, coarse, irregularly-shaped agglomerates at least 90% of which remained on a 100 mesh screen, i.e., >149 microns.

That the agglomerated gluten retains the chemical properties of the original powdered gluten and differs only in certain advantageous physical properties is demonstrated by the results given in the following table which compares relevant properties of the agglomerated gluten with those of the initial powdered gluten.

Comparison of Properties of Agglomerated Gluten and Powdered Gluten

Property	Agglomerated	Powdered
Moisture (percent)	8	5.
Protein (percent) dry basis	80	80.
Ash (percent) dry basis	0.8	0.8.
Colour	Cream	Off-white.
Bulk density (g./cc.) [1]	0.36	0.42.
Flowability (seconds) [1]	15	Does not flow.
Aqueous dispersion	Uniform	Lumpy; impossible to disperse.
Dispersion stability	Can be resuspended *	
Mesh analyses:		
on 25	2.2	
on 40	30.2	
on 60	37.4	
on 80	19.2	1.1.
on 100	3.2	30.5.
on 120	2.8	37.9.
on 140	2.2	10.9.
on 170	1.0	8.2.
on 200	1.8	10.1.
thru 200		1.3.

* With slight stirring after standing for 24 hours.

Frothable Wheat Gluten as Egg White Substitute

In producing bakery goods, pastry or meringues the white of egg used is frequently beaten up into a froth and added in this form to the material to be baked. Particularly when producing meringues or similar pastry it is necessary for the froth to be voluminous and stable so that it does not collapse before or during the baking. As it has not been possible to produce froth suitable for this purpose from vegetable albumin, animal albumin has been used exclusively.

It has been found by *J. Höhl; U.S. Patent 3,409,440; November 5, 1968* that a wheat protein suspension which is heat-stable and beatable can be produced in a very simple manner without separating individual albumin constituents by bringing denaturized wheat gluten into an aqueous suspension and establishing a pH of about 4 by means of a food acid. In this way a voluminous stable froth can be produced. Such a froth can be used as a completely equivalent replacement for animal albumin in the production of bakery goods with the further essential advantage of heat stability. Wheat of European origin, that is to say wheat with weak gluten, is particularly suitable as a starting product. Tartaric acid, citric acid or ascorbic acid can be used with great advantage for the acidulation, yielding particularly good results.

Example 1: 50 g of denaturized wheat gluten were placed in suspension in 250 cc of water and 2 g of tartaric acid and 30 g of glycerine added (pH—4.5). After beating at room temperature 3,300 cc of froth were obtained, which was very stable.

Example 2: A similar experiment to that in Example 1 was carried out with ascorbic acid instead of tartaric acid and brought to a pH value of 4.9, 3,100 cc of froth being obtained. Although the froth was not so compact and solid as that obtained using tartaric acid, nevertheless it was fully equivalent to that obtained from animal albumin.

Upgrading Odor and Taste of Gluten

A process for making a crude starch-containing gluten edible has been disclosed by *J.W. Knight; U.S. Patent 3,782,964; January 1, 1974; assigned to CPC International Inc.* The process comprises (1) subjecting the crude gluten to the hydrolytic action of alpha amylase to hydrolyze the starch; and (2) introducing heat into the amylase-treated gluten to heat the gluten to a temperature of at least 95°C for at least 15 minutes.

It has been found that the objectionable odor and taste associated with starch-containing crude gluten are bound into the starch-protein structure. The well-known absorptive properties of starch also contribute to the retention of this odor and taste.

It has been discovered that subjecting the crude gluten to the hydrolytic action of alpha amylase hydrolyzes the starch, disturbs the protein-starch structure, and frees the objectionable odor and taste bodies in such a manner that they are easily removed in the aqueous fraction; by subsequent heat treatment, preferably by steam distillation, and washing.

In a typical example, gluten cake from the string filters in a corn wet milling process is slurried to 13 to 14% solids and the pH adjusted to about 6. Alpha amylase is added in an appropriate amount and the mixture held at a temperature of about 80°C for approximately one hour. Live steam is then introduced for a period of about one hour. The resulting product is filtered and the protein-rich filter cake is washed with warm water. After drying, the filter cake yields a high protein product having substantially no odor or taste.

Suitable crude glutens for treatment by this process include corn gluten and any other starch-containing crude gluten materials, such as sorghum gluten, which have been isolated in the presence of sulfur dioxide.

The initial steps, i.e., subjecting the crude gluten to the hydrolytic action of alpha amylase, may be accomplished by slurrying the crude gluten in water to provide a slurry having a solids content from 10% or lower up to 30%. The alpha amylase is added to the crude gluten prior to or subsequent to heating to a suitable temperature for action by the alpha amylase. Any enzyme can be used which liquefies or drastically alters the starch structure. Malt enzymes are an excellent alternative to alpha amylase. The preferred enzyme for the conversion of the starch in the crude gluten is the type commonly referred to as alpha amylase. Suitable alpha amylases are derived from one of four sources; (1) pancreatic, (2) fungal, (3) bacterial, or (4) salivary. Although alpha amylases from any one of these is suitable for use in this process for expediency and brevity, bacterial alpha amylases will be particularly exemplified.

The pH of the conversion medium, when using a bacterial alpha amylase, is preferably that suitable for the optimum activity of the alpha amylase. Generally, this pH range is between 6.0 and 8.0. The most suitable temperature lies between that required for gelatinized starch which is at least 60°C and that at which the enzyme will lose a large portion of its activity which is about 95°C. It has been found that the preferred temperature range is between 70° and 92°C.

The quantity of the alpha amylase preparation required for hydrolyzing the starch will be dependent upon the activity of the alpha amylase preparation, the quantity of starch in the crude gluten, the conversion temperature, the pH of the medium, and the desired degree of hydrolysis of the starch present in the gluten. Suitable conditions are easily selected. For example, a bacterial alpha amylase having an activity equivalent to the HT-1000 product of Miles Chemical Laboratories would be used in an amount between 0.001 and 0.1% by weight of the crude gluten on a dry basis. The conversion conditions would include a temperature of 80°C and a pH of 6 for a period of time sufficient to hydrolyze the starch to the desired degree which generally will be about one hour.

After treatment with one or more of the enzymes and when the desired degree of hydrolysis is obtained, the temperature is raised to at least 95°C and preferably to 100° to 150°C for about one hour. Raising the temperature to a value above 95°C is necessary to terminate the hydrolytic action of the enzyme and assists in improving filtration rates. After the heat treatment which drives off the adverse odor and taste bodies by steam distillation or other heat methods suitable for driving off volatiles, the product may be subject to filtration, washing and drying to yield an edible protein-rich product having a moisture content less than about 15%.

Example: The crude corn gluten cake used in this example had the following analysis:

Moisture	60%
Protein	70% db
Starch	80% db
Reducing sugars	- -
Oil	3.2% db
Ash	1.0% db
pH	4.3
Crude fiber	1.0
Taste	Objectionable
Odor	Objectionable

The crude gluten cake was slurried to 13 to 14% solids and the pH adjusted to 6 with soda ash. Bacterial alpha amylase, (Kleistase) was added to the slurry in a quantity of 0.005% based on total dry solids. The mixture was held at 80°C for one hour to permit the alpha amylase to act on the starch.

After this time, the temperature was raised by the introduction of live steam to about 100°C for a period of one hour. The resulting mixture was filtered hot through a plate and frame press and the protein washed with warm water. The filter cake was diluted to 80% water and dried in a spray drier and had the following analysis:

Moisture	5.8%
Protein	88.4% db
Starch	Trace
Reducing sugars	0.53% db
Oil	4.5% db
Ash	0.54% db
pH	5.7
Crude fiber	0.8% db
Taste	Bland, satisfactory
Odor	Satisfactory

This product when tested was completely bland and had no adverse taste or odor. When subjected to shelf life storage tests for six months, the product showed no deterioration and was still substantially odorless and taste-free.

ZEIN

Zein, the alcohol-soluble protein from corn, is prepared commercially by extracting corn gluten with hot 80% isopropyl alcohol. It is commercially valuable because it can be spun into fibers and will form tough, adherent films. The uses of the film forming properties of zein have not been as great as originally hoped, due largely to the fact that, practically speaking, films must be cast from aqueous alcoholic solutions. This solvent is relatively expensive and thus adds considerably to the cost of zein films.

Numerous attempts have been made to utilized water as a zein solvent. If sufficient sodium hydroxide or other strong alkali is used, zein may be dissolved in water. The pH of these solutions must be above 11.5. Under these highly alkaline conditions useful films cannot be cast because strong alkali reacts to hydrolyze (degrade) the protein film. Further, these solutions are difficult and hazardous to handle because of their corrosive nature. Films may be cast from dispersions of zein in water but two serious disadvantages have hindered their commercial utilization. First, the films necessarily contain the agents used to disperse the zein, generally fatty acid salts. The presence of these products reduces the resistance of the film to water and other reagents and often impairs film toughness and adhesion. A second disadvantage is that stable emulsions are often difficult to prepare especially if a high zein content is desired.

Partially Deaminated Zein

A substantially undegraded zein soluble in water at a pH as low as 6.5 is prepared by *L.G. Unger and D.W. Howland; U.S. Patent 3,010,953; November 28, 1961; assigned to Corn Products Company*. They accomplished this by transforming the primary amide groups of zein into carboxyl groups (deamidation) without substantially hydrolyzing the peptide bonds and without esterifying the amide groups.

It was found that under carefully controlled conditions the deamidation reaction can be carried out without any substantial amount of peptide hydrolysis or esterification of amide groups. The desired reaction occurs if an aqueous solution containing 15 to 30% by weight of zein in 40 to 80% by weight of a solvent selected from aqueous secondary and tertiary aliphatic alcohols, acetone, and dioxane, is heated at 80° to 140°C for 5 to 60 minutes, the reaction being carried out at a pH less than 2.5 until it contains not more than about 18 new carboxyl groups per molecule of zein. The product thus obtained is soluble in water at a pH as low as 6.5.

Time, temperature and acid concentration are interdependent variables. At high acid concentration, i.e., 0.7 N sulfuric acid, short reaction times and/or moderate temperatures are required. At low concentrations, i.e., 0.1 N, longer reaction times and/or higher temperatures are necessary for the desired degree of reaction. The conditions preferred for use on a commercial scale are 6 min reaction time at 113°C in the presence of about 0.35 N sulfuric acid. Any strong mineral acid may be employed in this process. Most of the work has been with sulfuric acid since it is inexpensive and is relatively noncorrosive. However, either hydrochloric or phosphoric acid may be used satisfactorily. Although hydrochloric acid is more reactive than sulfuric acid, i.e., at equivalent concentrations more deamidation occurs with hydrochloric acid than with sulfuric, it is not sufficiently more reactive to offset its higher price. Furthermore, hydrochloric acid is more corrosive than sulfuric acid.

The unique feature of this process is the use of the specified solvent system. The solvents which produce the desired results are aqueous secondary and tertiary aliphatic alcohols, acetone, and dioxane. If a primary alcohol such as ethyl or methyl is used, esterification of the newly formed carboxyl groups occurs, and the resulting zein has a substantially reduced solubility in water. At solvent concentrations below 40% and above 80% the unmodified zein is not too soluble.

In addition to acetone and dioxane any secondary or tertiary alcohol, such as isopropyl, isobutyl or tertiary butyl, may be employed. However, isopropyl alcohol is preferred for practical as well as economical reasons; the main reason being that in the commercial manufacture of zein from corn gluten there is available a solution of zein in aqueous isopropyl alcohol.

This solvent system as the reaction medium for deamidation has another advantage over the colloidal suspension of protein in water used by previous investigators. The protein being in solution makes uniform attack on its molecule possible. This is important since one can stop the reaction at a low degree of deamidation and be certain that there is essentially no unreacted zein in the system.

Example: Preparation of Partially Deamidated Zein — Two hundred ml of flashed "heavy zein", a solution containing 56 grams of protein (N x 6.25) in 55% by weight isopropyl alcohol, was placed in a 500 ml 3-necked round bottom flask. Flashed heavy zein is produced during the commercial preparation of zein from corn gluten as described in an article on pages 226-229 of the September 1957 issue of *Chemical Engineering* entitled "Revamped Flow Sheet Wins Improved Zein". The flask was fitted with a thermometer, sealed mechanical stirrer and reflux condenser. Sulfuric acid (4 ml of 96%) was added to obtain a pH of about 0.5 and the solution heated at reflux (83°C) for 15 minutes. The flask and contents were cooled to below 50°C and the pH brought from about 1.0 to about 5.0 by the addition of 5.0 N sodium hydroxide solution. The acid treated zein was then precipitated by quenching the solution in 2 liters of cold water. The precipitated zein was filtered, slurried in additional cold water, filtered and then dried by lyophilization.

The product contained 7.2 new carboxyl groups per molecule of zein. Only 1.1% of the protein was degraded by this treatment, as measured by the amount solubilized in the precipitation and wash waters. Samples were run at various solvent concentrations. At solvent concentrations below about 40% and above about 80%, the unmodified zein was not too soluble.

Deaminated Zein by Alkaline Hydrolysis

F.C. Loew; U.S. Patent 3,370,054; February 20, 1968; assigned to Interchemical Corp. has developed a process of alkaline hydrolysis that only slightly degrades the zein. It has been found that deamidated zein dispersible in solutions having a pH of 6.50 or more can be prepared by alkaline hydrolysis in such a way that very little degradation takes place and therefore film properties are not significantly impaired. The process makes use of strong alkalies which can be neutralized to form precipitates so that only a negligible amount of salt is left in solution. The alkali and precipitant should of course be selected with the particular end use in mind.

The efficiency of precipitants varies somewhat with the solvent, of course. For instance lithium sulfate is fairly soluble in water (1 part in 4 of water) but is insoluble in ethyl alcohol. Where the end use is a pigmented film, as in an ink composition, the precipitate does not interfere and therefore there is no need to remove it from the reaction mixture. Examples of combinations are lithium or the alkaline earth hydroxides or oxides and as precipitates sulfuric acid, phosphoric acid, ammonium chromate, ammonium carbonate, etc. The economic aspects are of prime importance as far as commercial uses are concerned. From this aspect the use of barium hydroxide or oxide and sulfuric acid is the most attractive combination.

Hydrolysis may be carried out in monohydric aliphatic primary, secondary, or tertiary alcohols having 1 to 4 carbon atoms in the chain. Dihydric alcohols such as ethylene glycol, propylene glycol, diethylene glycol, or dipropylene glycol may also be used. In acid hydrolysis of zein, primary alcohols cannot be used without causing considerable esterification of the zein.

It is sometimes desirable to add stabilizers to counteract the tendency of the polymer to precipitate when the process is carried out near its limits of efficiency. Such stabilizers might be urea or carboxyl-containing rosin-derivatives. A wide range of materials may be used for this purpose; but this does not constitute a critical feature of the process. Any additive selected should of course be appropriate for the end use intended. Where the composition is to be stored for some time before using, it is advisable to add a suitable pesticide, as is well known in the art.

Example: A solution of 1.17 pounds of anhydrous barium hydroxide in 11.9 pounds of isopropanol and 31.6 pounds of water was warmed to 120°F and then zein (24 pounds) was added over a period of about 15 minutes while the batch was stirred rapidly. After the zein was practically all dissolved the temperature was raised to the reflux point of 184°F and held there 30 minutes. The charge was then allowed to cool below reflux temperature and $\frac{2}{3}$ pound of 98% sulfuric acid added, with due precautions. The batch was allowed to cool below 150°F and 0.34 pound of 28% aqueous ammonia was added. The resulting material had a pH of 8.5. Filtration was carried out in a plate and frame filter press. The barium sulfate passed through the filter press and remained in the zein solution. The ratio of alkali to zein in this example was 57 meq of barium hydroxide per 100 g of zein. The hydrolysis was conducted at 185°F for 30 minutes. Films of this zein were comparable to those from good acid process zein.

Solvent Extraction of Zein

It has been known that zein could be extracted from grain proteins such as corn gluten using an aqueous organic solvent such as aqueous ethyl or isopropyl alcohol, sometimes followed by reextraction of the extract with a fat solvent such as hexane or benzene, recovery of residual volatile hydrocarbon from the extract by distillation, and precipitation of the zein by mixing the remaining alcoholic solution with a large proportion of water. However, such prior processes have required a high capital investment and were characterized by high operating costs due to the complex solvent recovery systems required.

An improved process is disclosed by *R. Carter and D.R. Reck; U.S. Patent 3,535,305; October 20, 1970; assigned to Nutrilite Products, Inc.* for producing zein of high purity and light color from gluten, that is, from the mixed proteins in maize, wheat, rice, oats or barley. According to this process commercial gluten, such as corn gluten meal containing 40 to 70% protein on a dry basis, preferably 60% protein, is first extracted with a concentrated aqueous organic solvent such as 60 to 90% isopropyl alcohol, at a moderately elevated temperature. At least 3 parts of solvent, preferably 3.5 to 5 parts, per part of gluten can be conveniently used, the optimum ratio depending on whether a single stage or a multiple stage process is used, as well as on the moisture content of the gluten and the water content of the solvent.

Generally, it is desirable to select a strength and proportion of solvent to give an extracting liquid containing 85 to 90% ethyl or isopropyl alcohol and correspondingly 15 to 10% water. All proportions here are expressed on a weight basis unless otherwise indicated. This extraction is preferably carried out between 55° and 70°C. While aqueous isopropyl or ethyl alcohol are preferred as solvents, concentrated aqueous solutions of other oxygen-containing organic solvents known to dissolve zein can also be used.

Such solvents include acetone, methyl ethyl ketone, ethylene glycol, diethyl glycol, ethyl ether of ethylene glycol, furfuryl alcohol, tetrahydrofurfuryl alcohol and the like. If desired, the pH of the extracting liquid may be adjusted to between 6.5 and 7 by adding a small amount of an aqueous alkali metal hydroxide such as sodium hydroxide.

After completion of the extraction step, undissolved solids are separated from the liquid extract by mechanical means such as filtration or centrifuging. Typically, the extract may contain 5 to 15%, preferably 10 to 12% dissolved crude zein. The mechanical separation of solids from the liquid extract is preferably carried out at or near room temperature.

After separation of undissolved solid from the extract the latter is chilled to a temperature above the freezing point of the liquid solvent which causes the extract to separate into two distinct phases, e.g., to a temperature between +10° and -25°C., preferably to between -10° and -20°C, whereupon the resulting supernatant liquid is decanted from the heavy yellow taffy-like bottom phase. This taffy-like phase generally contains 20 to 40% zein with the remainder being solvent and a small amount of dissolved corn oil and oil solubles such as xanthophyll. When chilling to the preferred temperature range, the taffy-like phase contains 95% or more of the total zein extracted from the gluten feed and less than 20% of the extracted oil and oil solubles such as xanthophyll. By chilling to less deep temperatures, a relatively smaller proportion of the extracted zein is found in the bottom layer and relatively greater proportion remains in the liquid solvent phase.

The taffy-like phase or mass can then be dried in any convenient manner, e.g., in a vacuum tray drier at 50°C, preferably after first reducing its viscosity by allowing or causing its temperature to rise to at least 15°C, e.g., between 15° and 30°C. Drying temperatures preferably should not exceed 100°C, and temperatures below 70°C are preferred. In this way, dry, light yellow zein containing only 2% oil and oil solubles is readily produced. If a purer product is desired the taffy-like mass after being warmed to liquefy is diluted with a new portion of aqueous organic solvent, e.g., by adding a further amount of 85% isopropyl alcohol to produce a solution containing 2 to 15% zein, preferably 6 to 12%, and this dilute solution is then chilled again as indicated before, which reseparates the zein and removes most of the oil remaining after the first separation.

Example: The commercial corn gluten meal used as starting material contained 65% protein, 12% moisture as determined by weight loss on drying at 105°C, 2.5% oil and oil solubles, the remainder being various carbohydrates, fiber and ash. 100 parts of the corn gluten is placed in a stainless steel covered tank equipped with a mechanical stirrer. To this is added 400 parts of 88% isopropyl alcohol containing 1 part sodium hydroxide, preheated to 65°C. The resulting mixture is mildly agitated to maintain suspension of the meal particles in the liquid. After agitating for one hour, in which time the temperature of the mixture drops to 55°C, the mixture is centrifuged and the separated liquid filtered after cooling to room temperature (about 25°C). The cake from the centrifuge can be dried and used as animal feed.

The filtrate is chilled to −15°C by means of cooling coils, whereby it separates into a taffy-like zein-alcohol bottom layer (1 volume) and a supernatant liquid layer (4 volumes). The supernatant layer is decanted from the chilled zein-alcohol layer and the latter is allowed to warm up to room temperature. The zein-alcohol layer contains about 30% zein or about 24 parts zein per 100 parts of gluten feed.

One-half of the zein-alcohol solution (40 parts) is dried on a vacuum drum drier at 50°C and a pressure of 0.06 atm. The resulting dry zein sheet is milled to a fine, free-flowing powder which is light orange yellow and contains about 2% oil and oil solubles. To make a high-purity product, the other 40 parts of the zein-alcohol solution containing 30% zein is mixed with an additional 200 parts of 88% isopropyl alcohol and this diluted solution is again chilled to −15°C whereby it separates to give a taffy-like zein-alcohol bottom layer (1 volume) and a supernatant liquid layer (about 6 volumes). This supernatant layer when decanted may be used directly as solvent to extract fresh gluten meal.

The taffy-like zein-alcohol layer is again warmed to room temperature and dried and milled as previously described. The zein powder resulting from such a two-stage chill separation is light straw colored and contains only about 0.4% oil and oil solubles. The yield of zein from this two-stage chill separation process is about 22% based on initial gluten feed.

PROCESSING OTHER VEGETABLE PROTEINS

SUNFLOWER MEAL

When oil is expelled from sunflower seeds, a resulting by-product is the residue which is denoted sunflower meal. This meal can be a valuable source of protein for addition to food products such as breakfast cereals, meat products, peanut butter, imitation dairy products, pet foods, and snack foods to supplement the protein and/or to provide a protein constituent.

One method of isolating protein from sunflower meal involves alkali extraction of the meal to extract protein followed by acid precipitation of protein from the extract. The weight ratio of water to meal ordinarily ranges from 10:1 to 25:1. The pH of the water is adjusted to 9 to 11 by addition of a strong base, for example, sodium hydroxide. The meal/basified water mixture is then agitated from 15 to 90 minutes whereby protein and some nonprotein impurities are extracted from the meal to form a liquid extract and a solid residue of spent meal.

The extract which is a solution of water, protein and nonprotein impurities is separated from the spent meal, for example, by centrifugation. This extract is then treated in an acid precipitation step where the pH of the extract is adjusted from 3.5 to 6 by the addition of an acid, for example, hydrochloric or sulfuric acid. The precipitated protein is recovered from the acid-adjusted extract phase, for example, by centrifugation. The entire process is carried out without the application of external heating or cooling. However, higher or lower temperatures are sometimes utilized.

The recovered precipitated protein is denoted protein isolate. The protein isolate ordinarily has an intense green color. This is because the meal has as a constituent green color-forming precursors which in a conventional alkali extraction/acid precipitation process result in the formation of this green color in the protein isolate. Once it appears, the green color cannot be removed from the isolate product by dialysis or other conventional means of purification.

Purification by Acid Washing

A process has been developed by *D.E. O'Connor; U.S. Patent 3,586,662; June 22, 1971; assigned to The Procter & Gamble Company* which provides an acid washing step previous to alkali extraction and recovery of protein from the extract to produce a light-colored protein isolate which does not have an intense green color. More particularly, the process involves multistage washing of sunflower meal with water adjusted to acid pH, then subjecting the acid-washed meal to conventional alkali extraction followed by recovering protein from the extract by acid precipitation or other protein recovery method.

The sunflower meal is subjected to the multistage acid washing operation by washing the meal 2 to 7 times previous to its alkali extraction. In this acid washing step the meal is first mixed with water, the weight ratio of water to meal ranging from 3:1 to 20:1, preferably 5:1 to 15:1. Then the pH of the mixture is adjusted to 3.5 to 6.0 by the addition of an acid, for example, monobasic sodium phosphate, hydrochloric acid, sulfuric acid, or phosphoric acid.

The pH-adjusted meal/water mixture is then mixed for 10 to 60 minutes, to extract green color-forming precursors from the meal and provide a solid treated meal phase and a liquid aqueous acid phase into which green color-forming precursors have been extracted. The washing process is repeated on the treated meal phase at least one more time and up to six more times. Instead of multistage washing, a countercurrent washing operation can be employed which is the equivalent from a chemical engineering standpoint of a multistage process.

If pH's below 3.5 or above 6.0 are used in the acid washing step, protein losses into the wash water are excessive. If pH's above 7 are used, the final product will be characterized by a green color. If a weight ratio of water to meal less than 3:1 is used, an excessive number of washing steps may be necessary and green color-forming precursors may not be satisfactorily washed out of the meal. The weight ratio upper limit of 20:1 is a practical upper limit. Weight ratios much in excess of this limit can cause handling problems.

Temperatures less than ambient can be used but the extraction proceeds more slowly; there is no advantage to using the lower temperatures. If temperatures greater than 140°F are used, the protein can be denatured. At least 2 washing stages are necessary to remove sufficient green color precursors so that an acceptable food product can be achieved. The greater the number of washes used, the lighter will be the color. After seven washes the color improvement increment is so slight that extra washes are not of economic practicality. The wash water can be discarded. The acid-washed meal recovered from the last washing stage is then subjected to a conventional alkali extraction operation.

Example: Sunflower meal is obtained from dehulled sunflower seeds by grinding the seeds and extracting with hexane to remove oil. The hexane-washed meal is freed from solvent by allowing the solvent to evaporate from thin layers of the meal. A 10 gram sample of this meal is then added to 150 ml of water in a beaker. The weight ratio of water to meal is 15:1. The pH of the meal/water mixture is then adjusted to 4.5 by the addition of hydrochloric acid. This mixture is stirred 3 minutes in a household blender and then an additional 15 minutes with a mechanical stirrer whereby green color precursors are removed from the meal. The meal and liquid washing solution are separated by centrifugation, and the supernatant which contains removed green color-forming precursors is discarded.

The meal is washed five more times in exactly the same fashion. The resulting meal weighs 5.3 grams when dry. This meal is then subjected to a conventional alkali extraction/acid precipitation process. In this process the meal is first mixed with water, the weight ratio of water to meal being 15:1. To this mixture is added sodium hydroxide to adjust its pH to 10.5, whereby the meal is alkali-extracted to provide a liquid extract phase containing dissolved protein and a solid spent meal phase. These phases are separated by centrifugation.

The extract phase then has its pH adjusted to 4.5 by the addition of hydrochloric acid to precipitate the protein. The precipitated protein is then separated from the pH 4.5-adjusted extract phase by centrifugation. The precipitated protein is then slurried with 25 ml water, and the pH of this mixture adjusted to pH 7 by the addition of sodium hydroxide. The sample is then freeze dried to yield 2.36 grams of protein isolate. This protein isolate has a grayish white color. It does not become intensely green in color in a basic medium.

Purification by Membrane Ultrafiltration

D.E. O'Connor; U.S. Patent 3,622,556; November 23, 1971; assigned to The Procter & Gamble Company has also developed an improved process for obtaining light colored protein. The light colored protein is recovered from sunflower meal by alkali extracting the meal under an inert gas to form a solid spent meal and a liquid extract, separating the phases, and acid precipitating protein from the extract. In a preferred embodiment, the extract directly after being separated from the spent meal is subjected to membrane ultrafiltration to remove green color-forming precursors and protein is recovered from the retentate. The isolated protein is suitable for use as an additive for foods without imparting an unappetizing color.

In the first method of this process, the sunflower meal is treated in an alkali extraction step with conventional mixing of meal and water, water to meal weight ratios, basic pH adjustment, and agitation conditions. The sunflower meal can be mixed with water already under an inert gas blanket, followed by basic pH adjustment. Or, the water and meal can be mixed, then this combination blanketed with inert gas, and the pH then adjusted to the required basic level. What is required is that at all times when protein extract is at basic pH, this extract must be under an inert gas blanket. If inert gas blanketing is not used, the intense green color which cannot be removed by any ordinary purification method will appear.

Suitable inert gases for this method include, for example, nitrogen, argon, helium, fluorochloroalkanes having boiling points of less than 50°F and water vapor. As a result of this alkali extraction step using inert gas blanketing, a colorless liquid extract containing extracted protein is formed. Light-colored protein can then be recovered from this extract. In one method, the pH of this inert gas-blanketed extract is adjusted to acid precipitation pH ranges whereby protein is precipitated as a solid. As soon as the pH of the extract becomes acid, the inert gas blanketing can be removed without harmful color effect.

The isolated protein can be used in this state or it can have its pH adjusted as far as pH 7 by the addition of a base such as sodium hydroxide or sodium carbonate. The pH of the isolated protein should not be adjusted to the basic side, as the protein will take on an intensely green color. It is preferred that the isolated protein be dried by freeze drying or spray drying before it is used as a food additive. The isolated protein is characterized by a light color, that is, a white, gray, light brown, tan, light green-gray, or very light green color.

In the preferred embodiment of this process, where inert gas blanketing and ultrafiltration are used, sunflower meal is first alkali extracted under inert gas blanketing just as described above. The separated extract, still under inert gas blanketing is subjected to a membrane ultrafiltration step where green color-forming precursors are removed from the extract.

Membrane ultrafiltration is described in *Chemical Engineering Progress,* 64, No. 12, pages 31-43, December, 1968. In membrane ultrafiltration, hydraulic pressure activates separation of solutions into individual components by passage through a semipermeable membrane. It differs from reverse osmosis because that term is applied to the separation of low-molecular weight solutes rather than high-molecular weight solutes. The components which pass through the membrane are denoted the ultrafiltrate. The portion retained by the membrane is a concentrated solution and is denoted the retentate.

In the application of ultrafiltration to the removal of color-forming bodies from the protein-containing extract, the extract is passed over an ultrafiltration membrane under a pressure from 10 to 100 psig, preferably 15 to 40 psig. This pressure can be maintained by the inert gas blanket present above the extract or by pumping or other suitable means. The membrane has an average pore size radius ranging from 10 to 80 A, preferably from

15 to 70 A. This pore size allows molecules having a molecular weight of 100,000 to 10,000 and less to pass through while not permitting the passing of higher molecular weight molecules such as proteins. For efficiency it is preferred that the extract be passed over the membrane in a thin layer having a thickness of 0.01 to 0.05 inch. The ultrafiltration is conveniently carried out with the extract having a temperature from 35° to 120°F. The use of temperatures below 50°F is useful to retard or prevent bacterial growth in the extract.

Membranes suitable for the ultrafiltration process are commercially available. Any of the commercial membranes which have a pore size stated above and which are suitable for separating protein size molecules from smaller molecules are suitable. A number of such membranes are described in the previously referred to *Chemical Engineering Progress* article on page 32.

The retentate from the ultrafiltration step can then be subjected to acid precipitation to remove the protein. In other words, the pH of the retentate is adjusted to 4.5 by the addition of an acid. The protein is then separated from the liquid phase by any physical separation process, for example, centrifugation. The recovered protein can have its pH adjusted to be more or less neutral by the addition of a base and preferably dried by spray drying or freeze drying.

Protein isolate can also be recovered from the retentate by further ultrafiltration using the same equipment as was used to remove the green color-forming bodies. This further ultrafiltration concentrates the retentate by separating some of the water from the larger molecule protein. The remaining water can be removed by any conventional drying process, for example, freeze drying or spray drying to provide dried protein isolate.

Example 1: 300 ml of water in a 600 ml beaker was purged with nitrogen for 5 minutes. Then with the nitrogen blanket maintained, 20 grams of sunflower meal is added to the water. The weight ratio of water to meal is 15:1. The pH of the water is then adjusted to 10.0 by the addition of sodium hydroxide and the water meal mixture mixed for one hour by the use of a magnetic stirrer. The nitrogen blanket is maintained during the pH adjustment and mixing. The mixture is then centrifuged while the nitrogen blanket is still maintained, and the extract, that is, the liquid phase, is separated and recovered. To the extract, still under a nitrogen blanket is added hydrochloric acid to adjust the pH to 4.5. The nitrogen blanketing is now discontinued.

The precipitated protein is separated from the extract by centrifugation and dried in a freeze dryer. A yield of 4.0 grams is achieved. The color of the produced protein isolate is very light green. The protein isolate is incorporated in a breakfast cereal at a level of 25% by weight to supplement the protein therein. This food product has a slightly acid pH. The food product containing the protein isolate has an appetizing appearance without the intense green color normally associated with protein isolated by the alkali extraction/acid precipitation methods.

Example 2: 20 grams of sunflower meal is alkali-extracted under nitrogen blanket and then centrifuged under a nitrogen blanket just as in Example 1. The extract separated by centrifugation still under a nitrogen blanket is then subjected to ultrafiltration using a Diaflow XM-100 porous membrane estimated to have an average pore size radius of 60 A. The extract is placed in a nitrogen blanketed reservoir, then passed as a thin layer (0.02 inch thick) over the porous membrane with 16 psig pressure applied by a head of nitrogen used as an inert gas blanket.

Green color-forming precursors and some water pass through the membrane and are recovered as the ultrafiltrate. The protein in the extract and some of the water is retained above the membrane as the retentate. The retentate is recirculated to the reservoir containing the extract phase. The water lost through the membrane into the ultrafiltrate is continuously replaced by addition of water into the extract phase in the reservoir. This procedure is carried out until water having a volume of four times the original volume of the extract has been added to the reservoir. At all times during the procedure the extract and the retentate is kept under a nitrogen blanket. At the end of the procedure, the retentate in the reservoir contains 3 grams of dissolved solids. Nitrogen blanketing is now discontinued. The retentate then has its pH adjusted to 4.5 by the addition of hydrochloric acid. The precipitated protein is separated by centrifugation, has its pH adjusted to 7 by the addition of sodium hydroxide, and then freeze dried to provide a yield of 3 grams of protein isolate.

This protein isolate is light tan colored. It retains its light color in acid, neutral or basic medium; in other words, there is no intense green color appearance in basic medium. This protein isolate utilized as the protein in an imitation milk product retains its appetizing appearance, that is, it is not characterized by a green cast or a green color. When a Diaflow PM-10 membrane is utilized instead of a Diaflow XM-100 membrane, 5.5 grams of light tan protein is recovered instead of 3 grams. This protein isolate retains its light color in acid, neutral or basic medium.

SAFFLOWER SEEDS

Safflower seeds are grown primarily to extract their oil content of about 36%, for safflower oil is a valuable commodity with many desirable properties. After the oil is taken out, the residue, about 64% of the seed, is an intimate mixture of fractured hull particles and what is left of the meats. The hulls are very low in protein and are quite fibrous. The meat residue is very high in protein, being up to about 56% protein, and is an excellent feed material.

Both materials are useful in themselves and so is the mixture, but the mixture is a relatively low-grade feed of about 16 to 30% (typically 18 to 24%) protein content that brings only moderate prices and can be sold only in limited amounts. The problem is that it has been quite difficult and uneconomical to achieve sufficient discrimination between the hulls and the meats to give economic quantities of the high-protein meal. Hand-picking, classification by specific gravity and screening have not been satisfactory. Safflower products are used for production of protein fibers in U.S. Patent 3,175,909 (page 232).

Food Products from Oil-Free Safflower Seed Residue

A method has been developed by *G.A. Kopas and J.A. Kneeland; U.S. Patent 3,271,160; September 6, 1966; assigned to Pacific Vegetable Oil Corporation* to upgrade the residue left after oil removal from safflower seeds. The process provides for the production from this residue of at least three oil-free fractions: (1) a high-protein meal for human food and for poultry and swine feeds, (2) a medium-protein meal for cattle roughage feed, and (3) a low-protein fiber chaff for use as an adhesive extender and as an inert filler.

After the oil has been solvent-extracted from the seed, the solvent (e.g., hexane) may be removed from the residue by direct or indirect steam stripping or by superheated solvent stripping. If the remaining, substantially oil-free, mixture of meat and hull particles is too hot and moist (as it often is), it is then cooled and dried to a desired moisture content. For example, it may be cooled by dry air, to below 200°F, or preferably below 130°F, while being dried to a moisture content below 15%. This is the residue with which this process deals. Simple screening at this stage will accomplish very slight discrimination between the high and low protein materials.

This process includes the discovery that if this relatively dry residue is subjected to impact, either pneumatically by shooting it in an air stream through an air gun at a target, or mechanically, as in a vertical or horizontal impact mill (a hammer mill, entoliter, or attrition mill) or in a flour mill or similar device in which impact is the major element of force, this impact will both (1) detach most of the meat from the hulls and (2) fracture the meats into particles smaller than fibrous hulls and the meat particles are detached from the hulls. An important result of this discovery is that a classification by size alone can be used to divide the residue into three or more fractions of markedly different protein content. This makes classification into fractions of different protein content economical and practical. It is important for controlled impact to be the major element of the process, so that there is a breaking action due to impact but not a cutting action, for cutting would divide the hull particles as much as the meat particles. Similarly, over-impacting eventually breaks up the hull particles into small sizes that cannot be classified from the meat particles by sizing.

Furthermore, it was found that when the larger, low-protein fraction obtained from the classification of the once-impacted residue is reimpacted and again separated by size, a medium-to-high protein content meal fraction is obtained along with a lower-protein hull fraction. This enables a considerable saving and enables the process to be carried out quite economically, practically, and consistently.

Example: Dried residue of 21.6% protein content from prepressed and solvent-extracted safflower seeds was broken up by impact in a modified vertical hammermill. An air stream removed (1) a stream of fine particles 17% by weight of the residue and of 45.8% protein content, broken off the hulls and carried therethrough a $3/64$"-perforation screen. The coarse fraction, discharged through an opening in the bottom of the mill, was (2) a stream, 83% by weight of the residue and of 16.7% protein content.

The second stream (2) was subjected to a second mechanical force in an attrition mill, which broke away more protein from the hulls and also broke the meat residue more than it broke the hulls, for the ground secondary stream was separated on a shaking screen into (3) a stream, 32% by weight of the residue, of 30% protein content and (4) a stream, 51% by weight of the residue, of the larger, coarse hull particles, having a protein content of only 8.2%.

Considering the total effect of the two impacts, 49% of the meal was recovered in two very valuable fractions, 17% in a very-high protein prime-value product and 32% of a high-class middling fraction of far more value than the unclassified residue. Also, the 51% hull fraction had a protein content far lower than obtained from screening and more useful as an adhesive extender or filler than the original product.

Wet Processing of Safflower Seeds

A process which integrates the recovery of oil from safflower seeds with the production of high protein fractions is disclosed by *A.E. Goodban and G.O. Kohler; U.S. Patent 3,542,559; November 24, 1970; assigned to U.S. Secretary of Agriculture.*

One phase of the process concerns the application of wet milling to safflower seed materials. Another phase concerns the application of wet separation to such materials. At this point it is apropos to note that the term wet used here has no reference to water, but refers to the presence of an organic solvent, typically, hexane. Each of the steps in question provides important advantages over conventional (dry) operations, and it is within the process to use either one of these steps or both.

A particular advantage of the wet-milling step is that it exerts a differential size reduction, the kernel material (which is high in protein content) is reduced to smaller particle size than the hull material (largely fiber). This differential size reduction facilitates separation of the two fractions, that is, the proteinous material and the fibrous material. A particular advantage of the wet-separation technique is that the presence of hexane or other solvent in contact with the proteinous and fibrous fractions creates a difference in apparent density between these fractions so that one can be readily isolated from the other.

Moreover, by conducting either or both steps in the wet state, one can integrate oil recovery from the safflower material with recovery of high-protein fractions. The processes may be applied to safflower seeds themselves or to any of the materials derived therefrom in oil-removal systems. For example, the process may be applied to press cake, meal, or other residues from oil expressing and/or extracting procedures. Moreover, the process may be applied to meals which have been upgraded by conventional procedures. Since hexane is effective, readily available in oil-extraction plants, and relatively inexpensive, it is generally preferred as the solvent in both the wet-milling and wet-separation steps. However, it is by no means the only solvent which can be used. One may use any liquid hydrocarbon, hydrocarbon mixture, chlorinated hydrocarbon, fluorinated hydrocarbon, or other volatile fat solvent which has a density less than 1.2, preferably less than 1. Also suitable are oxygenated solvents such as methanol, ethanol, isopropanol, acetone, and the like. The use of these (oxygenated) solvents also aids in debittering the protein.

The following are descriptions of the two steps: wet milling and wet separation. However, other combinations such as wet milling with conventional dry separation and conventional dry milling with wet separation may also be covered by this process.

(A) Wet Milling — Safflower seeds, press cake, or meal, is milled in the presence of hexane or other solvent, using conventional comminuting apparatus such as a hammermill, press rolls, or the like. The milling is done in the presence of an amount of solvent sufficient to at least wet the safflower material. In general, at least 0.5 part of solvent is used per part of safflower material, and, where the next step (separation) is to be done in the wet state, an excess of the solvent may be used to provide a milled mixture which is directly ready for the next operation. The use of an excess of solvent in the milling step is also desirable in order to extract and recover oil from the starting material.

(B) Wet Separation — An example of this step involves a flotation or density fractionation to segregate the protein-rich and fiber-rich fractions contained in the mixture from the wet milling. The mixture at this point should contain an excess of the solvent, enough, to provide at least 2 parts of solvent per part of safflower material. A larger proportion of solvent can be used, and may be desirable where the overall purpose includes extraction and recovery of oil from the starting material.

In cases where the wet milling was carried out using a smaller proportion of hexane, more solvent would be added at the start of the separation step, whereas if the milling were carried out with an excess of hexane, the milled mixture would be ready for the separation procedure. In any event, the milled mixture is allowed to stand for a brief period, for example, 10 to 60 seconds, whereby the dense hull material will quickly settle out. The protein material, being less dense and in fine particle size, remains suspended in the solvent.

Accordingly, the solvent phase is decanted or otherwise removed, carrying with it the protein material. By centrifugation or filtration of the liquid, this protein material is isolated. It is then dried to remove residual solvent and is ready for use. The fractionation may be reapplied to the hull material to separate protein material which was not removed the first time.

In typical applications of the wet milling-wet separation procedure, products containing 50 to 65% protein and 2 to 8% fiber are obtained. Because of this high protein/fiber ratio the products are highly suitable for feeding poultry and other nonruminants. Indeed, the products, by application of a purification technique described below, can be useful for human use, e.g., for supplementing rice, wheat, and other low-protein foods.

Example: 70 grams of whole safflower seed and 500 ml of hexane were placed in an Osterizer. The blade was operated for about 5 minutes, then the liquid was decanted and filtered. Removal of hexane from the filtered liquid yielded 26.6 grams of oil. The material collected in the filter was removed and dried (Fraction a). The residue in the bowl was dried (Fraction b). The data obtained is tabulated below.

Fraction	% Yield	% Protein Content	% Fiber Content
a	17.3	52.0	7.6
b	42.0	8.0	54.3

SESAME SEEDS

Sesame seed (*Sesamum indicum* L.) is a member of the oil seed family and the exact analysis of an unhulled sample of the K-10 variety is as follows: 4.70% moisture, 23.85% protein, 53.10% oil, 4.52% fiber, 4.52% ash, 9.31% nitrogen free extract, and 13.83% carbohydrates. It is a high fat and a high protein seed which is very useful for nutritive purposes. The seed may be supplied in its natural form, partially dehulled and completely dehulled. It may be moisture free or the moisture content may be higher, for example, up to 8% at which seed can safely be stored. Wet seed from the dehulling operation or partial dehulling operation can be used directly in the mix containing up to 30% moisture content.

Sesame Seed-Flour Food Product

L. O'Neal; U.S. Patent 2,990,285; June 27, 1961 has developed an improved mix or blend of ingredients including sesame seed which can be readily extruded to form a composition which will retain its physical shape in a hot frying liquid. This is accomplished by first preparing a substantially dry mix containing the following ingredients all in percent by weight: 10 to 60% sesame seed, 10 to 60% cracker flour, 10 to 60% cereal flour, 3 to 6% salt, and 0 to 5% shortening.

In the above composition the dehulled sesame seed is prepared by conventional methods where the hull is removed completely or partially, and the seed is washed and dried. The seed can also be toasted prior to incorporating it into the mix. Natural seed when used should be clean and free of extraneous matter. The cracker flour is a composition which is also sometimes referred to as cracker meal or cracker crumbs made by baking a cracker-forming composition and grinding the resultant product.

The cereal flour is one containing a substantial proportion of gluten and wheat flour is preferred. The amount of cereal flour used should be sufficient so that the gluten thereon enhances the cohesiveness of the product for extrusion. Water should be referred to as being used to form a dough necessary for extrusion.

To prepare the dough for extruding the following specific formula is used containing: 30% sesame seed, 33% cracker flour, 33% flour, and 4% salt (NaCl). With each 100 pounds of this mixture add 53 pounds of water together in a dough mixer of common usage in normal bakery operation, using an upright mixer or horizontal self dumping mixer. Blend until the moisture is incorporated into the dough, usually 2 to 3 minutes is all that is necessary. Remove the dough and put into an extruder which utilizes hydraulic pressure against a ram within a tube or barrel to push the dough through dies that form the pieces in a predetermined shape.

To push the dough through these slots, pressure of approximately 1,000 pounds per square inch is exerted on the hydraulic ram. The thickness of the slot in this specific instance is $^{40}/_{1,000}$ inch, $^{5}/_{8}$ inch long and when extruded the pieces of dough can be cut to any length desired, usually 1" to 1½" in length. The extruded pieces may be deep fat fried to form a snack type food.

Sesame-Whey Combinations

J.H. Kraft; U.S. Patent 3,669,678; June 13, 1972; assigned to John Kraft Sesame Corporation has found that a useful food composition can be prepared by mixing comminuted sesame and whey. The composition can be used as a liquid, or a semisolid, with or without added flavoring ingredients, but is preferably converted to a powder, for example, by spray or drum drying. The term comminuted sesame as used here includes ground dehulled whole sesame seed, sesame seed cake containing 10 to 20% oil and extracted sesame seed cake containing less than 10% sesame oil, usually around 0.5% sesame oil (i.e., the cake which remains after the sesame oil has been solvent extracted). Mixtures of any two or more types of comminuted sesame are also included. The term whey as used here includes ordinary whey; delactosed whey (which is higher in protein); demineralized, delactosed whey; and demineralized whey as well as mixtures of two or more of the forms of whey. When a dried composition is prepared, a liquid comminuted sesame-whey mixture is first heated. The heating is believed to produce some interaction between the components of the ingredients. A preferred method is to prewarm the mixture at 145° to 205°F for 30 minutes or at a higher temperature for a shorter period of time. The mixture is then evaporated under 18 to 27 inches of mercury at 105° to 165°F to a concentration of 30 to 55% by weight solids and the concentrate spray dried. The ingredients could also be dry blended. However, it is more difficult to produce intimate association and less interaction of the components is likely.

The dried compositions can be used as such or they can be blended with other ingredients, e.g., flavoring substances and/or taste intensifiers, and/or agents to increase water wettability, and/or suspending agents, e.g., salt, vanilla, sugar, cocoa, lemon, lime, orange, sodium glutamate, lecithin and/or carrageenin.

The weight ratio of comminuted sesame to whey solids is usually within the range of 2:1 to 1:2. Water can be present in various amounts depending upon the desired consistency of the product. Thus, a liquid product might contain 80 to 99% by weight water, a semisolid product might contain 40 to 60% by weight water and a powdered product would contain less than 15% by weight water.

Usually, a dried or powdered product will contain some moisture because it is not practical to remove all of the water and even when a product is thoroughly dried some atmospheric absorption of moisture will occur unless the product is immediately packaged in containers that are sealed from the atmosphere.

In general, a drink mix will contain for each part of sesame 0.5 to 2 parts by weight of whey solids, plus the flavoring ingredients (usually 1 to 50% of the total solids). Lecithin or other edible water wetting agent and carrageenin or other edible suspending agent will each constitute 0.5 to 1% by weight of the total solids.

In addition to the uses previously mentioned, the compositions can be used in making salad dressings, pudding, candy, cakes, ice cream, and other bakery products and confections. They can be used as such in liquid, semsolid or solid state or they can be mixed with other substances of the type previously decribed. For some purposes, for instance, bread making, it may be desirable to add sodium diacetate, calcium propionate or other edible mold and rope inhibitors in mold and rope inhibiting amounts.

Example 1: Eight parts by weight of ground dehulled sesame seed are mixed with 100 parts of whey having a pH of about 6.0 and containing approximately 6% by weight whey solids and 94% water. The mixture is heated to 185°F for 30 minutes and evaporated to 35% total solids in a double effect evaporator at 27 inches of mercury vacuum in the second effect. The resultant mixture is then spray dried in an apparatus of the type used for making powdered milk by spray drying. A white powder is obtained which resembles powdered milk. The weight ratio of sesame to whey solids is approximately 4:3. The taste of the product is excellent.

Example 2: Eight parts by weight of ground dehulled sesame seed are mixed with 100 parts by weight of whey having a pH of about 4.6 and containing approximately 6% by weight whey solids and 94% water. The mixture is heated to 350°F (177°C) under pressure with removal of water until a mixture suitable for spray drying is obtained. The resultant mixture is then spray dried in an apparatus of the type used for making powdered milk by spray drying. A white powder is obtained which resembled powdered milk. The weight ratio of sesame to whey solids is approximately 4:3. The taste of the product is excellent.

Separation of Oil and Protein from Sesame Seeds

In a process disclosed by *J.J. Liggett; U.S. Patent 3,476,739; November 4, 1969* oil-containing seeds, especially sesame, are treated in an aqueous system containing saturated calcium hydroxide to recover both protein and oil. Generally, in commercial processes for recovering protein and oil from oil-containing seeds, the oil is first extracted with an organic solvent or the seeds are treated in some other manner to remove the oil before any attempt is made to recover the protein solids. This is done because the protein solids are normally recovered in the presence of water and the presence of the oil during the processing of the protein solids with water complicates the recovery of the protein solids. One of the difficulties is the formation of an emulsion between the oil and the water used in processing the protein solids.

A schematic diagram of this combined process is given in Figure 10.1. As shown there the process is carried out by mixing dehulled, cracked or crushed oil-containing seeds **1** with water containing dissolved calcium hydroxide in a contactor tank **4** to produce a mixture **5** containing oil, water, dissolved protein and fiber. The fiber **7** is separated from this mixture and the oil and protein are then separated from the residual mixture **8**. The separation of the fiber can be done by passing the initial mixture **5** through a suitable extractor **6**, e.g., a centrifuge or expeller. The fiber content of the raw product will depend upon the seed being treated, in the case of sesame seed being approximately 4 to 8%.

FIGURE 10.1: SEPARATION OF OIL AND PROTEIN FROM SESAME SEEDS

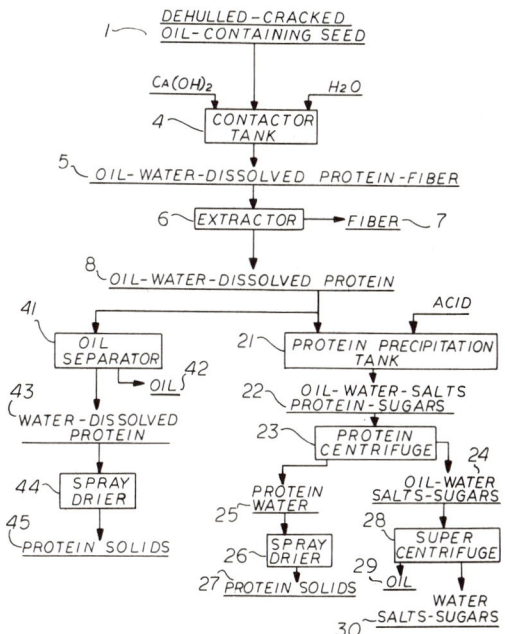

Source: J.J. Liggett; U.S. Patent 3,476,739; November 4, 1969

The separation of the oil and protein can be carried out in one of two ways. In the first method, the mixture **8** from which the fiber has been separated is passed into a protein precipitation tank **21**. At this point, due to the presence of the calcium hydroxide, the pH of the mixture is around 10 or 11. In order to separate the protein, an acid is added to lower the pH to the isoelectric point of the protein which is usually around pH 4 to 5. At the isoelectric point the protein precipitates and the resultant solution **22** contains oil, water, salts, protein and sugars. The precipitate can be separated by filtration or by passing it through a centrifuge **23**. The protein-water mixture **25** can be converted to a dry solid, preferably by spray drying in a conventional spray drier **26**. The resultant product comprising protein solids **27** is useful in food and for many other purposes.

After the removal of the protein precipitate the residue **24** contains oil, water, salts and sugars. The oil can be separated from this residue, for example, by passing the residue through a centrifuge **28**. In this way, a highly purified oil **29** is obtained. The aqueous liquid **30** from the centrifuge can be discarded or used in other processing or subjected to drying, for example, spray drying, to recover the salts and sugars.

In a second type of process, the liquid mixture **8** containing oil, water and dissolved protein after removal of the fiber is passed to an oil separator **41** where the oil **42** is removed. The residual liquid mixture **43** consists of water and dissolved protein. The dissolved protein in this case is calcium proteinate which can be recovered as protein solids **45** by drying in a spray drier **44** or in any other suitable manner. It can be used as calcium proteinate in foods and in industrial applications.

Example: 250 parts by weight of crushed wet sesame seed, based on the dry weight of the seed, are mixed thoroughly in a stainless steel container equipped with a sweep-type agitator for 5 to 30 minutes with 740 parts of water containing 10 parts calcium hydroxide USP grade [a saturated $Ca(OH)_2$ solution], the water being heated to 180°F. The mixture was then pumped through a centrifuge with only slight pressure. The liquid phase of the mixture from the centrifuge was then passed to a separator tank where it was allowed to settle and the oil was separated by decantation. The aqueous phase containing water and calcium proteinate (basic salt) was spray dried to recover the protein solids as calcium proteinate.

CASTOR BEANS

The high content of protein and the presence of an extremely toxic substance, ricin, a specific form of protein, stimulated at an early date unusually great interest in the investigation and possible uses of the castor bean proteins. The seeds are composed of about 25% husk and 75% kernel. The castor bean in the past years, has been raised almost exclusively for its oil which is extensively used for medicinal and industrial and lubricant purposes and which in later years, in dehydrogenated state, has been found valuable for use in paints.

Ricin comprises about 1.5% of the oil-free meal. Ricin has been found to be an albumin and to constitute a part of the natural proteins present in the kernel of the castor bean. Like all alubumins, pure ricin remains soluble in water and remains in solution after all of the globulin has been separated from a saline extract of the meal. From past scientific work, the fact seems now undisputed that ricin as other toxalbumins differ in some pronounced manner from other proteins and that their toxicity is a peculiar property of the protein itself.

The toxic effects of ricin produces after several hours, a paralysis of the respiratory and vasomotor systems. Consumption by mouth causes diarrhea, general prostration and hemorrhagic condition of the intestines, renal congestion and hyperemia of the spinal medulla and brain. It agglutinates the red blood corpuscles, this agglutination being more pronounced in mammals than in fowl.

Autodigestion to Remove Ricin from Castor Beans

The process of *E. Darzins; U.S. Patent 2,920,963; January 12, 1960* relates to the utilization of the proteins of the castor bean and particularly to the production of a reconstituted castor cake product, high in nutritious proteins and amino acids and free of all toxic substance and particularly adapted as a feed for animal production and for high protein, industrial use. This process gives a cake which when pulverized or granulated, is substantially white (when dry) and has very desirable taste to domestic mammals and fowl. Its protein content may exceed 34% and includes a relatively very high amino acid content. The natural castor beans are preferably first peeled or skinned in any of the accepted commercial methods which removes large quantities of carotenoid pigments and some of the superficial layer of kernel containing poisonous ricin cells. After the peeling, the protein content of the beans will vary from 35 to 40% determined in extracted meal, after the oil is substantially completely extracted.

The castor bean meal obtained either from peeled or unpeeled beans is comminuted and dried in any conventional manner. The dried meal is preferably comminuted again to a second fine milling forming a flour. The oil may be directly extracted by hydraulic press or the solvent method of extraction may be utilized, imposing as the chemical solvent sulfuric or petrol ether, light benzine or commercial hexane in the approximate proportion of one part castor cake flour to two parts of the extract. The extraction is carried out in the requisite time (usually about 3 hours) and the solvent is then drained and the resulting flour dried.

In this process, the dry flour is then mixed with water in the proportion of one to three by volume although this proportion may be varied considerably with successful results. The mixture is stirred and thoroughly agitated, remaining for about 3 hours, and the water then drained off. This draining removes more soluble carotenoid pigments, considerable portion of water-soluble allergens and a part of the water-soluble, poisonous albumin, but most of the ricin albumins and globulins remain in the flour or cake. In the next step, the wet flour is mixed for the second time with water, preferably in the proportion of substantially one to three by volume.

A strong proteolytic bacteria culture is preferably added at this time to the mixture, although it may be added subsequently during part of the procedure of fermentation or autodigestion. Various proteolytic bacteria may be used such as Pseudomonas or other bacteria such as those obtained from decomposed organic substances, including those isolated from sewage (e.g., Clostridium). It is important that the culture selected is strong, virile and has strong proteolytic activity. From the culture selected, a bacterial broth is made in conventional manner. The culture is added to the aqueous mix in the proportion of 20 ml of the culture to form 4 or 5 kg of the meal actually utilized in the wet mix.

The wet mix with the strong proteolytic bacteria culture preferably added at this time, is then submitted to fermentation and auto-digestion caused by such bacteria and by the natural enzymes of beans, for preferably a period of 72 hours, at a temperature varying from 20° to 45°C, (preferably between 25° and 35°C). In the fermentation, the enzymes are liberated from the cells and the culture, the cells are split and the proteins become soluble in water with the result that most, but not all of the toxic ricin is destroyed or converted in auto-digestion.

To ferment and produce proper auto-digestion and conversion of all toxic ricin, the substantially liquid mass of castor bean flour and water must be maintained throughout the auto-digestion at a pH from 7.5 to 9. To this end, the step of auto-digestion includes the addition of an amount of alkali of the earth metals (preferably calcium hydroxide) in the proportion of one half to one percent of the mass by weight. Salts of sodium, magnesium and strontium or potassium may be substituted for the calcium hydroxide. Sodium or potassium hydroxides, however, subsequently produce small quantities of water-soluble soaps. It is an advantage, but not essential, to add from one half to one percent by weight of sodium chloride to the wet mixture. The addition of this chloride facilitates the solubility of the auto-digestion of protein.

After auto-digestion and/or fermentation for approximately 76 hours, the digested mass is transferred into an autoclave and heated for 1 hour at a temperature from 100° to 120°C (at a pressure of one atmosphere gauge). In this step, the minute quantities of castor oil which may remain in the flour and which may be liberated through auto-digestion are converted into insoluble, nontoxic lime soap. In the auto-digestion at specified pH, and subsequent heating under pressure, the poisonous ricin albumins and other proteins coagulate and convert into proteins which have no toxicity. The last mentioned step in the heating under pressure (autoclave) at a temperature between 100° and 120°C, converts and detoxicates the slightest remaining, poisonous albumin. After autoclaving the digested mass the water is drained off and the mass is dried to leave a granulated or pulverized material containing less than 6% moisture.

The product is white, having a grayish tinge depending on the amount of husk present in the meal and this product by tests on red blood cells and upon animals, is found to have no toxicity and no ricin present. The products may be obtained as a cake or the mass may be milled or comminuted to granular or pulverized state. The dried product almost always contains up to 32% nontoxic, half-digested and boiled proteins, the proteins decreasing in proportion according to the amount of husk present in the original meals. This product has been proved to have very high feeding value with desirable taste qualities for fowl and domestic animals.

Deallergenizing Castor Beans with Ammonium Hydroxide

In addition to the toxins in castor bean pomace, there are also allergens which cause sensitization from handling the materials. In general, the toxic materials offer less of a problem as they are readily destroyed by heating the castor pomace. However, the allergens are a particular problem because they are heat-stable.

According to the process developed by *L.L. Layton and F.C. Greene; U.S. Patent 3,294,776; December 27, 1966; assigned to U.S. Secretary of Agriculture* deallergenization is accomplished by mxing the pomace with aqueous ammonium hydroxide and heating the mixture under pressure. The ammonium hydroxide is applied as such when the treatment is initiated but it is evident that ammonia will be formed during the heating and thus the treatment can be considered as a heating of the pomace in the presence of water, ammonium hydroxide, and ammonia. The temperature of the treatment may be varied from 100° to 150°C and, as in other chemical reactions, the destruction of the allergenic principles occurs more rapidly at the higher temperatures in the range.

To prevent loss of water and ammonia, and to maintain the water largely in the liquid phase, the treatment is conducted in a pressure-tight vessel such as an autoclave. The pressure generated will depend on the temperature selected and may be referred to as an autogenous pressure, that is, one created by the conditions applied. A critical factor in the treatment is that there be present during the entire course of the process an adequate amount of water, namely, at least one part of water per part of pomace.

It has been demonstrated that ammonia treatments applied to pomace containing relatively low proportions of water, i.e., up to 50%, yield ineffective results in that the allergens are not inactivated or only inactivated to a minor extent. Where, however, a high proportion of liquid water is present as mentioned, the deallergenization is essentially complete.

The amount of ammonium hydroxide used may be varied with the proviso that an amount is added so that the water in the system has a molar concentration of NH_4OH of at least one. Faster deallergenization is attained with higher concentrations and in general one can economically use a molar concentration as high as four.

Following the treatment as described above, the product may be treated to recover the ammonia. This may be accomplished by heating the product in a conventional evaporation apparatus equipped with a condenser or absorption unit for collecting the ammonia from the evolved vapors. The product may then be dried in the same apparatus or in a conventional dehydrator with a current of hot air or it may be dried under vacuum.

Although the deallergenization procedure of this process is usually applied to castor pomace, it can be applied to other castor bean materials, for example, castor bean flour, nondefatted castor materials, etc. In the following example, the products were tested for allergenicity by the passive cutaneous anaphylaxis (PCA) test.

Example: 5 grams of castor pomace and 10 ml of 1.0 molar aqueous ammonium hydroxide solution were heated in an autoclave at 120°C (15 psig) for 1 hour. The product was then removed from the autoclave and subjected to vacuum to remove ammonia and water. Two other runs were carried out as described above but using different concentrations of ammonium hydroxide, namely, 0.75 molar and 0.5 molar. The three products were tested by the procedure described above. The results are tabulated below.

Run	Concentration of NH_4OH, molar	H_2O in reaction mixture,* percent	P.C.A. test
1	1.0	65.4	Negative.
2	0.75	65.7	Positive.
3	0.50	66.0	Do.

*Exclusive of moisture content of pomace.

CRYSTALLIZATION OF BEAN PROTEIN FROM DILUTE SALT SOLUTIONS

A process of isolating protein from beans using dilute aqueous salt solutions of carboxylic acids has been disclosed by *P. Melnychyn; U.S. Patent 3,450,688; June 17, 1969.* The protein is obtained in a crystalline form and in high yields directly from lima beans, pinto beans, kidney beans and white beans.

The process comprises the steps of mixing ground beans with a dilute aqueous solution of a salt of a mono-, di- or tricarboxylic acid having a salt normality between 0.1 and 0.8 and an acid pH, separating undissolved material from the resulting solution, allowing the solution to stand while keeping the solution refrigerated until the protein crystallizes out of the solution, and then separating the crystalline protein from the solution.

A primary advantage of this process is that the protein is not isolated from solution by any precipitation techniques which may occlude materials from the solution but is isolated directly by crystallization. In direct crystallization, little if any occlusion takes place so that the protein isolated is highly refined and may be used for many applications. The aftertreatments required with the usual precipitation procedures are thus eliminated or minimized.

Another factor which contributes to the purity of the products is that they are crystallized from solutions containing very low concentrations of water-soluble salts (specifically salts of monocarboxylic, dicarboxylic and tricarboxylic acids). This means that the opportunity for contamination of the product with salts is reduced simply because there is not much of the salts in the environment.

The protein products exhibit a crystalline structure and are white or essentially colorless. They are soluble in acid and in alkaline solutions. The products obtained from lima beans have a bipyramidal crystal habit. Proteins from other varieties of beans will also exhibit a bipyramidal crystal habit when crystallized from extracts of water-soluble salts of dicarboxylic and tricarboxylic acids whereas they may assume other forms if crystallized from water-soluble salts of monocarboxylic acids.

A sample of crystalline protein derived from lima beans was found to have a molecular weight of about 180,000 by light scattering measurements; about 100,000 to 120,000 by ultracentrifugal measurements. Analysis of the protein indicates the presence of a relatively high proportion of mannose, i.e., about 9 to 10% of the protein, by weight. In essence, the process involves the extraction of beans with a dilute acidified aqueous solution of a water-soluble salt of preferably unsubstituted or hydroxy substituted aliphatic straight or branched chain mono-, di- and tricarboxylic acid such as acetic, propionic, butyric, lactic, glycolic, glyoxylic, malonic, fumaric, succinic, glutaric, adipic, malic, tartaric, and citric.

It has been found that the water-soluble salts of the di- and tricarboxylic acids are preferred over the same salts of the monocarboxylic acids because lower concentrations of these polybasic acids can be employed to effect isolation and crystallization of protein. The resulting extract or solution is then allowed to stand, whereby the protein crystallizes out of solution.

In addition to the types of beans listed previously, the process may also be applied to varieties of *Glycine soja*, that is, soybeans, but in this case the products may not be crystalline. Ordinarily, the beans to which the process is applied are the usual dry beans available in commerce. The products are often referred to as dry shell beans because the product includes only the mature seeds, the pods having been removed. To enhance contact between the bean material and the extracting medium, the beans are ground in conventional apparatus to the form of a powder or flour. It is sometimes desirable to defat the beans prior to contact with the protein extracting medium. The defatting step can be conducted in known manner by extraction with suitable fat solvents.

As noted above, the extracting medium applied to the ground beans is a dilute aqueous solution of a water-soluble salt of the desired acid. The pH of the extracting medium should be on the acid side, i.e., 1.0 to 6.0. Usually excellent results are obtained with a pH of 3.5 to 5.0 and this degree of acidity is preferred. A convenient way to prepare the extracting medium involves diluting the acid with water to the approximate desired normality, adjusting the pH by addition of a suitable alkali, e.g., sodium hydroxide, and then adding more water to adjust the normality to the level desired for the extraction procedure.

In general, the concentration of the extracting medium may vary from 0.05 to 0.8 normal. A concentration within this range enables a high yield of crystalline protein to be obtained, and the concentration chosen depends upon the extracting medium and upon the type of bean to which the process is applied. For example, with lima beans, if one selects a solution containing a salt of acetic acid, it is preferable to use a 0.1 normal sodium acetate solution; with pinto beans or soybeans, 0.6 normal sodium acetate solution; with kidney beans, 0.6 to 0.8 normal sodium acetate solution; with large white beans, 0.4 to 0.8 normal sodium acetate solutions.

If one selects a solution containing a salt of a dicarboxylic acid, such as fumaric acid, or a salt of a tricarboxylic acid, such as citric acid, in the extraction of protein from kidney beans, the extraction and crystallization may be effected with a solution which is 0.05 to 0.1 normal. This illustrates the greater effect of the polybasic acids, as previously mentioned.

Example 1: A 0.1 normal acetic acid solution was adjusted to pH 4 by addition of sodium hydroxide. 100 ml of this solution was mixed with powdered defatted lima beans, using 10 ml of the acetate solution per gram of beans. The slurry was allowed to stand at room temperature for 30 minutes, then centrifuged. The supernatant solution (extract) was placed in a refrigerator (about 5°C) for 24 hours to enhance crystallization of protein. The protein crystals were separated from the remaining liquid. Nitrogen analyses indicated that of the total amount of protein in the beans 27% was recovered as the crystalline product.

Example 2: 20 grams of ground or crushed pinto beans (nondefatted) were extracted for 0.5 hour with 200 ml of 0.1 N sodium fumarate solution (pH 4.0) at elevated temperature (38°C). The mother liquor was removed and the protein was permitted to crystallize under refrigeration. The protein was harvested after 12 to 18 hours. The nitrogen precipitated in crystalline form (protein) corresponded to 21% of the total protein originally present in the bean.

Example 3: 200 ml of a 0.1 N solution of sodium tartrate (pH 4.0) was employed to extract protein from 20 grams of undefatted, ground lima beans in accordance with the procedure of Example 2. Insoluble residue was removed and protein crystallization was facilitated by refrigerating the clarified supernatant phase. The estimated yield of crystalline protein obtained was about 20% of the total originally present in the bean.

PEANUTS OR GROUNDNUTS

Most processes for peanuts deal with the preparation of peanut butter which is traditionally made by shelling peanuts, roasting them and after blanching, grinding the product to a pasty consistency. Since the contents average about 47% oil, 20% carbohydrate, and 26% protein, the peanut butter obtained is one of the most popular spreads.

Another product made from peanuts is peanut flour. This is distinguished from peanut butter by the lower oil content since in the preparation of flour, an oil extraction takes place for the purpose of making a free-flowing product by grinding the defatted nuts. Peanut flour has many uses as an addition to other foodstuffs, because though poorer in oil, it is high in protein and carbohydrate.

Other processes relating to processing of peanuts include: U.S. Patents 3,090,779, page 16 and 2,928,821, page 14 on hammermill release of cellular protein; U.S. Patent 3,809,767, page 95 on preparation of protein concentrates; U.S. Patent 2,830,902, page 259 on textured proteins; and, U.S. Patent 2,919,192, page 277 on preparation of cheese-like products from peanut protein.

Process for Making Peanut Flakes

This process by *J.H. Mitchell, Jr.; U.S. Patent 3,800,056; March 26, 1974* relates to a method for converting peanuts into precooked full fat flakes which are thermostable, bland flavored and pathogen free. The process includes the steps of steaming the unground peanuts, drying to a moisture level of 2 to 4%, removing the skins and hearts, grinding the peanuts to a fine consistency, cooking the finely divided peanuts with water, and then drum drying the product.

Although the flakes may contain 46 to 54% oil, depending upon the composition of the peanuts used, they do not present an oily appearance unless they are ground to a fine state. If this is done, free oil is released and the product assumes the characteristics of a paste. If materials, such as low-fat peanut flour, precooked potatoes, rice, or other substances of low oil content are incorporated with the cooked finely ground peanut-water mixture just prior to drum drying, the resultant flakes have a reduced oil content, depending upon the relative quantities of each component. Such flakes may be reduced to a flour instead of a paste by grinding.

This process involves precooking the raw ground peanuts with moist heat prior to drum drying. This precooking brings about a number of desirable changes relative to flavor in that it releases compounds having a somewhat unpleasant aroma. These compounds are removed during subsequent drum drying since they are volatile with the steam issuing from the thin layer of material coating the surfaces of the drums.

This results in a flaked product of bland taste, free from raw, beany, or other objectionable flavor. The precooking also prevents oil leakage during drum drying, contributes to stability of the finished product during storage, and produces a desired texture of the flake. It further develops an optimum consistency for drum drying and destroys pathogenic organisms.

The material after drum drying may be used in formulating a wide variety of foods. The compounds responsible for objectionable flavors when raw ground peanuts are heated with water have already been released when the precooked frum-dried flakes are used in food formulation and are not reformed even though such foods are heat processed.

The precooking prior to drum drying results in stabilization of the oil in the peanut flakes. Since peanuts have an average oil composition of about 46%, it might reasonably have been expected that rancidity would develop quickly because of the large amount of exposed surface in the flakes. This does not occur. This stability, however, was not achieved in flakes prepared from peanut material which was not precooked, as shown in the following table.

The acceptability rating used is one based on a 9 point scale ranging from extreme dislike at (1) to extreme like (9). A rating of 3.0 to 4.0 indicates rancidity.

Treatment	Storage at 100°F Peroxide*			Storage at 100°F TBA**		Storage at 145°F TBA**		Storage at 100°F Acceptability***	
	0 mo.	3 mo.	12 mo.	0 mo.	12 mo.	0 wk.	2 wk.	0 mo.	12 mo.
None	3.0	20	300	92	14	-	-	8.0	3.0
None	3.0	-	289	94	18	94	50	8.3	3.0
20 minutes at 240°F	2.5	-	82	90	87	93	85	8.0	6.6
20 minutes at 240°F	3.0	-	40***	94	90	-	-	8.2	7.0
None + 0.017% BHA + 0.01% citric acid	3.0	-	149	90	87	-	-	-	4.0

*Peroxide as milliequivalents of peroxide oxygen/1,000 grams of extracted oil.
**Thiobarbituric acid value as percent transmission at 530 mg.
***Ten months storage.

Example: Steam raw shelled peanuts with live steam at atmospheric pressure to expose the peanuts to a temperature of 212°F for 15 minutes. The peanuts are then dehydrated to a moisture content of 2 to 4% in a mechanical dehydrator. After the steamed peanuts have been dried to a moisture level of 2 to 4%, red skins and hearts, the latter a source of bitter flavor, are removed, and the cotyledons are ground to a fine state which releases essentially all of the oil from the individual cells which compose the cotyledons.

As an indication of the degree of fineness of grind which is desired, it is preferred that less than 5% of the oil-free solids be retained by a 200-mesh screen. To achieve the desired type of particle, it is necessary that the moisture level of the peanuts be in the range of 2 to 4% at the time of grinding.

Mix finely ground peanuts with 2 to 3 parts of water and heat to 203°F (95°C) for 2 to 3 minutes with flowing steam. The exact amount of water used depends upon the cooking time and temperature, and should be adjusted to give a consistency of the cooked material which yields the best results in the drum drying step.

The preferred consistency was a flow rate of between 2 and 12 units in 10 seconds, as measured on a Bostwick consistometer. This consistency was satisfactory for a small drum dryer. It will be understood that the greater the Bostwick unit, the more fluid the material. Cooking for 2 to 3 minutes at 203°F also effectively destroys pathogenic organisms.

Drum dry the precooked peanut material. The surfaces of the drums should be chrome plated or made of a material such that contamination of the product with copper, iron or other heavy metal does not occur. The space between the drums is adjusted to about 0.016 inch and the revolving speed to about 1 revolution per 18 seconds. A steam pressure of 80 to 90 pounds per square inch gave satisfactory results. Conditions of operation would need to be adjusted for commercial size equipment.

PROTEINS FROM SEEDS OF BRASSICACEAE FAMILY

Proteins from Mustard and Rape Seeds

Mustard seed, rape seed, and other seeds belonging to the genus Brassica are regularly processed to obtain a vegetable oil which is used in oriental countries as a substitute for soybean oil. Such processes furnish a by-product, an oil meal cake, which because of its unpalatable taste and odor is either not usable or has only very limited uses. Only small quantities of the proteinaceous oil meal cake of some of these seeds, for instance, of rape seed, may be added to a feed without making the feed unpalatable or even toxic. It is the object of the method developed by *K. Goering; U.S. Patent 2,987,399; June 6, 1961; assigned to Oil Seed Products Inc.* to provide a process by which the unpalatable taste and odor of these meal cakes is removed and a proteinaceous feed material is produced which, either alone or in combination with other feed materials, produces an excellent animal feed. Further, some of the oil residue meals contain and permit the recovery or extraction of substances which are highly valuable in themselves. This is especially the case in connection with mustard seed.

The process is based on the fact that the seeds such as mustard seed, rape seed, etc., contain thioglucosides, for instance, sinigrin, which is responsible for the unpalatable taste and in some instances for the toxic properties. It has been found that the sinigrin may be hydrolyzed by the enzyme myrosin (myrosinase, sinigrinase) contained in the seeds themselves and the products of this hydrolysis are, in the case of mustard seed, for instance, glucose, potassium acid sulfate, and allyl isothiocyanate, the latter being responsible for the unpalatable taste. This last-named product is a volatile oil which may be removed by distillation.

The process is illustrated in the accompanying flow diagram Figure 10.2 which shows several different possible modifications of the process by way of example. The optional addition of mustard or rape seed is indicated in the flow diagram by dotted lines. Otherwise, the modifications are indicated by the inscriptions.

FIGURE 10.2: ANIMAL FEED PROCESS FROM MUSTARD OR RAPE SEED

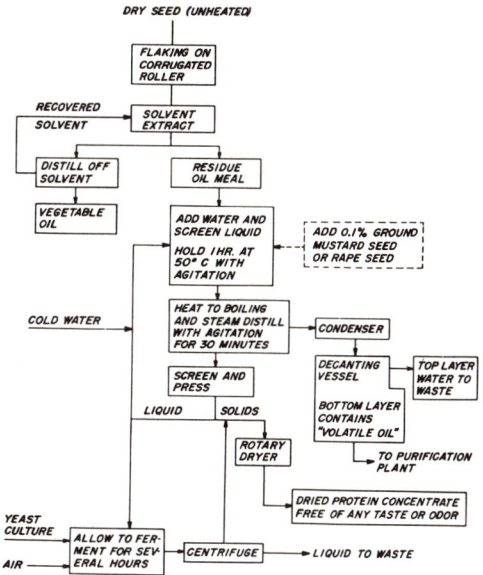

Source: K. Goering; U.S. Patent 2,987,399; June 6, 1961

As shown in the flow sheet, the dry seed is first flaked, but is not heated either before or during this step. The flaking may be carried out by corrugated rollers. The flaked seed is then subjected to a conventional extraction process carried out by a solvent, e.g., hexane, for removing the oil. For economy, the solvent may be distilled off for reuse and reintroduced into the cycle in a subsequent extraction step. The vegetable oil thus obtained is similar to and may be used as a substitute for the soybean oil. What remains after extraction is a residue oil meal which may be, for instance, mustard oil residue meal or rape seed oil residue meal. To prepare from this meal a concentrate usable as a feed substance, the residue oil meal is placed into a jacketed vessel and is mixed with tap water, preferably warmed to about 50°C. Parts of the fresh water which must be warmed may be replaced by the screen liquid from the screen which is described below and which has a higher temperature. The residue oil meal and water is gently agitated in a closed system, such as the jacketed vessel, connected with a condenser. The pH of the mixture will be between 5.1 and 5.5, which is an optimum pH for the action of the myrosin on the sinigrin.

After the previous temperature has been maintained for 1 hour, the temperature is raised as rapidly as possible to the boiling point by heating the jacket of the vessel in which the meal and water is kept. As soon as the boiling point is reached, live steam is injected into the bottom of the vessel and agitation is continued, the steam distillation being continued for about 30 minutes. The mixture is then screened and pressed to recover the solids which are subsequently dried in an oven.

The screen liquid may be reused by mixing with cold tap water as mentioned before, which is then added to another batch of the cake. The addition of cold tap water cools the screen liquid so that the temperature of the mixture of tap water and screen liquid is between 45° and 50°C, the volume of screen liquid and tap water being selected so that the optimum temperature for the hydrolysis by the enzyme is obtained. It was observed that after the screen liquid was used several times, the specific gravity of the filtrate remained constant. This indicates that solids adhering to the screened portion equal those dissolved from the fresh cake.

Example: 1,000 grams of dry mustard seed which is unheated is flaked by means of corrugated rollers and is treated with a solvent such as hexane. For each 100 grams of residue, 600 ml of water (either tap water or a mixture of screen liquid and tap water) warmed to 50°C is used and is agitated for 1 hour. Due to the buffering action of mustard seed, this will produce a pH of 5.1 to 5.5.

The temperature is then rapidly raised to 100°C and live steam is injected into the vessel containing the mixture for about 30 minutes. The condensate from the steam distillation is collected in a receiving vessel where it is separated into layers. The volatile oil, which is in this case allyl isothiocyanate (specific gravity 1.013), settles on the bottom of the receiving vessel, while the top layer is formed by water which may be discharged and go to waste.

The volatile oil can be separated from the water by drawing it off from the bottom. The allyl isothiocyanate may then be further purified by extraction with ether, separating the ether layer from the water layer and evaporating the ether from the allyl isothiocyanate. From 1,000 grams of mustard seed, 5 grams of allyl isothiocyanate may be recovered. The recovery of the allyl isothiocyanate is in itself sufficient to economically justify the above described process as it is valuable enough to more than pay for the process of purification of the mustard seed protein.

The remaining material is screened and pressed and passed through a rotary drier. This yields 380 grams of a dried protein concentrate which is free of any taste or odor and nontoxic. If the liquid fraction is concentrated and dried and added to the solids, the total recovery is 570 grams.

Protein from Rape Seed

Brassica napus and other closely related species sometimes referred to as raps or rapes have acquired some importance as a source of edible oils and have been raised for this purpose. The plants are first pressed and then solvent extracted to remove substantially all oils, leaving a protein-containing residue that has considerable biological value. The extraction process typically leaves a by-product presscake that contains about 40% protein having a net protein utilization of 74. Analysis of this protein for amino acid content has indicated that it can be a substantial source of proteins for animal consumption and possibly for human consumption. The proteins, however, are contaminated with certain toxic substances which render then unsuitable for animal or human consumption. These substances produce goiter, hepatic fibrosis and atrophy. Rape seeds contain thioglycosides which remain in the presscake during the oil extraction steps and hydrolyze in the presence of an enzyme known as mirosinase to liberate 5-vinyl thiooxazolidone and isothiocyanates.

F.M. Barros; U.S. Patent 3,615,648; October 26, 1971; assigned to the University of Chile has disclosed a process for extracting the toxic substances from presscake formed by the pressing and extraction of oils from *Brassica napus* and rendering the extracted presscake suitable for animal consumption.

The presscake remaining after oils have been removed from plants in the genus Brassica can be substantially improved in biological value by this process. The residue is first macerated in water substantially in excess of the weight of the residue for a period of up to 15 hours at ambient temperature. The water and macerated residue are separated and the macerated residue is agitated in a second quantity of water for a period of up to 3 hours. The second quantity of water and extracted residue are separated and the residue is dried at a temperature not above 60°C.

Example: A presscake remaining after *Brassica napus* is subjected to pressing and solvent extraction to remove oil, is analyzed for toxic substances. It contains 8.25 mg/g of 5-vinylthiooxazolidone and 5.95 mg/g of isothiocyanates. The biological quality of this rape presscake, measured as net protein utilization, is only 40%. 100 grams of this presscake are added to 500 grams of water and macerated at 20°C for 15 hours. The presscake is then filtered from the water, added to a second 500 grams of water and agitated at 20°C for 3 hours. The extracted presscake and water are separated by filtration. The presscake is dried at ambient temperature and finely ground.

The processed presscake contains 0.80 mg/g and 0.28 mg/g of 5-vinyl-thiooxazolidone and 5.95 mg/g of isothiocyanates respectively. The biological quality, measured as net protein utilization is 68%. It is thus comparable to milk, in which casein has a net protein utilization of 72%. Portions of original presscake, processed presscake and powdered milk, as a control, are fed to three groups of rats in order to determine the relative protein value of the original and processed presscakes. The rats are fed at a level of 20% protein calories (about 55% by weight). The results are given in the table below.

Weight in Grams of Rats at Different Ages

Study Time, weeks	Original Rape Presscake, 16 animals	Treated Rape Presscake, 60 animals	Powdered Milk (Control), 30 animals
0	34 ± 1.6	34 ± 1.8	39 ± 1.4
1	28 ± 4.5	52 ± 2.9	57 ± 2.1
2	30 ± 4.2	71 ± 4.1	77 ± 2.5
3	34 ± 3.2	89 ± 4.6	97 ± 2.9
4	38 ± 4.6	110 ± 5.5	118 ± 3.4
5	39 ± 5.6	126 ± 7.5	136 ± 4.0
6	42 ± 6.1	151 ± 7.4	154 ± 4.1

The weight gain of rats fed on processed presscake is but slightly less than that of rats fed on powdered milk and very substantially improved over that of rats fed on the original presscake. A histological study of the thyroid, liver, and kidneys of rats fed on processed presscake shows no damage or changes.

Crambé Proteins

The seeds of *Crambé abyssinica* in common with other members of the mustard family (Brassicaceae) genus Cruciferae contain enzyme-susceptible thioglucosides which are also hydrolyzed by enzymes of the gastrointestinal tract to free the volatile isothiocyanate oils. In addition the seeds of *Crambé abyssinica* and those of rape seed also contain another thioglucoside that upon hydrolysis yields a nonvolatile cyclic compound, 5-vinyl-2-thiooxazolidone (goitrin), which has marked goitrogenic activity that is not overcome by feeding supplemental iodine. Three processes have been developed by workers at the Department of Agriculture Laboratories for the purification of hexane defatted crambe meals.

In the first process by G.C. Mustakas and L.D. Kirk; U.S. Patent 3,173,792; March 16, 1965; assigned to U.S. Secretary of Agriculture enzymatic hydrolysis is used to detoxify the seed. By this process, substantially total recovery of a toxicant-free, bland, high protein content animal feed with corresponding recoveries of bland triglyceride oil and volatile isothiocyanate can be achieved. *Crambé abyssinica* seeds are first dehulled to improve the efficiency of the subsequent enzymatic hydrolysis and also to raise the content (proportion) of triglyceride oil from a value of about 40% to a value of about 46% (moisture-free basis) and the protein from the level of 40% to a value of 51% and then further treating the dehulled seeds as follows.

The dehulled seeds (endosperm) are tempered to 10% moisture to permit flaking between rollers. Additional tempering moisture is then added in a mixing chamber to provide a total moisture content of 30% and an enzyme-activating temperature of 50°C is applied for about 15 minutes to permit enzymatic hydrolysis of the thioglucosides. Then the temperature in the mixing chamber is raised to 100°C by indirect steam while injecting sparge steam at the bottom of the chamber. After about 30 minutes the sparge steam is cut off and the indirect heating continued for a few minutes to lower the moisture content.

The hot meal containing about 19% moisture is rapidly air-cooled, forming crisps having a residual moisture content of 12 to 14%, a zero thiooxazolidone content, and from zero to 0.005% residual isothiocyanate. The readily frangible crisps are screen-reduced, slurried in hexane to remove the triglyceride oils, and the high quality protein product having a residual lipid content of 1 to 3% is recovered by simple filtration. The seed oil is recovered from the solvent by stripping and then freed of any trace of contaminant isothiocyanates by conventional bleaching and deodorization steps.

Example: *Crambé abyssinica* seed (5,200 grams) was cracked between corrugated rolls, dehulled by aspiration, tempered to 9% moisture, and rolled between smooth rolls to produce flakes having an average thickness of 0.002 to 0.005 inch. About 4,000 grams of the flakes were charged to a converter-cooker equipped with an agitator to provide thorough mixing. While stirring the charge at room temperature, moisture was added to provide a total moisture content of 30%. The moist flakes were heated to 50°C (130°F) and held at this temperature for 15 minutes to promote enzymatic hydrolysis of the thioglycosides to thiooxazolidone and isothiocyanates.

At the end of this period, open steam was added to the charge through a perforated coil sparger and at the same time heat was applied indirectly through a jacket containing steam at 30 psig. The charge was steamed for 30 minutes and then dried to approximately 19% moisture using indirect heat only. After removing the charge and cooling with air, the crisped material was passed through smooth rolls twice at 0.001 inch clearance and was then subjected to a miscella extraction with hexane for removal of the triglycerides.

The slurry was then filtered and the cake was washed successively with hexane washes containing 5, 1 and 0% oil content, each weighing 5,200 grams. After addition of the final wash the cake was drained under vacuum for 10 seconds and air-dried to produce approximately 2,200 grams of meal. Whereas the untreated raw meal had analyzed 0.058% of isothiocyanate (calculated as allyl isothiocyanate) and 0.897% of 5-vinyl-2-thiooxazolidone, the detoxified meal analyzed 0.005 and 0.000%, respectively.

The second process by G. Mustakas and L. Kirk; U.S. Patent 3,392,026; July 9, 1968; assigned to U.S. Secretary of Agriculture involves steaming the seed material at 200° to 215°F to inactivate enzymes, contacting the seed material with reactants ammonia gas and aqueous NH_4OH, subjecting the ammoniated seed material to live steam and finally drying under reduced pressure.

Example: A commercial processor of vegetable oils subjected a 3,000 pound batch of hexane-extracted meal from dehulled *Crambé abyssinica* seeds to toasting for 30 minutes in a steam-emitting commercial desolventizer-toaster that almost instantaneously provided temperatures of about 230°F, i.e., well above the myrosinase-inactivation temperature of 175°F. The crambe oil was found to be free of contamination with sulfur, and beyond this, forms no part of this process.

A 250 pound aliquot of the above enzyme-deactivated meal analyzed by weight, 6.18% nitrogen, 8.6% thioglucoside, as well as a considerable amount of the butter alkaloid sinapine as shown by an intense fluorescence under ultraviolet light. The meal was introduced into a steam-jacketed spherical cooker rotating on a horizontal axis so as to tumble the meal. Gaseous ammonia was admitted until the pressure in the cooker reached 10 psig.

Following a period of 5 minutes for absorption and reaction, ammonia gas was again admitted to the same pressure, a sequence of seven such admissions being given. Immediately after the seventh admission the double-walled jacket of the cooker was heated by live steam at 230°F, thereby increasing the pressure in the cooker to about 25 psig.

After maintaining the heating for 30 minutes the cooker was vented and the ammoniated meal was transferred to a ribbon blender having steam admitting means, where the steam-moistened meal was further mixed for 30 minutes before lowering the moisture content with dry heat and slight vacuum. Chromatographic analysis of the air-cooled meal showed that it contained zero percent thioglucoside, zero percent thiooxazolidone, and zero percent sinapine (confirmed by absence of fluorescence under ultraviolet). The Kjeldahl analysis showed 7.44% nitrogen.

In the third process by *G.C. Mustakas and L.D. Kirk; U.S. Patent 3,391,000; July 2, 1968; assigned to U.S. Secretary of Agriculture* use is made of alkaline compounds such as sodium hydroxide and sodium carbonate to detoxify the seeds. This overcomes the need for use of specialized equipment needed to handle the pressurized ammonia used in U.S. Patent 3,392,026.

It was found that the reaction at moderately elevated temperatures of about 2% based on the dry weight of meal of powdered sodium carbonate or other alkaline salt with heat-inactivated moist crambe flakes or meal effectively debitters and detoxifies the seed material so that it is usable as a palatable supplement in livestock feeds and a nontoxic supplement in poultry feeds.

Example: Several hundred pounds of *Crambé abyssinica* seed were dehulled, the hulls aspirated, the dehulled seeds cracked between corrugated rolls gapped at 0.060 inch, and the cracked seeds flaked by a single passage between smooth rolls. An intense fluorescence under ultraviolet light of the chromatogram spots (principal spot R_f 0.4) from a 4:1:4 n-butanol-ethanol-water solvent system confirmed the presence of about 0.5% of sinapine and several unidentified minor alkaloids.

A Wetter analysis of the flakes after myrosinase activation showed a 1.66% content of 5-vinyloxazolidine-2-thione, and a gravimetric sulfate analysis on the defatted flakes indicated a total thioglucoside content of 9.48%. 200 pounds of the untreated flakes were placed in a steam-jacketed cooker equipped with a ribbon agitator. 4 pounds of finely powdered sodium carbonate were intimately mixed with the flakes and the dry mixture was rapidly heated to approximately 220°F by indirect steam to destroy the myrosinase and thereby prevent any deleterious hydrolysis of the thioglucosides.

Then the moisture content of the flakes was raised to 22% by the addition of 53 pounds of water, and the contents directly heated for 40 minutes at 220° to 230°F by the addition of sparge steam through the agitator shaft. Then the direct steam was stopped and the flakes were subjected to indirect heat for 45 minutes. The partially dried flakes as discharged from the cooker contained 16% moisture and were air-dried. Analysis of the meal showed a nitrogen value of 6.79%, a zero content of thioglucoside and a zero thiooxazolidone value, and an 0.04% content of sinapine. The treated seed is acceptable in cattle feed.

ALFALFA PROTEIN BY ALKALINE-PANCREATIN TREATMENT

Food products acceptable for use by humans are prepared from alfalfa or clover in the process disclosed by *G.W. Edwards and A.W. Edwards; U.S. Patent 3,780,183; Dec. 18, 1973.* The alfalfa or clover is first mixed with an aqueous solution of a base which will not leave a residue which may affect the suitability of the product or leave it unpalatable. The aqueous basic solution digests the plant material and yields an aqueous extract.

The aqueous extract, with or without a prior separation of undissolved material, is then treated with pancreatin to convert starches into soluble carbohydrates, digest lipids and release protein hydrolysates. The resulting aqueous solution, advisably after separation of lipids and lipoproteins, can be added to other foods without prior concentration, or it can be concentrated by removal of water before being used.

The product can be added to fruit and vegetable juices, starch products, blancmange, chocolate foods of all types, ice cream, ground and comminuted meats, other plant protein foods, soups, cookies and candy using products which have an appropriate solids content for use in the foodstuff. The product can function as a wetting agent for less soluble protein foods and as a water-retaining material in foods. It also constitutes a complete growth medium for *Lactobacillus arabinosus, L. fermenti, L. casei* and other bacteria. It can be used as a carrageen extender in foods such as ice milk, ice cream and candy. It is also a good acid buffer.

In the first step of the process, alfalfa or clover containing a significant amount of protein can be used. The materials presently considered most suitable are cut alfalfa and dehydrated alfalfa with dehydrated alfalfa being the starting material of choice. While a number of basic materials can be used for the extraction, it is advisable to use an alkali metal hydroxide or carbonate, or an alkaline earth hydroxide or carbonate, or any combination of these bases in aqueous solution.

Sodium hydroxide, because of its low cost, effectiveness, suitability for use in food processing and ready availability, is usually used. The concentration of the basic solution used in digesting the plant material is not narrowly critical. However, it should have at least a pH of 8 or higher. It is considered that an aqueous basic solution having a pH of 8 to 14 gives the best results. Particularly useful is a 0.33% by weight solution of sodium hydroxide in water having a pH of 10.5.

The ratio of the amount of plant material by weight to the volume of the aqueous basic solution used for the extraction is not narrowly critical. However, from 1 to 5 liters of aqueous basic solution can be suitably employed to extract one pound (454 grams) of plant material. The extraction is done at an elevated temperature up to the boiling temperature of the solution. The aqueous basic solution can be preheated before being combined with the plant material or the slurry of the plant material can be first formed and then heated. Stirring the slurry during the extraction is advisable. The extraction is considered completed after 30 minutes to 2 hours, depending on the temperature of the extraction. When the slurry is boiled, an extraction time of 30 minutes to 1 hour is generally suitable.

The slurry is then cooled to 38° to 75°C without separating the liquid from the dispersed solids but with adjustment of the pH to 14 if necessary, and pancreatin is added. When a strong basic solution is used for the digestion, the resulting extract will be more basic than suitable for pancreatin effectiveness. An acid may be added to lower the extract pH to a level which favors pancreatic activity. Generally, enough acid is added to bring the pH to 6 to 14. Any suitable acid may be used but phosphoric acid is advisably employed since it exerts a buffering action.

The most suitable temperature for the pancreatin treatment appears to be 40° to 50°C. In general, 0.2 to 5 grams of pancreatin can be added per pound of plant material used as starting material. The most suitable range, however, appears to be 0.5 to 1 gram of pancreatin per pound of plant material, particularly when alfalfa is the starting material. The inclusion of 0.5 to 1 gram of corn and potato starch amylase per pound of starting material, with the pancreatin, is beneficial and aids in reducing bitterness in the final product. Usually from 12 to 48 hours of incubation is sufficient although a longer period does no appreciable harm. After digestion with pancreatin for 24 hours, the slurry can be expected to have an odor somewhat like grapefruit juice. As a result of the pancreatin treatment, the slurry pH may drop to 5 to 6.

Following the pancreatin incubation, the slurry is separated into a solid and a liquid phase by suitable means, such as filtration, centrifugation or decantation. When the product is filtered, the extract may develop a greenish black substance comprising decomposed chlorophyll and lipids which settles out. This substance can be removed by centrifugation or gravity settling. The liquid extract constitutes a human and animal food product rich in digestible protein, carbohydrates, essential amino acids, vitamins and minerals. The liquid extract, containing 4 to 5% solids, can be added to other foods. It can be concentrated, however, by a number of means to raise the solids concentration up to 50 to 100%.

Example: 113 grams (0.25 pound) of alfalfa meal is added to 850 ml of an aqueous 0.33% sodium hydroxide solution. The mixture is boiled for 1 hour. The solution is cooled to 45° to 50°C and then 0.125 gram of pancreatin and 0.25 gram of baker's yeast are added. The mixture is incubated for 48 hours. The greenish black lipoprotein was separated by decantation and the liquid extract containing 5% solids was evaporated to a product containing 50% solids. The extract contained 28% of the 113 grams of starting alfalfa meal. The extract solids analyzed: crude protein, 19.20%; fat, 0.231%; crude fiber, 0.037%; nitrogen free extract carbohydrates, 80% and ash, 0.37%.

COCONUT PROTEIN

K.F. Mattil of Texas A and M University has studied fresh coconuts as a source of low cost protein for foods and beverages. The work was described in PB 213 594, September 1, 1970. Although no final process was developed, results of the work in the following areas are reported: the ultrastructure of coconut meats; physical processing methods for maximum extraction of fat and protein; the amount of heat-treatment coconut meat and meal can tolerate without adverse effects on product quality; identification of major classes of protein in coconut meats; and characteristics of each class of protein. Their results are given in the following.

In structure studies, electron microscopy of the stained sections showed that the cells are more or less polygonal in cross-section and long and fiber-like in longitudinal section. The long axis of the cells is normal to the inner surface of the endocarp. Large intercellular spaces run parallel to the long axis of the cells. The intercellular spaces appear to be empty and apparently can be filled with air easily. These observations make the fibrous quality of the endosperm understandable in terms of cell structure and cell arrangement in the tissue.

Results suggest that the triglycerides and protein of solid coconut endosperm are not intimately mixed in the cell. It seems reasonable to suppose that unnecessarily vigorous grinding of the fresh tissue may create emulsions of protein and fat that would be difficult to separate once formed. The fact that most of the protein and fat is inside the cell points up the need for effective cell rupture in developing economical procedure for recovering protein and fat from coconuts. Improved cell rupture might be effected by slicing coconut tissue parallel with the surface of the endocarp. This would cause the fibrous cells of the solid endosperm to be sectioned at nearly right angles to their long axis. This would improve the chances for cutting each cell thus permitting the recovery of more cellular fat and protein.

Physical processings studies have shown that the Urschel mill did an effective job of disrupting the cellular matrix so that both the fat and the protein were available for extraction by the laboratory procedures used. In heat-treatment studies on coconut meat and coconut meal, the meal was subjected to various temperatures (60°, 90°, 105°, 120° and

150°C) for varying lengths of time (5, 15, 30 and 60 minutes). Properties evaluated were browning (by visual observation). odor; lysine destruction (by comparing amino acid analyses of the heated meal samples with that of the untreated meal and by examining the changes in unavailable lysine); protein solubility at pH 2, 8 and 10.5; and nitrogen solubility index (NSI, AOCS Tentative Method Ba 11-65). Inspection of the data indicates that coconut meal can tolerate 105°C for at least 60 minutes with no significant loss in protein solubility.

On the other hand, 150°C obviously is almost immediately detrimental to protein solubility. The breaking point appears to occur at about 120°C. Coconut meal is evidently tolerant to 120°C for a short period of time but after 15 minutes heating, there are indications of loss of protein solubility.

The classical Osborne classification of the proteins of coconut meal were determined. These data are shown in the table below. They indicate that over 90% of the proteins would be classed as albumins and globulins.

Classification of Coconut Meal Proteins According to Solubility

Fractions	Extraction Conditions	% Protein Nitrogen
Albumin	CO_2-free water pH 6.7	30.6
Globulin	1 M NaCl solution pH 7.0	61.9
Prolamine	70% aqueous ethanol	1.1
Glutelin	0.2% NaOH solution	4.7
Insoluble residue	–	1.8
Nonprotein nitrogen	12% sodium tungstate to precipitate proteins	0.1

Various results on preparing protein isolates are also reported in this study.

MISTLETOE PROTEIN

Mistletoe has been used in medicine for hundred of years. Its treatment of epilepsy and vertigo had been established by Pliny in the second century A.D. Nowadays mistletoe preparations are used to treat high blood pressure, vascular diseases and arteriosclerosis. The preparations used are crude, or so-called total extracts. The potential anti-neoplastic action of certain mistletoe preparations during the 1920's led to the manufacture of crude extracts for treatment of cancerous diseases.

It was obvious to isolate active principles from plant material of *Viscum album*. Apart from choline and acetylcholine, urson and certain resin alcohols, it was viscotoxin (K. Winterfeld and M. Leiner, *Naturwissenschaften,* 42, page 487, 1955) which was isolated in this connection. This product is a polypeptide distinguished by its cardiac effect and with very strong irritant and necrotizing properties. The shrivelling of vaccination tumors in mice on intratumoral administration of mistletoe total extracts has been attributed to the necrotizing properties of viscotoxin. The extremely toxic properties of viscotoxin, however, makes the intravenous administration of the substance or of total extracts very difficult.

Extraction of Mistletoe Protein

The process of *F. Vester; U.S. Patent 3,394,120; July 23, 1968; assigned to Ciba Corp.* is based on the observation that from plant material of *Viscum album* L. and Loranthus species, especially of mistletoe growing on deciduous trees such as oaks, apple, or poplar, another protein can be isolated in purified form. This protein was accorded the references Nx1. It can be resolved into individual fractions which have been given the references explained below. This protein, or its individual fractions or mixture have about 100 times the antitumoral activity of the clear juice and are comparatively little toxic. The antitumoral action was examined, for example, on ascitic tumors, such as sarcoma 180 on Swiss mice and also on cell cultures (for example, Hela or sarcoma 180 cultures, and Chang liver).

Purifying Mistletoe Proteins by Ultracentrifugation

F. Vester, H. Majer and J. Mueller; U.S. Patent 3,472,831; October 14, 1969; assigned to Ciba Corporation have found that the ballast substances and especially the atypically toxic portion can be removed from protein fractions obtainable by the process of U.S. Patent 3,394,120 by ultracentrifuging an aqueous solution of these protein fractions and isolating the tumor-inhibiting protein fraction in a highly purified form from the sediment.

The ultracentrifugation is carried out in a known manner, at as high a rotational speed as possible, advantageously, at over 20,000 revolutions per minute, while subjecting the solution used as starting material to a gravitational field of over 100,000 g, for example, about 170,000 g. This field is maintained for several hours, e.g., for about 15 to 30 hours, working at a low temperature, e.g., between 0° and +10°C, to conserve the protein fractions. If desired, the treatment on the ultracentrifuge may be repeated and/or the resulting, purified protein fraction further purified by filtration, e.g., filtration under sterile conditions, by which an additional removal of ballast can be achieved.

The resulting, highly purified protein fraction can be stored in the lyophilized form; it is, however, advantageous to free it prior to the lyophilization operation by dialysis against water from inorganic constituents, especially sodium chloride.

Example: 22.0 grams of the ammonium sulfate precipitate from the eluate $Nx12_{0.35}$ I (as prepared in U.S. Patent 3,394,120) are dissolved by careful addition of water (final volume, 20 ml) and then redialyzed against an 0.9% aqueous sodium chloride solution. The above solution is diluted to a concentration of 1.7% by adding more 0.9% aqueous sodium chloride solution and then exposed in an ultracentrifuge (preparative ultracentrifuge, model L, using a preparative rotor, type 30 (Beckman/Spinco) for 24 hours at 5°C and 30,000 revolutions per minute to a gravitational field of 105,000 g. The electrophoresis of the starting material, of the sediment and of the decantate reveals the following values.

	Starting material, $Nx\,12_{0.35}$ I	Decantate, $Nx\,12_{0.35}$ I UZD	Sediment, $Nx\,12_{0.35}$ I UZS
Protein content:			
In mg	579	259	320
In percent	100	44.7	55.3
Sediment:			
Area B_1+C	21.1	34.3	7.3
Area B_2	18.7	18.8	22.0
Active fraction, area A	60.2	46.8	70.7

The areas A, (B_1+C) and B_2 are shown in percent of the total area.

Solution of the sediment, dialysis against plain water and lyophilization converts the enriched protein fraction $Nx12_{0.35}$ I UZS into a storable dry form.

Electrophoresis of Mistletoe Protein

F. Vester; U.S. Patent 3,475,402; October 28, 1969; assigned to Ciba Corporation further found that it is possible to obtain in a simple and surprising manner a highly active tumor inhibiting protein freed from ballast matter and from atypically toxic impurities when press juice from mistletoes or purified tumor-inhibiting protein fractions obtained therefrom, are separated by a carrier-free deflection electrophoresis and the tumor-inhibiting protein is isolated. The principle of carrier-free deflection electrophoresis is known (cf. Hannig, Z., *Anal. Chem.*, vol. 181, page 244, 1961; Hannig, Eine Neuentwicklung der trägerfreien Ablenkungselektrophorese und ihre Anwendung auf cytologische Probleme, postgraduate thesis, Munich, 1964). This process can be shown by the following example.

Example: The dialysis retentate designated as Nx1 in the Example given in U.S. Patent 3,394,120, is salted-in with the aid of an 0.03 molar aqueous 2-amino-2-hydroxymethyl-1,3-propanediol buffer. The centrifuged solution is adjusted to the substance concentration of 1% by addition of the same buffer solution. 20 ml of this solution, containing 0.2 gram of substance that can be compared with one of the protein fractions designated by Nx12 are supplied at a rate of 1 ml/hour at a distance of 12.5 cm from the cathode to the 0.5mm thick buffer film of the deflection electrophoresis apparatus (Hannig, Z., *Anal. Chem.*, vol. 181, page 244, 1961), traveling in a width of 50 cm across the electric field.

The buffer current flows at a speed of 80 ml/hr, corresponding to a migration speed of the buffer front of about 30 cm/hour. The electric field is kept constant at 1,900 volts, and the current intensity is of the order of 180 ma. When 20 ml of the substance solution have been consumed, more of it can be poured in and the separation may be continued for as long as is desired.

The buffer front issuing from the bottom end of the separating apparatus is collected in 48 contiguous fractions; these are classified by measuring their ultraviolet absorptions at 280 mμ (2 hours after starting the electrophoresis, that is to say when the substance begins to run out) and continually combined to collective fractions. The fractions thus obtained may be salted out, for example, by addition of aqueous ammonium sulfate up to a molarity of 3 and centrifuged off as a precipitate and, if desired, liquefied, dialyzed and then cautiously lyophilized; alternatively, they may be used as they result from the deflection electrophoresis.

EXTRUDED FIBER PROCESSING

As natural polymers, proteins, including vegetable proteins, can be processed to form filaments or other fibrous products. It is very desirable to form fibers which simulate the texture and chewiness of meat fibers. Such fibers can be used as extenders in meat dishes or can be processed forming simulated meat products.

Protein filaments are generally produced by dispersing the proteins which are the starting material in a dispersing medium such as an alkaline aqueous solution. Depending upon the material dispersed, and the dispersing agent used, the dispersion may amount to a colloidal solution. This dispersion is then forced through a porous membrane such as a spinneret used in the production of rayon, and passes into a coagulating bath which is generally an acid salt solution. The streamlets coming through the spinneret are precipitated in the form of filaments. One spinneret will produce several thousand filaments which are very fine (about 20 microns in diameter). These several thousand filaments will constitute a group of filaments or a bundle of filaments which group or bundle may have a diameter of perhaps ¼ inch.

If a battery of spinnerets or dies is provided, there will be a considerable number of bundles or groups of filaments which, when assembled together, constitute a tow of filaments, which tow, depending upon the number of dies or spinnerets used, may be 3 to 4 inches in diameter. If the filaments emerging from the spinneret or die are not stretched at all, they will be weak, tender, inelastic and kinky. On the other hand, if the filaments are pulled away from the spinneret or die at a speed sufficiently great to keep them straight and prevent kinking, a further orientation of the molecules occurs and the filaments become stronger, attain elasticity, and usually will be smaller in diameter than the orifice from which they emerged.

A synthetic product made from the kinky or unstretched fibers will lack chewiness, whereas the product made from more highly oriented fibers will have improved chewiness and a more meat-like texture. Coarser filaments can be produced by starting with the proteins as powdered material plasticized with about 25% of alkaline water and then extruding the plasticized protein material through dies. The filaments produced by this process may be of much greater thickness than those produced by a spinneret. These coarser filaments may be of a thickness on the order of paint brush bristles and even though they are stretched afterwards, the final filament will be relatively coarse.

The early work frequently referred to in many of the patents reviewed in this volume includes three patents by R.A. Boyer and coworkers which will be covered here only briefly.

BOYER PROCESS

R.A. Boyer; U.S. Patent 2,682,466; June 29, 1954 used the technique of forming filaments in tows, except for carrying out the insolubilizing treatment of the filaments. The process is designed to produce synthetically from vegetable protein or animal protein or combinations thereof a product which closely resembles natural meat as to its appearance, as to its fibrous qualities, as to its flavor, as to its nutritive value, and as to its chewiness.

When the formed filaments are placed in a salt solution to prevent them from redissolving, the pH of this salt solution may be adjusted to any desired level and it will be obvious that the fibers upon immersion in it will soon attain the same pH as the solution. Since most meats have a pH range of 5.6 to 6.4, the pH of the fibers is frequently adjusted to these levels. However, satisfactory results are obtained over a wide pH range.

In order to convert the filaments which have been produced into a meat product they should be freed from excess salt solution by squeezing or centrifuging. They need not be dried but they should not be dripping wet. The groups of filaments thus treated should then be treated with edible binders.

Finally the groups of filaments are assembled by pressing together to form a tow, and are cut into suitable lengths convenient for handling and sale. The pressure in assembling can be light or heavy depending upon the density and texture desired in the final product.

In a second process, *R.A. Boyer; U.S. Patent 2,730,447; January 10, 1956; assigned to Swift & Company* provides a method for introducing and incorporating modifiers into protein fibers or filaments, which represents an improvement over U.S. Patent 2,682,466.

The modifiers or additives may be added, and preferably are added, to the protein material in dispersion. It has been found that by so doing a uniform distribution of the modifier or additive in the resulting filament is obtained. It is essential when following this procedure that the additive or modifier be thoroughly mixed into the protein dispersion preliminary to forcing the same through the spinneret.

Examples of suitable additives or modifying agents which may be incorporated into the edible protein fibers are cod-liver oil, salad oil, spice oils, skim milk solids, sugar, starch, hydrogenated vegetable or animal oil shortening, lard, cottonseed oil, butter, monosodium glutamate, protein hydrolysates, spices, gums, binders, flavoring agents, pigments and dyes.

In the third process *R.A. Boyer and H.E. Saewert; U.S. Patent 2,730,448; January 10, 1956; assigned to Swift & Company* provide a method for controlling and adjusting the degree of toughness of edible protein fibers. The process comprises the subjection of edible protein fibers, to the action of an alkaline medium for a sufficient period of time to increase the pH of the fibers to within the range 4.0 to 7.0.

The protein fibers or filaments are preferably produced by dispersing the proteins in an alkaline aqueous solution. Depending upon the material dispersed and the dispersing agent used, the dispersion may amount to a colloidal solution. This dispersion is then forced through a porous membrane, such as a spinneret, into a coagulating bath which is generally an acid salt solution. The filaments issuing from the spinneret, which actually is a small die having 5,000 to 15,000 holes each on the order of 0.003 inch in diameter, will be of a diameter of about 0.003 inch. When these latter filaments are stretched, they are elongated and reduced in diameter until the average thickness is on the order of 20 microns.

Following the formation of the protein material into the fibers, the fibers are subjected to a stretching operation. The fibers leaving this stretching operation will have a pH in the neighborhood of 3.0. Such a pH has proven undesirable for the subsequent use of such fibers in a food product both from a taste standpoint and also because of the toughness of such fibers. In adjusting the degree of toughness of the fibers, the tow is immersed in a salt solution of ½ to 12% concentration. The solution is continuously adjusted to a pH within the range 4.0 to 7.0 by the addition of an alkaline solution prepared from any alkali or buffering agent which can be considered suitable for use in food products in order to compensate for the decrease in pH of the salt solution due to the acidity of the fiber.

Through control of the pH of the fibers, it is possible to eliminate the sour taste of the fibers without requiring a washing operation. Variations in the pH control of the fibers permit a variation in the moisture content of the final product, the higher pH values permitting the fiber to hold substantially more moisture.

PROCESSES FOR EXTRUDED FIBERS

Continuous Fiber Preparation from Protein Slurry

R.W. Westeen and S. Kuramoto; U.S. Patent 3,118,959; January 21, 1964; assigned to General Mills, Inc. provide a process for the continuous production of protein fibers having a high degree of uniformity where the step of preparing a stock protein solution of increased pH could be eliminated with savings in time and equipment. This is a continuous process for the preparation of shaped protein products, particularly filaments, from an aqueous protein slurry having a pH of less than 7.0.

The shaped protein products can be prepared continuously by: (1) charging an aqueous protein slurry having a pH of less than 7.0 and an aqueous solution of an alkaline material into a mixing device to yield a mixture having an initial pH of at least 10.5; (2) intimately blending the aqueous protein slurry and the aqueous solution of the alkaline material while the slurry and the solution are being advanced to the discharge orifice of the mixing device; (3) discharging the resulting spinning solution from the discharge orifice of the mixing device; and (4) extruding the stream of spinning solution through an extrusion device immersed in an acid coagulating bath to form the shaped protein product.

A wide variety of edible protein materials can be used in preparing the shaped protein products. Representative of such materials are soybean, safflower, corn, peanut and pea proteins as well as various animal proteins such as casein and keratin. Generally, the proteins are used in relatively pure form. Thus, for example, soybeans may be dehulled and solvent extracted, preferably with hexane, to remove the oil therefrom.

An important feature of the process is the screw pump used for mixing the protein slurry and alkaline material. This mixer is illustrated in Figure 11.1.

FIGURE 11.1: SCREW PUMP

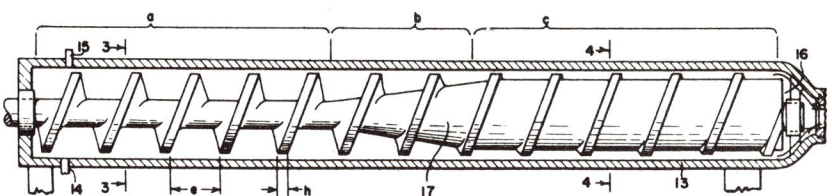

Source: R.W. Westeen and S. Kuramoto; U.S. Patent 3,118,959; January 21, 1964

The screw pump **7**, comprises a cylindrical housing **13**, a helix **17** having a mixing section **a** of channel depth **f**, a transitor section **b** of decreasing channel depth and a metering section **c** of channel depth **g**, inlets **14** and **15** for introduction of the protein slurry and alkali solutions and a discharge orifice **16**. The dimensions of the various sections of the helix, based on the diameter D thereof are preferably as follows: Length mixing section a, 4 to 12 x D; length metering section c, 4 to 12 x D; length transitor section b, ½ to 6 x D; channel depth f of mixing section, ⅓ to ⅙ x D; channel depth g of metering section, 1/10 to 1/20 x D; pitch e, ¼ to 1 X D; and width h flat edge of helix, ⅙ to 1/15 x D.

The channel depth of the mixing and metering sections is preferably uniform while the channel depth of the transition section increases from that of the mixing section to that of the metering section, the increase being relatively uniform over the length of the transition section. The ratio of the lengths of the mixing and metering sections is in the range of 0.3:3.0 to 3.0:0.3, respectively. The clearance between the cylindrical housing and the flat edge of the helix should be less than 1/150 of the diameter D of the helix. The pitch e can be constant or variable, decreasing from the mixing section to the metering section. The diameter D of the helix can vary considerably but is preferably within the range of 1 to 4 inches. The following example serves to illustrate the process.

Example: 300 pounds of hexane extracted soybean flakes were charged into a solution of 1.5 pounds Na_2SO_4 in 3,000 pounds water at 40°C. The pH of the resulting slurry was adjusted to about 8.0 by the addition of 15.2 pounds of a 10% by weight aqueous NaOH solution. After 35 minutes of agitation, the slurry was centrifuged to separate the extract from the spent flakes and then recentrifuged to remove traces of insolubles. The pH of the concentrate was adjusted to 4.5 with sulfur dioxide and the precipitated protein concentrated by centrifugation. Traces of solubles were removed by reslurrying the protein in 5 volumes of water followed by reconcentration. The solids content of the slurry was adjusted to 17% by weight.

The slurry was charged into a makeup vessel equipped with a stirrer. This slurry was pumped into the inlet end of a screw pump at a rate of 254 gal/min. A 10% by weight NaOH aqueous solution was pumped from a makeup vessel into the inlet end of a screw pump at a rate of 33 gal/min to yield an initial mixture in the pump having a solids content of 15% by weight protein. The screw pump had the following dimensions: overall length, 24"; length mixing section, 6"; length metering section, 12"; length transition section, 6"; channel depth of mixing section, 0.5"; channel depth of metering section, 0.1"; diameter of helix, 1.50"; pitch, 0.75"; width flat edge of helix, 0.10"; and clearance helix to housing, 0.07".

The screw was rotated at 210 rpm which developed a pressure of 25 psi at the discharge end of the screw pump. The spinning solution was discharged into a line containing a filter and a metering pump. It was pumped to the extrusion device containing a spinneret having 15,000 holes (diameter of 0.003 inch) immersed in an acid coagulating bath (12% by weight NaCl and about 1.8% by weight acetic acid). The average time between the first contact of the protein slurry and alkali solution and the extrusion of the resulting spinning dope through the spinneret was about 5 minutes. There was obtained a continuous tow of protein filaments having excellent texture, color and odor. The fibers also had the desired strength for use in the production of simulated meats. After containing the spinning for 4 hours, the fibers still had the same outstanding properties.

Gaseous Coagulation of Protein Fibers

Protein fibers are prepared by contacting an extruded protein with a fast acting acid gas traveling at a velocity greater than the protein in the process developed by *R.D. Dannert and M.E. Manwaring; U.S. Patent 3,794,731; February 26, 1974; assigned to General Mills, Inc.*

The starting protein solution can be made from a variety of functional protein materials, that is, not heat or otherwise denatured. Additionally the protein materials must be capable of forming a pumpable solution and of forming fibers upon being contacted with the acid gas. Preferred proteins for use in this process are the oilseed proteins and casein. Of the oilseed proteins, soy protein isolate is particularly preferred.

The most effective fast acting acid gas is HCl, preferably anhydrous HCl. Mixtures of air and anhydrous HCl have also given good results. The acid gas is used in an amount sufficient to react with the alkaline protein stream to form fibers. The various proteins normally have different isoelectric points or ranges and thus the quantity of gas varies in relation to the protein material and also the pH of the protein solution. The alkalinity of the solution is preferably obtained by using an alkali metal hydroxide, particularly sodium hydroxide. The quantity of base needed depends somewhat on the particular protein or combination of protein source materials. Preferably, a large excess of base is not used since this requires greater quantities of acid gas to coagulate and form fibers from the alkaline protein streams.

The protein solution is forced through an orifice having a diameter of 6 to 50 mils. A multiple of such orifices can be used provided that the acid gas stream comes into intimate contact with all of the issuing protein dope streams. The acid gas pressure is not critical, it being sufficient to provide a gas stream having a velocity greater than that of the protein stream.

The extrusion nozzle used in this process can be illustrated by the following drawing in Figure 11.2, although other modifications of such extruders can be used.

FIGURE 11.2: EXTRUSION NOZZLE

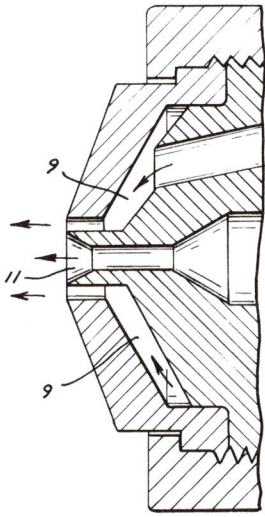

Source: R.D. Dannert and M.E. Manwaring; U.S. Patent 3,794,731; February 26, 1974

The orifice **11** for the exiting protein solution has a cross-section or diameter of from 6 to 50 mils as set forth above. Figure 11.2 shows the orifice which is fluted and, in such nozzle construction, the acid gas stream **9** leaves the atomizing nozzle in a generally parallel relationship with the exiting protein solution stream.

Example 1: A protein solution was prepared from the following ingredients: 440 grams soy isolate (90 to 95% protein Promine R), 1,380 ml H_2O, and 180 ml NaOH (10 N). One-half of the soy isolate was slurried with the water and then two-thirds of the NaOH was added. The mixture was heated to 50°C and then the remainder of the NaOH and soy isolate were slowly added with continued mixing. The solution temperature was raised to 60°C and the same was centrifuged. The solution was sprayed at a pressure of 10 psig and an anhydrous HCl pressure of 15 psig (the solution temperature and the spray apparatus were maintained at 60°C). Very fine fibers which were fairly strong were obtained.

Example 2: Example 1 was essentially repeated using the following protein containing solution: 100 grams soy isolate (Promine R), 281 grams H_2O, 23 ml NaOH (10 N), and 1 gram ammonium sulfite. The resulting solution was smoother than the solution of Example 1 and the resulting fibers were white in color (in comparison to the fibers of Example 1 which had a greenish hue). Thus the ammonium sulfite improves solution viscosity and fiber color.

Heat Binding Extruded Fibers

R.K. Dudman; U.S. Patent 2,785,069; March 12, 1957; assigned to Swift & Company provides a method for preparing food products from man-made, edible protein fibers without the use of extraneous binders. The process comprises the treatment of edible protein fibers, which have been adjusted to the proper pH, salt concentration, and moisture content, with controlled heat to obtain a binding action attributable solely to the protein fibers themselves.

The protein fibers or filaments are preferably produced by dispersing the proteins in a suitable dispersing medium such as an alkaline solution. It is often desirable to incorporate additives uniformly within the protein fibers or filaments. This dispersion, with or without modifying agents incorporated therein, is then forced through a porous membrane, such as a spinneret, into a coagulating bath which is generally an acid salt bath solution.

The filaments issuing from the spinneret, which actually is a small die having from perhaps 5,000 to 15,000 holes each on the order of 0.003 inch in diameter, will be of a diameter of 0.003 inch. Following the formation of the protein material into filaments or fibers, the fibers may be subjected to a stretching operation. Additionally, it may be desirable to subject the fibers to a mild heat treatment during or after such stretching. The stretched fibers or filaments are then placed in a salt solution (such as sodium chloride) having a concentration of 2 to 12%. It may well be desirable at this point to adjust the pH of the fibers to 4.0 to 7.0 by the use of an alkaline medium.

Groups of these fibers are then formed into bundles or tows and freed from excess liquid by squeezing, centrifuging or the like. At this point it has previously been considered necessary to add a suitable binder to the bundles of man-made, protein fibers, whereupon the bundles or tows were passed through a bath of melted fat or the like and then incorporated into the final product. In this process, the use of extraneous binders can be dispensed with if the fibers are subjected to a heat treatment suitably high to cause fusion thereof. Upon cooling, the fibers will be found to be bound together in a desirable fashion.

The temperature of the binding operation will vary depending upon the particular type of protein being treated. For example, fibers prepared from casein must be heated to an internal temperature of 100°F or higher for from 5 to 60 minutes. The temperatures required for the fusion of soybean fibers are somewhat higher, i.e., about 110°F or higher, and generally a longer period of heating is required.

Example: Soybean fibers, i.e., with a pH of 5.2 and a salt concentration of 3%, were squeezed to remove excess moisture. 100 grams of these fibers were shaped into a 1 inch diameter bundle and placed on a rack in an uncovered pan in a 185°F oven. Fusion required 2 hours of heating at this temperature. A similar run in a 240°F oven required 1 hour of heating to produce fusion.

ADDITIVES FOR FIBER PROCESSES

Acid Coagulating Bath Containing Sulfur Dioxide

S. Kuramoto, R.W. Westeen and J.L. Keen; U.S. Patent 3,177,079; April 6, 1965; assigned to General Mills, Inc found that the flavor of synthetic protein fibers can be greatly improved by treating the precipitated filaments or fibers with sulfur dioxide.

Generally, the process comprises: (1) forming a dispersion of edible protein material in aqueous alkali; (2) forming filaments from the dispersion by precipitation in an acid coagulating bath; and (3) treating the precipitated filaments with sulfur dioxide. The treated filaments may also be (4) neutralized to pH 4.0 to 7.0 to raise the pH thereof. Simulated food products may be prepared by impregnating the filaments with additive materials such as binders, flavoring agents and the like. Also, the impregnated filaments can be allowed to set up by heating.

The original protein dispersion and the filaments or fibers can be produced by known methods from a wide variety of edible protein materials. Representative of such materials are soybean, safflower, corn, peanut and pea proteins as well as various animal proteins such as casein.

The flavor of the fibers is materially improved if they are treated before neutralization with sulfur dioxide or a sulfur dioxide producing material. This can be accomplished by bubbling sulfur dioxide through the fiber-containing bath or by adding a sulfur dioxide producing material to the coagulating bath. Such materials include sulfurous acid, alkali metal hydrosulfites, such as sodium hydrosulfite ($Na_2S_2O_4 \cdot 2H_2O$), alkali metal bisulfites, such as sodium bisulfite ($NaHSO_3$) and the like. Any material which produces sulfur dioxide on addition to the fiber suspension can be used in the process.

The amount of sulfur dioxide added will vary considerably but should be sufficient to effect the improvement in flavor. Amounts of a few to about 10,000 ppm based on the dry weight of the fibers are preferred. An especially preferred range is 300 to 1,000 ppm. The filaments or bundles thereof (tows) are stretched by pulling them from the coagulating bath over a take-away reel. The pH of the filaments or tows after treatment with the sulfur dioxide, is acidic which is undesirable since food products prepared therefrom would have a sour taste and would normally be too dry and tough. The filaments can be neutralized by passing them through a neutralizing bath or by spraying them with a neutralizing solution while they are being stretched or being conveyed to the next step in the process. The neutralization is carried out until the pH of the fibers is 5.0 to 6.4 which is the pH of most natural meats.

Examples 1 and 2: A spinning solution comprising water, 18% by weight soy protein and 7% by weight NaOH and having a viscosity of about 20,000 cp and a temperature of 40°C was used to prepare fibers. The solution was forced through a rayon spinneret having 5,000 holes (diameters of 0.004 inch) into a coagulating bath composed of water containing 10% NaCl and 1.5% lactic acid. In Example 1, the tow of precipitated fibers was taken out of the bath by a take-away reel adjusted so that stretch amounting to 100 to 200% was exerted on the fibers. The fibers were water washed to remove excess acid and to raise the pH of the fibers to 5.0. In Example 2, sulfur dioxide was added to the fibers in the precipitation bath so that the fibers, containing 65% by weight water, had an SO_2 content after treatment of about 1,000 ppm. The fibers were then processed as in Example 1.

Analysis of the fibers before the water washing showed that the peroxide number of those of Example 1 was 2.4 and those of Example 2 was 0.0 (milliequivalents of O_2/1,000 grams total sample, i.e., fibers containing 65% water). The fibers of Example 2, in addition to the greatly reduced peroxide number, were very bland while those of Example 1 had the characteristic and objectionable flavor of soybeans and a distinct paint flavor. Although sulfur dioxide was used in Example 2, equally good results are obtained using sulfurous acid, sodium hydrosulfate, sodium bisulfite, and other sulfur dioxide producing materials. The sulfur dioxide treated fibers can be stored or sold as such or used in the preparation of simulated food products, particularly meats.

After the neutralization step, the filaments are freed from excess neutralizing solution and impregnated with binders, flavoring agents and the like. The binder preferably consists of, or contains a substantial proportion of a heat coagulable protein such as albumen. Various meat flavors which are available commercially can be added.

Sulfites in Protein Coagulation Baths

In the process described by *R.A. Boyer, A.A. Schulz and E.A. Schatzman; U.S. Patent 3,468,669; September 23, 1969; assigned to Ralston Purina Company* spun protein products are prepared by mixing a sulfite, other sulfur dioxide producing material or sulfur dioxide itself, with an edible protein to improve the flavor and texture of the product and to facilitate its preparation.

The treatment of the edible protein may take place at various stages of the production. For example, the sulfite may be mixed with the dry protein or the wet protein curd prior to formation of the spinning solution, it may be added to the spinning solution, it may be added to the coagulating bath or it may be added to a supplemental bath employed after the coagulating bath. If desired, the sulfite treatment may be applied to the protein at more than one location. The sulfite may be formed in situ in addition to or in lieu of adding it as such.

Either sodium sulfite, potassium sulfite or ammonium sulfite, or mixtures of two or more of the foregoing, may be utilized. Other useful sulfites are hyposulfites, hydrosulfites, bisulfites and metabisulfites exemplified by the sodium salts of these various sulfites. Sulfites other than sodium sulfite may likewise be employed as long as they form sulfur dioxide in solution. If preferred, sulfur dioxide per se or sulfurous acid may be utilized in carrying out the process.

Not less than 0.25% sodium sulfite equivalence by weight of the dry protein should be employed. Higher proportions of sulfite may be uitlized if desired but the proportion should not be increased to the point where the protein spinning solution tends to set up to gel form. In this connection, 1.5% of sulfite gives excellent results while 16% of sodium sulfite causes the spinning solution to set up to gel form. If sulfur dioxide or a sulfur dioxide producing material other than a sulfite is employed, the amount added to the protein is preferably between as few as 200 and 10,000 ppm SO_2 based upon the dry protein. Preferred protein is soy protein or casein.

Example: Edible soy protein which has been isolated at the isoelectric point is spray dried. To the dry product is added 1.5% by weight of sodium sulfite. The combination is then dispersed in an aqueous solution at 80°F so that the solution contains 14.6% total dry solids. The solution is agitated and sodium hydroxide is added at a rate of 7% of the weight of the dry protein. The combination is mixed until all of the protein is dissolved.

The solution of protein and sulfite in the sodium hydroxide solution is then spun into filaments using a spinneret in the manner described in U.S. Patent 2,682,466. After spinning, the filaments are formed into bundles to constitute tows of filaments and subsequently processed as taught in U. S. Patent 2,682,466. Among the additives which may be incorporated are binders, flavoring materials, coloring materials, oils, fats, emulsifiers and mixtures thereof.

The protein product formed has an improved flavor and structure over a product formed in the same manner but without the sulfite and, when chewed, provides the effect of tenderness. In addition the viscosity of the spinning solution is lower than if no sulfite were included. The product exhibits increased resistance to bacterial and fungal growth.

Sulfur Compounds in Soy Fibers Extruded at High Temperatures and Pressures

S.L. Jenkins; U.S. Patent 3,496,858; February 24, 1970; assigned to Ralston Purina Company describes a method of producing a meat simulating product from soybean meal, comprising the steps of: forming a mixture of soybean having an oil content of less than 5% by weight, sufficient moisture to form a mixture of moisture content between 19 and 41% by weight, and a sulfur reagent selected from elemental sulfur, potassium sulfide, and sodium sulfide; mechanically working the mixture under an elevated pressure and at a temperature above

212°F to convert the mixture into a flowable substance; and forming a porous, expanded, fibrous structure by extruding the converted mixture through a two stage restricted orifice means, into an environment of substantially lower pressure.

Experimentation has shown that of the above three sulfur reagents, by far the most effective is elemental sulfur, with the second most effective being potassium sulfide, and the third most effective being sodium sulfide. The potassium sulfide produces a product which is considerably better than the sodium sulfide, with the elemental sulfur producing a product which is far better than either of the other two.

These reagents have a profound effect upon this molecular structure of the soybean protein causing apparent breakdown and reorganization from the original molecular formation under the pressures and temperatures used in this extrusion process. An extremely desirable fibrous characteristic is obtained when the product expands as it exits from the restricted extruder outlet.

The additives should be used in the following amounts: sulfur, 100 to 5,000 ppm; potassium sulfide, 400 to 20,000 ppm; and sodium sulfide, at least 1,000 ppm or more.

The extruder used is illustrated by the following Figure 11.3.

FIGURE 11.3: EXTRUDER ASSEMBLY

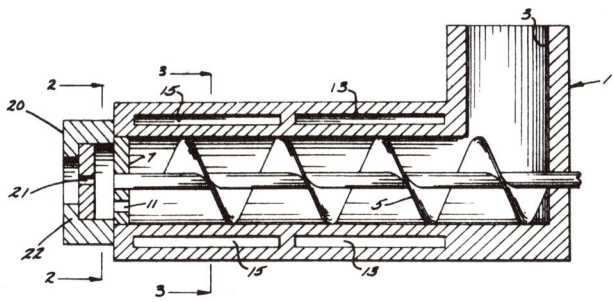

Source: S.L. Jenkins; U.S. Patent 3,496,858; February 24, 1970

With reference to Figure 11.3, the extruder assembly **1** comprises a housing of elongated configuration with a meal mixture inlet **3**, a rotationally driven pressure screw **5** extending through an elongated cylindrical extruder chamber in the housing, and product outlet means on the end of the housing opposite the inlet. The outlet means is provided and controlled to prevent premature ejection of the product prior to formation of the fibrous structure. Specifically, a restrictor or retention pressure plate **7** closes off the end of the extruder chamber, except that it has a restricted port such as one or more small orifices **11** therein. This orifice may be of any selected configuration, but normally is circular. If only one orifice is present, it normally has a diameter of about ¼ inch or so. If more than one is used, the diameter of each is smaller so that the total cross sectional area of the orifices is approximately the same.

The pressures in the extruder are believed to fall within the range of 300 to 600 psig. Part of the pressure is caused by the screw and retainer plate combination, and part by the high temperatures which result both from the friction between the flowing products and the components of the extruder, and from external heat that is purposely added during operation. This added heat is obtained by passing steam through the forward or front annular jacket **15** within the extruder housing and around the chamber but separated from it.

The temperatures reached by the material in the extruder must be above 212°F and actually should be considerably higher to form a meat simulating product with good fibrous structure. This temperature varies with variations of the other mixture characteristics of which the most significant is moisture, but normally shall be between 270° and 310°F. As the moisture content increases from 20 to 40%, the temperatures may decrease from 310° to 270°F.

An annular cooling jacket **13** surrounds the rear portion of the extruder chamber as it has been found desirable to maintain lower temperatures in the initial stages of the extruder to prevent the product from overheating before it exits and yet to enable sufficient heat to be added in the latter stages of the extruder. Again, the amount of cooling water and its temperature to cause the desired cooling effect will vary, but can be readily determined be trial and error. The product outlet means also includes a smaller secondary chamber into which the material discharges from restricted orifice **11**. The output from this chamber is restricted also by a die plate **20**, containing a restricted outlet **21**. This plate, which can be supported by a suitable retention collar **22** or the like, provides the second stage of a double stage restriction set up, the first stage being achieved by restrictor plate **11**. This double stage restriction has been found necessary to obtain top quality fibrous formation in the product.

The expanded product emerging is very porous, and has a fibrous network structure somewhat resembling that of actual meat. The product can be kept moist and used directly for food materials, or can be dried easily by passing it through a conventional drying chamber and then packed in a convenient fashion for later use.

Example: Elemental Sulfur — Seventeen pounds of soybean meal having a 50% protein content and a fat content of 0.5% were mixed with 2,600 cc of water to which 0.007 pound of finely divided elemental sulfur has been added. After mixing to obtain a homogeneous mixture, the mixture was added to the extruder, with steam at 20 psig circulated through jacket **15**, and room temperature cooling water at 60 psig passed through cooling jacket **13**. The orifice **11** and pressure plate **7** were $\frac{1}{2}$ inch in diameter, with the orifice **21** in the die plate being $\frac{3}{8}$" x $\frac{1}{8}$".

The temperature of the mixture at the downstream or outlet end of the extruder reached 280°F, at a pressure of about 400 psi. It was extruded through the double orifice series and ejected into the atmosphere in a continuous length of expanding structure ovular in cross section with a major diameter of approximately $\frac{3}{4}$ inch. It had an excellent porous, fibrous structure.

EXTRUDED FIBERS FROM PROTEIN BLENDS

Fibers from Blends of Casein and Soy Proteins

A process has been developed by *A.S. Szczesniak and E. Engel; U.S. Patent 2,952,543; September 13, 1960; assigned to General Foods Corporation* to produce fibers having the combined advantages of casein fibers and soy fibers. In addition the process makes use of a relatively inexpensive defatted soy flour.

The process comprises preparing an acidic casein aqueous dispersion having a pH of 5.3 to 5.8; preparing an alkaline dispersion of essentially undenatured soy protein having a pH of 11 to 11.5; blending the casein and soy protein dispersions to provide a pH in the neighborhood of 8.5 to 9.3 at which the dope is of such viscosity and tackiness that it can be readily spun into filaments. The protein of these filaments is next precipitated in an acid bath in order to produce a protein fiber having a uniform texture throughout its cross-section and characterized by a freedom from either alkaline or acid centers as well as a freedom from an overly dense or firm curd skin around the fiber.

By initially providing a casein dispersion or dope moiety having an acidic pH in the neighborhood of 5.3 to 5.8 the casein dispersion is provided with such acidity as to allow its being blended with a cooperating soy dispersion or dope moiety at a high alkaline pH such that the blends of these moieties provide a spinning dope solution having maximal spinning properties. It has been found that the optimal range of ratios of soy to casein dope moieties ranges between 40:60 and 60:40 parts by weight. In following the foregoing relationships between soy and casein dope moieties the alkalinity of the separate moieties is automatically adjusted so as to provide a spinning dope when the moieties are blended having a pH generally in the range of 8.5 to 9.3 for maximal spinnability.

When a pH substantially above 9.3 is practiced there is a reduction in dope viscosity and a loss in tackiness such that the filaments when spun suffer from a loss in drawability or stretchability. Where the pH of the eventual composite spinning dope is too low (say below a pH of 8.0) the viscosity of the dope is increased to the point of gelation such that it is difficult to spin.

The protein fiber formed upon precipitation can be dried to a moisture content of less than 10% and to as low as 3% and can be rehydrated in any compounded meat-like product in the presence of other modifying ingredients such as fat, starches, flavoring agents and the like to resemble meat fibers having satisfactory tensile strength and substance as manifested by the mastication required to disintegrate the fibers. In addition, the fibers are notable for their heat stability since they can be formulated with the foregoing modifying fats, starches, gums and like ingredients into most readily cookable forms and will not disintegrate on boiling in water, baking in an oven at temperatures up to 450°F, or deep fat frying up to temperatures of 450°F.

Example: 600 grams of casein were slurried in a large Waring Blender with 1,800 cc room temperature tap water. This slurry was then placed in a hot water bath of 55°C and agitated with slow stirring while 120 cc of 5% sodium hydroxide solution was added. After holding the slurry for 30 minutes with gentle agitation it has a pH in the range of 5.3 to 5.8.

A slurry of 600 grams of finely ground, to a size passing a 200 mesh screen, defatted soybean meal and 1,800 grams of tap water at room temperature was produced by blending in a large Waring Blender. The slurry was placed in a water bath at 55°C with slow agitation and 500 cc of 5% sodium hydroxide solution was slowly added to bring the pH to between 11 and 11.5, the slurry was stirred for at least 30 minutes at 55°C.

The soy slurry was gradually added to the casein slurry with continuous slow agitation to avoid mixing air into the spinning dough produced by such addition. Stirring was continued for 1 hour at 55°C. To improve the spinnability and strength of the dough thus produced it was held at a temperature of 5°C for 12 hours, during which time the soy and casein became fully hydrated, and entrapped air was permitted to rise to the surface. The dope was brought back to a temperature of 55°C with agitation and 120 grams of melted hydrogenated cottonseed oil was added to improve the spinnability of the dough as well as enhance the texture of the final product. Blending was continued for 30 minutes at 55°C with slow agitation and the blend had a total solids content ranging from 23 to 25%.

The blended spinning dope held at 35° to 45°C is extruded through a 100 hole 0.015 inch diameter spinneret and dropped into a coagulating bath containing 8 liters of water at room temperature having 520 cc 85% lactic acid and 960 grams table salt (sodium chloride) to produce the fibers. The fibers pass through the coagulating bath into a rinse bath containing 7 liters of H_2O and 420 grams of table salt. The fibers are held and stored in 4% table salt solution at 5°C. When it is desired to utilize the fibers in a meat-like product, the fibers are brought to room temperature and excess salt solution removed by squeezing the fibers, to obtain fibers of approximately 35% total solids.

Fibers of Wheat Gluten and Soy Flour Blends

The bite or chewiness of simulated meat products prepared from gluten can be improved by the addition of a heat coagulable protein prior to oven-setting and/or by the inclusion of hardened gluten strips prepared by cooking a gluten composition under steam pressure.

These food products have good chewiness and good variations in bite, however no distinct fibers or fibrous texture is evident in the final products.

N.A. Kjelson; U.S. Patent 3,197,310; July 27, 1965; assigned to General Mills, Inc. found that simulated meat products having improved eating qualities and a fibrous texture can be prepared by simple inexpensive processing from gluten by including a defatted oilseed flour or meal, intimately blending the mixture and then allowing the whole composition to set up by heat.

The gluten used to prepare the food products is preferably freshly prepared wheat gluten. However, dried gluten which is then reconstituted with 50 to 75% and preferably 60 to 70% by weight water can also be used. The defatted desolventized oilseed flour or meal is derived from oilseeds such as soybeans, peanuts, castor beans, safflower seeds, and the like. The flour is preferably prepared from soybeans.

The compositions also preferably contain a heat coagulable protein such as egg albumen or dried egg whites. The amount thereof can be in the range of 5 to 40% by weight based on the weight of the wet gluten. Thus, flavoring agents, colorants, oils and fats can be added, preferably after the gluten and defatted oilseed flour have been intimately blended to form the fibrous structure. The desired amount of defatted oilseed flour is added to the gluten and then the mixture is intimately blended until a fibrous body results. Such blending may be accomplished in a Waring Blender in 5 to 15 minutes, for example. It is preferable to add a heat coagulable protein to the gluten and defatted oilseed flour prior to the fiber formation.

After the above described fiber forming step, the composition is set up by oven heating. The temperatures and times of the oven heating are not critical, but will generally be in the range of 250° to 450°F and a few minutes to 50 minutes or more, respectively. While the process relates generally to a homogeneous meat product prepared from the above described ingredients, it also includes products which contain portions or areas having variations in bite or chewiness. Such products may be prepared by first forming fibrous compositions having different gluten, defatted oilseed flour, or heat coagulable protein contents and then mixing or layering the compositions in a suitable pan to provide areas or portions composed of different compositions. The resulting mixture is then oven set.

Additionally, the food products may include hardened strips or shreds of gluten containing no heat coagulable protein or defatted oilseed flour. Such strips or shreds are prepared by mixing the gluten containing 50 to 75% by weight water with the desired coloring agents, flavoring agents, oils or fats, and then subjecting the mixture to cooking under steam.

Example: A mixture of 100 parts fresh wheat gluten, 10 parts dried egg whites and 20 parts toasted, defatted soybean flour (200 mesh) was intimately mixed in a Waring Blender. Distinct fibers were produced after blending for about 10 minutes. The fibrous composition was then oven set at a temperature of 240°F for 10 minutes. The fibers were still apparent in the resulting product giving it a meat-like texture and appearance.

Fibers from Soy-Keratin Mixtures

Although the process disclosed by *C.A. Anker and P.I. Burchill; U.S. Patent 3,684,522; August 15, 1972* covers primarily the use of keratin in preparing fibrous protein, it also covers keratin mixtures with vegetable proteins such as soybeans. The fibrous proteins are prepared by extruding plastic masses comprising keratin protein or keratin soy mixture directly into a gaseous medium and then elongating the resulting extrudate. The products are edible and serve as substitutes for natural meats.

The keratin used in the process is prepared by the following steps. The first step is to extract the keratin source material with an aqueous solution of an alkali metal sulfide. Representative sulfides are sodium sulfide and potassium sulfide. The protein containing solution is separated from the insoluble residue by conventional means such as decantation, filtration or the like. The protein containing solution is next treated with the alkali metal sulfite. Representative sulfites are sodium and potassium sulfite and bisulfite. A preferred treating agent is sodium sulfite (Na_2SO_3). After the described sulfite treating step, the protein is precipitated by the addition of acid in the conventional manner. Any of a variety of inorganic acids can be used. The precipitated protein is separated from the protein barren liquid by conventional techniques, i.e., decantation, filtration and the like. In all of the steps of the process the temperature is not critical but is preferably in the range of 20° to 50°C. The precipitated and separated protein can be dried if desired. Any conventional drying technique can be used, i.e., spray, drum, tray, freeze or the like. The resulting keratin protein is a high quality product having good odor, color and flavor characteristics. The following examples illustrate the preparation of mixed soy-keratin fibers.

Example 1: A solution containing 100 ml water, 20 ml 6 N NH_4OH, 50 grams glycerol and 3 grams Na_2SO_3 was mixed in a Waring Blender with 40 grams soy isolate (approximately 95% by weight protein) and 60 grams feather keratin. The resulting mixture was formed into a plastic mass and extruded using the Brabender extruder equipped with a 1" wide ribbon die having a 5 mil gap. The temperatures in the three heating zones of the extruder were 80°, 100° and 95°C, respectively. As the extruded ribbon left the die, it was stretched so that it had a width of approximately ½". The extruded product was very fibrous.

Examples 2 through 4: Example 1 was essentially repeated except that the ratio of soy protein to feather keratin was changed as follows: Example 2, 50 grams soy isolate and 50 grams feather keratin; Example 3, 60 grams soy isolate and 40 grams feather keratin; and Example 4, 70 grams soy isolate and 30 grams feather keratin. In all instances a fibrous product was obtained on stretching of the extruded ribbon. As the level of soy isolate was increased, the fibrous texture of the product was reduced. Thus in a similar run using 80 grams soy isolate and 20 grams feather keratin, the extruded and stretched ribbons contained no noticeable fibers.

FIBERS FROM PROTEIN-CARBOHYDRATE POLYMER BLENDS

Mixed Fibers Having a Polymeric Carbohydrate Gel Precursor

In a process developed by *N.H. Ishler, R.V. MacAllister, A.S. Szczesniak and E. Engel; U.S. Patent 3,093,483; June 11, 1963; assigned to General Foods Corporation* food products having fiber-like textures are produced by forming a sol of a polymeric carbohydrate gel precursor and the sol is formed into oriented, thermostable, gelled fibers. The fibers are gelled with an alkaline earth metal ion, typically a calcium ion.

The term thermostable gels refers to those gels which will not reverse into a sol upon heating at temperatures required to cook or heat the food product, such heat treatment ranging from gentle warming to deep fat frying.

Desirable products can also be produced by forming fibers from a sol of a polymeric carbohydrate gel precursor containing a modifying agent. Such fibers can also be aggregated to form fibrous compositions having unique textures and eating qualities. The term modifying agent includes materials capable of actually modifying the gel-forming characteristics of the gel precursors, as well as additives which would simply be carried by such gel precursors. Included in the former class of compounds are such compositions as raw farinaceous doughs like raw wheat flour dough, raw corn flour dough, raw starch doughs, etc., calcium caseinate, plasticizers, fats, etc. Included in the latter class of compounds and considered as additives are such materials as relatively inert filler materials, colors and flavors, inert proteins, such as various seed meals, cottonseed meal, soybean meal and farinaceous materials such as cooked wheat flour, corn flour, starches, etc.

The combined sol thus produced is formed into fibers, preferably by extrusion and using many of the fiber-orienting procedures of the man-made textile prior art, to wit: a spinneret capable of producing fibers having a thickness in the order of 0.003 to 0.050 inch in mean diameter (although even higher cross-sectional dimensions are contemplated). While it is preferred to stretch or orient the fibers after extrusion it is possible in some instances to obtain suitable oriented fibers merely by forcing the sol through a spinneret followed by setting the extruded sol in a coagulating bath as has heretofore been described. The following example shows the formation of fibers wherein the additive is soy flour.

Example: (a) An aqueous colloidal solution containing 3% sodium alginate (900 centipoises) and 7% dry solvent extracted soybean meal was extruded through a 50 hole spinneret having 0.008 inch diameter holes into a coagulating vat containing 10% by weight of calcium acetate and 1% by weight of acetic acid. The treated filaments which formed in the coagulating vat were taken up on a collecting reel and were then washed with pure water. A nutritious, high protein fiber having desirable chewing and eating properties was obtained.

(b) An aqueous colloidal solution containing 3% sodium alginate (900 centipoises) and 5% commercial casein was extruded through a 50 hole spinneret having 0.008 inch diameter holes into a coagulating vat containing 10% by weight of calcium acetate and 1% by weight of acetic acid.

The treated filaments which formed in the coagulating vat were taken up on a collecting reel and were then washed with pure water. A nutritious, high protein fiber having desirable chewing and eating properties was obtained.

Aluminum Modified Protein-Alginate Fibers

Synthetic protein-carbohydrate fibers which can withstand boiling in salt solutions are produced by *W.T. Atkinson; U.S. Patent 3,455,697; July 15, 1969;* assigned to *Archer Daniels Midland Company.* His process comprises spinning a protein-carbohydrate gel mixture into an aqueous coagulating solution containing an alkaline earth metal salt and a water-soluble aluminum salt.

Preferably, the process comprises: (a) forming a dispersion of a protein material (e.g., an inert or a modifiable protein material) and a solution of a polymeric carbohydrate gel precursor capable of producing a continuous thermostable gel; (b) forming the resulting dispersion into fibers, e.g., by extrusion through a spinnerette; and (c) treating the fibers in a slightly acidic, aqueous bath containing an alkaline earth metal cation and aluminum to gel the precursor and transform the protein, if of a modifiable nature, from its solution state into a substantially solid state. The final fibers are usually arranged into tows and bound together by use of an edible binder to form a meat-like product.

The gelled fiber is comprised of aluminum and an alkaline earth metal carbohydrate fiber. It has been found that the presence of aluminum in a carbohydrate fiber protected the system from deteriorating ion exchange which can occur if the fiber is heated in an alkali metal salt solution. When, for example, calcium alginate fiber is heated in the presence of sodium or potassium salts, the calcium alginate reverts to the sodium or potassium alginate with the resultant loss in fiber strength. The presence of aluminum in the fiber prevents the reversion.

The aluminum salt which is to be added to the gelling bath can be any of the ionizable aluminum salts, such as aluminum chloride, aluminum sulfate, aluminum nitrate, aluminum chlorate, and the like, including mixtures. The aluminum salt should be added from 0.10 to 5.0% by weight of the bath depending upon the type of fiber desired, i.e., large amounts of aluminum will give a hard and tough fiber while small amounts will give a fiber which is much more susceptible to ion exchange.

The carbohydrate material used is selected from the thermostable gel precursors such as those derived from algins or pectins and like polymeric carbohydrates, e.g., alkali salts of alginic acid and low methoxyl pectin, which generally have a high order of sensitivity to reaction with polyvalent cations. The gel-forming agents in the gelling bath are alkaline earth metal ions such as calcium and the like. These alkaline earth metal cations serve to gel the precursor and also to coagulate or to precipitate the protein, if the protein which is used is of a modifiable (i.e., precipitable or coagulable) nature. Ordinarily, the alkaline earth metal ion which is used is calcium acetate, and it is present in an aqueous solution.

Among the protein materials which can be used are modifiable extracts of soybean meal, isolated proteins, peanut meal, fish protein meal, albumins, such as egg albumin and soy albumin, gelatin, caseins, globulins, glutelins, and the like. Other color, flavor, and texture modifying ingredients can also be added.

The temperature of the spinning dispersion is usually held at about room temperature, 20° to 25°C. The extrusion step is usually conducted at slightly elevated temperatures and pressures, e.g., about 20° to 60°C and 0 to 100 psi. The pH of the spinning dispersion should be at least slightly acidic, e.g., 5.0 to 7.0, preferably 6.0 to 7.0. As stated above, the gel precursor is gelled by treatment of the fibers immediately after formation with an aqueous solution of a polyvalent cation. A preferred bath is comprised of calcium acetate at slightly acidic conditions, e.g., at a pH of 4.5 to 6.0, preferably 5.0 to 5.5. The protein, if of a modifiable form, is either precipitated by lowering the pH of the bath to the isoelectric pH of the protein, or it is coagulated by heating the fibers after the alkaline earth metal cation solution treatment to gel the precursors. Alternatively, the protein can be in a nonmodifiable form, and the fibers would require no treatment other than gellation of the polymeric carbohydrate gel precursor.

Example: An aqueous colloidal solution having the following composition with all parts by weight, 1,800 isolated soy protein, 19,500 water, 100 antifoam agent, 178 sodium alginate, and 90 fat (chicken), was extruded through a 204 hole spinnerette having holes 0.015" in diameter into a coagulating bath containing 1.0% by weight aluminum chloride, 1.0% by weight calcium oxide, 2.5% by weight glacial acetic acid, and 95.5% water.

The coagulated fibers were washed with water, after which a small amount of the fibers was boiled for 30 minutes in an aqueous 1.0% solution of sodium chloride. The boiled material was quite tough and was considerably tougher than a control sample which was made by the same method as above except that the coagulating bath contained 3.5% by weight calcium acetate and no aluminum chloride. The rest of the washed fibers were arranged into tows and bound together by use of a small amount of an edible protein binder. These tows were autoclaved and canned in a 0.5% sodium chloride solution. On standing, there was no noticeable discoloration nor was there a loss of toughness. The control sample suffered discoloration and loss of strength when canned in a similar fashion.

Preservation of Spun Protein-Alginate Fibers

Bacterial growth is controlled in protein fiber production using the process developed by W.T. Atkinson; U.S. Patent 3,645,746; February 29, 1972; assigned to Archer Daniels Midland Company.

Briefly, the process accomplished the preservation of protein materials by the steps of:

(1) Heating the protein material to a temperature above 140°F, and
(2) treating a mixture of the heat-treated protein, a water soluble alginate, and water with peroxide.

The treated protein materials are especially useful in the preparation of edible spun fibers. It is desirable to include carrageenin prior to spinning. Accordingly, the process also involves the subsequent steps of:

(3) Adding carrageenin to the protein-alginate mixture before or after the heat treatment step (1), and then
(4) spinning the mixture into a coagulating bath.

Although any suitable means can be employed for heating the protein, in a preferred embodiment the protein is mixed with water to form a slurry which is then admixed with a solution of the soluble alginate to form a mixture. This mixture is then heated to a temperature above 140°F and below the boiling point of the material, and preferably at a temperature of 140° to 180°F. The heating is continued until the mixture no longer decomposes peroxide, which generally occurs after heating periods of from 5 to 60 minutes. Heating periods of about 10 to 30 minutes have been found to be satisfactory and are preferred. Heating periods in excess of 4 hours are to be avoided because of resultant undesirable darkening of the protein and a reduction in toughness and tensile strength.

Any type of edible protein of vegetable, fish, or animal origin may be used in this process. The preferred proteins are the oilseed proteins such as peanut, cotton seed, sesame seed, or soybean proteins. The protein may be used in substantially pure or water-soluble form, or, as is preferred, in the form of flakes or flour, generically referred to as meal, obtained on solvent extraction of oils and other fatty material.

The compositions utilized in the process can have widely varying weight ratios of protein to soluble alginate, but these ratios are generally between 1:100 and 100:1, and preferably between 2:1 and 20:1. The water present in the mixture can comprise up to 95 weight percent, based on the combined weight of protein, soluble alginate, and water. The peroxide is generally added as an aqueous solution. The peroxide solution is added to the protein materials in amounts providing up to 2.0 and preferably 0.01 to 1.0 weight percent peroxide, based on the combined weight of protein and soluble alginate. The peroxide can be any water-soluble peroxide capable of decomposition to release oxygen in the presence of catalase. Inorganic peroxides, especially alkali metal peroxides such as sodium peroxide and potassium peroxide, can be used.

Example: The following are combined as indicated:

	Ingredients	Quantity
(1)	Soy flour	5,400 g
(2)	Water at 100°F	15,750 ml
(3)	Sodium alginate (Kelco HJ)	518 g
(4)	Water at 212°F	27,000 ml
(5)	Hydrogen peroxide (50 wt % aqueous solution)	98 ml
(6)	Dytol (60% lauryl alcohol)	5 ml

Ingredients (1) and (2) are thoroughly mixed in a first vessel, (3) and (4) are thoroughly mixed in a second vessel. The contents of the first vessel are then poured into the second vessel to form an alginate/soy flour slurry having a temperature of 135°F, which is then heated (for approximately 10 minutes) to 170°F, and held at a temperature of 170°F for a period of 0 to 5 minutes. At the end of this period ingredients (5) and (6) are added. The slurry exhibits no bacterial count even after 20 hours.

The protein-alginate composition is extruded through a spinnerette having a plurality of holes into a coagulating solution comprising:

Ingredients	Quantity
$CaCl_2 \cdot 2H_2O$	1,005 g
Water	15,000 ml
HCl (37 wt % aqueous solution)	75 ml

The ingredients are mixed together to form a coagulating bath having an initial pH of 1.3. A hank of spun tow was washed in water before storage to prevent hydrolysis of the fiber at low pH's. The washed fiber had a pH of 3.8. Optionally, 98 grams of carrageenin may be added and thoroughly mixed with the above slurry. The resulting composition can then be extruded into edible fibers in the same manner described above.

Protein-Thermogelable Polysaccharide Fibers

Fibers with improved texture and mouthfeel are produced by *K. Sawada, S. Moritaka,*

Y. Nakao and K. Yasumatsu; U.S. Patent 3,806,611; April 23, 1974; assigned to Takeda Chemical Industries, Ltd., Japan by adding a thermogelable polysaccharide to the fiber forming solution.

The thermogelable polysaccharide used in this process is a white or off-white powder, with a characteristic absorption band showing the β-glucosidic bond at 890 cm^{-1} by infrared analysis. In addition, this polysaccharide swells on the addition of water and its aqueous suspension whose concentration is not less that 1% (weight/volume) is gelled into a jelly or agar-like mass when heated. This change is irreversible, and the properties are not affected by cooling, nor can it be dissolved in water.

The strength of the gel obtained by heating a 2% aqueous suspension of the thermogelable polysaccharide in boiling water for 10 minutes is between 470×10^3 to $1,300 \times 10^3$ dyne/cm^2. The thermogelable polysaccharides are, for example, a thermally gelable β-1,3-glucan named curdlan, a thermogelable β-1,3-glucan-type polysaccharide named PS. The thermogelable polysaccharide is produced by aerobic cultivation of a thermogelable polysaccharide-producing microorganism belonging to the genus Alcaligenes or Agrobacterium.

For production of the polysaccharide, these microorganisms are incubated in a medium containing assimilable carbon sources (e.g., glucose, sucrose, sorbitol, dextrin, starch hydrolyzates, organic acids) digestible nitrogen sources (e.g., inorganic ammonium salts, nitrates, organic nitrogen sources such as yeast extract, cornsteep liquor, corn gluten, soybean meal), inorganic salts (e.g., salts of manganese, iron, magnesium, calcium, zinc and cobalt). If desired, trace growth promoters as vitamins, etc. may be added.

Although the preferred conditions vary with different microorganisms, cultivation of the main culture for the production of the polysaccharide is generally done at a pH of 5 to 8 at 20° to 35°C for 2 to 4 days, using a shake culture or a submerged culture method. As the polysaccharide produced occurs mainly extracellularly, advantage may be taken, in order to recover it, of a combination of known techniques for separating and purifying the polysaccharide, e.g., dissolution, filtration, precipitation, desalting, liquid-solid separation, drying, powdering etc. Details of the preparation of these polysaccharides are given in the patent.

High quality edible fibers can then be prepared by this process by (a) preparing an edible protein fiber spinning solution having a pH of 9 to 13.5 and containing (1) the protein and (2) the thermogelable polysaccharide in a (1) to (2) ratio of 100:1 to 100 (on a weight basis), the combined amount of (1) and (2) being 7 to 20 weight percent relative to the above spinning solution, (b) extruding the spinning solution through a spinneret into an acid coagulating bath thereby coagulating the solution in the form of filaments, (c) neutralizing the same and (d) dehydrating them.

The protein used as the raw material in this method includes such proteins as soybean, peanut, casein, and the like. Other vegetable proteins, animal proteins such as fish flesh, proteins obtainable by the cultivation of microorganisms including the so-called petroleum proteins may also be used effectively. The fiber forming process is shown in detail in the following example.

Example: To 5 kg of defatted soyflour is added 50 kg of water and the mixture stirred at 40°C for 1 hour. The mixture is centrifuged to obtain a soybean protein extract. The extract is adjusted to pH 4.5 with hydrochloric acid and centrifuged to obtain 4.5 kg of acid-precipitated soybean protein curd having a moisture content of 65%. Then, the additives indicated in the table below are added to aliquot portions of the above curd in the protein-additive ratio of 13:1 (weight basis). After the addition of water, each mixture is evenly stirred in a homogenizer to prepare a slurry sample. To each of those slurries, a 10% aqueous solution of sodium hydroxide is added, followed by stirring to homogeneity. The procedure yields various alkali-solubilized solution samples, each having a pH of 12.3 and a combined soybean protein-additive concentration of 14 weight percent.

Each solution is aged under gentle stirring at 40°C for 20 min and defoamed to form the spinning solution. This spinning solution is extruded through a spinneret having orifices 0.12 mm in diameter into a coagulating bath comprising a 3% aqueous solution of acetic acid containing 10% of sodium chloride. The filaments are neutralized to a pH of 5.5 by immersing them in an aqueous solution of sodium bicarbonate, washed thoroughly with water and dehydrated by squeezing over rubber rollers.

A sensory test was made on the resulting edible protein fibers by the scoring method (on a 5-point grade) using a panel of 20 experts. For hardness, a higher value represents an increased degree of hardness; for difficulty to bite off, a higher value represents an increased difficulty to bite off, for smoothness, a higher value represents an increased degree of smoothness; and for whiteness, a higher value means that the sample is nearer to black. The hardness of each sample was measured with a texturometer. The results of the above tests are shown in the table below. Compared with the control samples, the edible protein fibers by this process have excellent qualities, including soft tenacious mouthfeel and ease of biting off.

Sensory Test Scores

Number	Additive	Hardness	Difficulty to Bite Off	Smoothness	Whiteness	Texturometer Hardness (texturo-units)
1	None	3.8	3.5	1.9	3.7	7.5
2	Polysaccharide PS-B	2.1*	2.3*	3.0**	3.3	5.6
3	Shortening	3.3	3.1	2.4	3.6	6.9
4	Starch	3.1	3.0	2.2	3.2	6.8
5	Casein	3.9	3.7	2.0	3.1	7.4
6	Gelatin	3.3	3.3	2.3	3.3	6.8

*The sample is significantly different from the control (no additive) at the 1% level of significance.
**Same as above except that the level of significance is 5%.

MISCELLANEOUS FIBER-FORMING COMPOSITIONS

Protein Fibers from Safflower Seed Meal

L.F. Elmquist; U.S. Patent 3,175,909; March 30, 1965; assigned to General Mills, Inc. has found that fibers can be produced directly from safflower seed meal without first having to isolate the protein. Prior processes have used isolates from such vegetable proteins as soy, corn, peanuts, etc.

The safflower seed meal used as the starting material is obtained by dehulling and deoiling naturally occurring safflower seeds by conventional procedures. Preferably, the oil is removed by solvent extraction. After the extraction, the dehulled and substantially oil-free seeds are desolventized by any suitable method to remove residual solvent(s) and then preferably ground to a relatively fine particle size to facilitate the preparation of the spinning solution or dope.

The spinning dope can be prepared by dispersing the safflower seed meal in a dispersing medium in varying amounts. Preferably the amount of the meal in the spinning dope will be from about 20 to 30% by weight. If desired, the dope can be prepared from mixtures of the safflower seed meal and other edible protein materials. A suitable dispersing medium is water containing an alkaline material, i.e., about 1 to 10% by weight alkali metal hydroxide. The pH of the spinning solution or dispersion can vary within relatively wide limits, but will generally be in the range of about 9 to 13.5. The viscosity and temperature of such dispersions will generally be within the range of about 5,000 to 100,000 cp and about 20° to 45°C, respectively.

After formation of the spinning dope, it is forced through a small orifice into a coagulating bath which can be an acid salt solution. If a single filament is desired, the solution may be forced through a device having a single opening, the diameter of which can be varied according to the size of the filament desired. A porous membrane, such as a spinneret used in the production of rayon, containing a plurality of openings can be used where more than one, i.e., a tow of filaments is to be produced. The streamlets coming through the spinneret are precipitated in the acid-salt bath in the form of a tow of filaments. It is also possible to have a series of spinnerets producing filaments from the safflower seed meal dispersion. Such spinnerets may have the same or different number of holes making it possible to directly produce tows of filaments having the same or different diameters.

The coagulating bath contains an acid constituent and is preferably an aqueous solution of salt and an acid. The salt (i.e., NaCl, for example) can be used in widely varying concentrations, such as from 0.5 to 15% by weight. Representative acidic compounds are acetic acid, lactic acid, citric acid, adipic acid, hydrochloric acid and the like. The concentration of the acids in the bath is not critical and may vary between about 0.5 to 10% by weight.

Example: To a dispersion of 12.5 grams safflower seed meal in 37.5 grams water was added 4.92 grams of a 10% aqueous solution of NaOH. The safflower seed meal had the following analysis: 58.25% protein, 6.10% oil, and 8.6% water. The meal was prepared by hexane extraction of dehulled safflower seed meats. The blend was stirred at 40°C until smooth. The spinning dope was cooled to 25°C and a portion thereof placed in a syringe attached to a Number 23 needle (0.013 inch diameter bore). A streamlet of the dope was forced through the needle into an aqueous precipitating bath consisting of 0.1 N HCl and 10% by weight NaCl. The filament formed in the bath was almost white, bland and had good strength and elasticity.

Polyacrylic Acid Additive for Protein Fibers

Fibers are spun from solutions having a low protein content in a process disclosed by *T. Sakita, M. Ebisawa and H. Mimoto; U.S. Patent 3,749,581; July 31, 1973; assigned to The Nisshin Oil Mills, Ltd, Japan.* The fibers have good chewing properties desired in a foodstuff. This is accomplished by using an additive selected from polyacrylic acid and the alkali salt and ammonium salt thereof in any desirable steps of preparing the protein fiber.

The proteins usable in the process include both vegetable and animal proteins. The vegetable proteins include proteins from oil-bearing seeds such as soybean, sun flower, safflower, peanuts, cottonseed and the like and wheat protein, and the animal proteins include casein and gelatin, etc. The polyacrylic acids useful in the process are the water-soluble long linear macromolecular electrolytes having moecular weights of several millions, which have plural carboxylic groups along their hydrophobic vinyl chains. The molecular weight of the polyacrylic acids and salts should be such that the degree of polymerization of the polymers is at least 15,000 and preferably 30,000 or more. The higher the molecular weight the larger the spinning effect.

The process has the following advantages:
- (1) The protein fiber can be directly and continuously produced from any extract without any steps of preparation, dispersion and peptization of the protein curds. As a result equipment, time, labor, and chemicals can be saved and mechanical loss of the product can be reduced.
- (2) The dispersed protein solution can be directly spun into tough protein fiber without need of removing oil and nonprotein substances in the edible protein. The yield of the fiber is very high and the resultant fiber is extremely strong in spite of containing nonprotein matter in the resulting fiber.
- (3) The concentration of the dispersed protein solution can be selected arbitrarily, and any protein solution of lower protein concentration, for example, as low as 2 weight percent can be spun to useful fiber.
- (4) The spinning can be conducted at lower pH range such as at pH 5.5 to 11.
- (5) The amount of additives is very low, below 2 weight percent, with respect to the

protein of the solution, or below 0.5 weight percent with respect to acidic coagulating solution of polyacrylic acid or the salt thereof being required for the spinning.

(6) The spinning solution after the addition of polyacrylic acid or its salt has fairly good spinning ability, in spite of extremely low viscosity.

(7) The spinning solution can be prepared in a very short time; the time required including the pH regulation is merely about 10 minutes.

(8) Polyacrylic acid or its salt exerts an antifoaming effect when added to the protein solution prior to the spinning, the antifoaming effect is particular marked at pH 8 to 10, and the conventional defoaming step can be avoided.

(9) A neutral salt such as sodium chloride, sodium sulfate, sodium acetate and the like is not required in the acid coagulation bath. The acid to be used in the coagulation bath includes hydrochloric acid, sulfuric acid and acetic acid, etc. which are conventionally used in acid coagulation baths.

(10) A fine and tough protein fiber with more favorable color and taste can be obtained as compared with the products obtained by the conventional methods.

Example: A dilute or concentrated alkaline extract of soybean protein obtained conventionally (pH 9.0) was added to polyacrylic acid and the resulting spinning solution was extruded from nozzles of 0.08 mm in diameter in a coagulation bath containing 5% sulfuric acid. The protein concentration of the extract during coagulation into protein fiber and the effective amount of polyacrylic acid required were investigated. The results are shown in the table below.

Required Amount of Polyacrylic Acid to Effect Spinning

Protein concentration of the extract (%)	2	4	6	8	10	12
Amount of polyacrylic acid required*	4	2	0.8	0.4	0.2	0.1

*Weight percent based on the amount of protein

The alkaline extract of soybean protein obtained by the conventional method contains more than 4% of protein, then addition of only 2 weight percent or more (based on the protein) of polyacrylic acid gives useful effect to the spinning.

Fibers from Protein-Lipid Extracts

B.G. Newsom and M.P. Tombs; U.S. Patent 3,794,735; February 26, 1974; assigned to Lever Brothers Company provide a process for the extraction of protein from protein-bearing seed which avoids the low-temperature extraction of seed with a lipid solvent, and yields a fluid lipid-containing protein material which can be used in the manufacture of food products.

In the first stage of the process, the seed is finely subdivided, and an aqueous emulsion of protein and lipid is formed which contains in suspension the water-insoluble carbohydrate of the seed. In carrying out the first stage, the seeds are split or the hulls removed and leave the cotyledons exposed, and these are finely subdivided in the presence of water, conveniently by passing them with water through a homogenizer equipped with cutters. Alternatively, whole (unhulled) seeds can be finely ground in the presence of water in a mill. Disintegration of the cotyledons releases powerful emulsifying agents naturally occurring in them, and under the action of these agents the protein and lipid of the seeds become emulsified.

It is desirable to include in the water used from 0.01 to 5% by weight of a dissolved agent (such as a water-soluble sulfite, bisulfite or dithionite) that prevents the linking of protein —SH groups to form intermolecular —S—S— linkages. By this means one prevents the aggregation of protein molecules that tends to occur in aqueous protein systems and often results in gelation. The aqueous emulsion is then treated to remove the water-insoluble carbohydrate that it contains in suspension. This is best done by centrifugation.

The supernatant protein-lipid emulsion is then acidified to reduce its pH to the isoelectric pH of the protein component of the emulsion. The use of hydrochloric or sulfuric acid as precipitant is preferred. The protein of the emulsion is not precipitated alone, but brings down with it a fairly high proportion of the lipid, that is to say, a coprecipitate of protein and lipid is formed. This precipitated material is separated, suitably by centrifugation, and is obtained in the form of a cake, usually of solids content in the range 30 to 60% by weight.

The separated coprecipitate is then mixed with an edible salt; this, when used in at least the minimum proportion defined in the next paragraph, has the effect of converting the solid protein in the coprecipitate into a fluid aqueous preparation of protein concentration at least 15% by weight and usually in the range 15 to 50%. This fluid aqueous preparation is rather viscous and contains distributed in it the lipid present in the coprecipitate, and it can be used in the manufacture of food products.

The edible salt used is preferably sodium chloride, but other edible salts can be used, for example sodium phosphate, potassium chloride and calcium chloride. The salt is used in an amount such that the fluid aqueous preparation that is formed from it and from the separated coprecipitate and associated water contains dissolved edible salt in a concentration equivalent to an ionic strength of at least 0.2.

Example 1: Unhulled soybeans (20 kg) were mixed with aqueous 0.1% sodium sulfite solution (200 liters) and milled in a Hobart MCV 12 mill, so as to pass a 0.2 mm gap. The aqueous protein-lipid emulsion formed was passed twice through a supercentrifuge to remove water-insoluble carbohydrate, and was then acidified to pH 4.9 with hydrochloric acid. The resulting coprecipitate of protein and lipid was centrifuged off, and the separated cake (53.7% solids content) was analyzed and found to contain 28.7% protein and 19.6% lipid. 4% by weight of solid sodium chloride was mixed with the cake, and a viscous aqueous preparation formed containing 20.9% protein and 14.4% lipid, and having an ionic strength of 1.16.

Example 2: This example illustrates setting treatments of a fluid aqueous protein-lipid preparation such as is obtained by the procedure of Example 1. A preparation obtained following the procedure of Example 1 and of the percent composition of water, 65.4; solids, 34.6; protein, 20.7; lipid, 7.1; and ash 3.6; was filtered through a 100 mesh (BSS) sieve to remove traces of solid carbohydrate and was extruded through spinnerets of diameter 0.2 mm into a setting bath maintained at 75° to 80°C. In four separate procedures, the setting baths had the compositions: (1) isopropanol, (2) 75% isopropanol, 25% water, (3) 50% isopropanol, 50% water, and (4) 25% isopropanol, 75% water. The composition of the fibers produced is shown below, and compared with that of the aqueous preparation from which they were obtained by extrusion.

Composition of Fibers

	Water	Total solids	Protein	Lipid	Ash	Protein/ lipid (%)
Fibers 1	29.7	70.3	41.3	11.1	6.4	3.64
Fibers 2	53.5	46.5	31.5	8.9	1.7	3.54
Fibers 3	55.3	44.7	31.2	8.8	0.9	3.54
Fibers 4	56.0	44.0	31.2	9.6	0.5	3.24

The fibers obtained as described above can be used as a foodstuff, for example as part-replacement for the meat ordinarily used in the preparation of beefburgers or canned reformed meat. The flavor of soybean in material thus made is only slight, and hardly detectable at all with use of the fibers made in setting bath (1).

OTHER PROTEIN FIBER PRODUCTION METHODS

COAGULATION OR PRECIPITATION OF FIBERS

Coagulation of Alginate-Protein Suspensions

An acid coagulation-accelerating solution is used by *T. Arima and Y. Harada; U.S. Patent 3,627,536; December 14, 1971; assigned to General Foods Corporation* in a process for preparing fibers from vegetable and animal protein products. This process comprises the steps:

(1) Stirring an aqueous suspension of protein containing 10 to 60% by weight, based on total solids, of a soluble alginate salt in a blender and gradually adding an aqueous solution of coagulant selected from the group of calcium chloride, calcium phosphate, calcium lactate, calcium gluconate, and calcium sulfate to form a semicoagulated protein curd. The curd is simultaneously sliced by the rapidly rotating blender blades to form soft fibers of random short length in flake form.

(2) Reducing the pH to 3.2 to 2.4 with a coagulation-accelerating solution selected from gluconic acid, sulfuric acid, acetic acid, hydrochloric acid, lactic acid and phosphoric acid whereby the semicoagulated fibers become harder and simultaneously partially dewatered.

(3) Boiling the proteinaceous fibers with the coagulating and coagulation-accelerating solutions to thermally coagulate the protein in the flake type fibers.

(4) Washing the fibers with water to remove the excess coagulating solution and coagulation-acceleration agent.

(5) Boiling the fibers in excess water to leach out any remaining unreacted materials, the excess coagulating solution and the excess coagulation-accelerating solution.

(6) Washing the fibers further with water so that the fibers become hard, bland to taste and odorless, having a tendon-like external appearance.

(7) Partially dehydrating the fibers.

(8) Finish drying the fibers.

Proteins found useful in this process may be from any source, for example, soybeans, cottonseed, wheat, peanuts, corn, egg albumen and milk casein. Mixtures may be used also, and such mixtures have been found to result in excellent products.

Other Protein Fiber Production Methods

It was unexpectedly discovered that different ratios of various of these proteins mixed with alginate solutions of varying amounts result in fibers having properties approximating those of certain kinds of meats. Listed in the following table are ratios of various proteins found suitable for simulation of beef, pork and chicken protein fibers.

Fiber Composition (Percent)

	Beef-Like Fibers	Pork-Like Fibers	Chicken-Like Fibers
Soybean protein	20:50	50:80	40:60
Peanut protein	5:20	5:20	10:20
Egg albumen	5:20	5:10	10:40
Milk casein	–	5:10	5:10
Sodium alginate	40:60	10:20	20:40

The protein fibers of this process are preferably made by utilizing a continuous process in a series of tanks, but may be also advantageously manufactured by using a batch technique. While any of the well-known apparatus for continuous type tank treating may be used, the box-type blender designed by Satake Chemical Machine Seisoku-Sho, Morigughi City, Osaka, Japan, has been found to be particularly useful. A total treating time of 30 minutes has been found preferable. The following example will illustrate the process in more detail.

Example: Protein Fiber Production—Utilizing the process and the apparatus as described above, 3 kg of soybean protein, 1 kg of peanut protein, 3 kg of egg albumen and 3 kg of sodium alginate were stirred in a blender for 5 minutes, after which 100 liters of water were added and stirred for 20 minutes. The mixture was introduced into the first tank at a rate of 2 kg/min for a period of 55 min. 555 liters of an aqueous solution of 10% calcium chloride was concurrently added at a rate of 10 l/min while the mixture of protein and sodium alginate was being stirred. The curd formed was stirred and sliced for 15 min.

After stirring and slicing to form soft flake type fibers, 55 liters of glacial acetic acid was added, at a rate of about 1 l/min, to harden the fibers. The flakes (or fibers) obtained were passed into the second tank, where the flakes were boiled for about 5 minutes to set the heat coagulable protein content of the flakes. After boiling, the aqueous solution of calcium chloride, acetic acid, and the flakes were transferred to the third tank.

In the third tank, the coagulating solution was drained from the flakes (or fibers) and the flakes were then washed with water for about 1 minute using a shower-type washer and thereafter transferred to the fourth tank, where the flakes (or fibers) were further boiled for about 5 minutes and, thereafter, transferred to the fifth tank where they were rewashed with water again for about 1 minute.

The flakes (or fibers) from the fifth tank were drum-dewatered to a water content of about 60%; then, on a belt conveyer, passed through an air drying box until the water content was reduced to about 50%. The yield was approximately 20 kg of fibers having a width of about 1 to 4 mm and a length of about 8 to 20 mm.

Protein Coagulation with Divalent Cations

T. Nagasawa, M. Tomita, T. Obayashi, Y. Tamura and Y. Kenmotsu; U.S. Patent 3,674,500; July 4, 1972; assigned to Morinaga Milk Industry Co. Ltd., Japan have developed a process for preparing foodstuffs having a porous and elastic meat-like texture by reacting a protein solution with a divalent cation in a highly alkaline region (pH 9.0 or above) and then neutralizing, washing with water and removing the reaction product.

The porous, elastic protein coagulate obtained by this method is light colored, tasteless and odorless. The protein coagulate contains over 96% of the available protein. When fat and carbohydrate are incorporated homogeneously in the protein solution, the yield of fat and protein in the coagulate is over 96%. Carbohydrate recovery is not as good, but the carbohydrate loss is surprisingly low (12%).

(1) Preparation of the Protein Solution: Casein, soybean protein, wheat gluten or their mixture, or protein-fat mixtures and protein-fat-carbohydrate mixtures may be used as the raw material. When employing soybean protein powder as a protein source, it is preferably dispersed in hot water and dissolved by using an alkaline agent to prepare a solution containing more than 2%, preferably 5 to 8%, of protein. The alkaline agents which can be used include the hydroxides, phosphates, hydrogen phosphates and carbonates of sodium and potassium, and in particular, the phosphate is preferable. The presence of the phosphate radical serves to form a calcium caseinate phosphatide complex and improves the texture of the coagulate.

(2) Preparation of the Reaction Liquid: In order to react a divalent cation with a protein, which exists in the free state in a high alkaline region (pH above 9.0), and, therefore, in order to maintain the pH constant during the course of reaction, calcium hydroxide, magnesium hydroxide, etc., may be used, with calcium hydroxide being preferable. Other divalent cations to be used will include, in addition to calcium and magnesium, ferrous ion, cupric ion, zinc ion, etc. For the preparation of reaction liquid, it is necessary to form a supersaturated solution of, e.g., calcium hydroxide, add hydrochloric acid or calcium chloride to make the pH above 9.0, preferably 11.0 to 12.0, control the concentration of free calcium ions to above 0.25 millimol per 1 gram of protein, particularly in the range of 0.6 to 3.75 millimol per 1 gram of protein, and maintain the temperature of liquid above 40°C, particularly 85° to 95°C.

(3) The Reaction Step: The protein solution described in (1) and the reaction liquid described in (2) were heated to a temperature of above 40°C, more exactly, to 85° to 90°C, separately, and were reacted. The properties of the meat-like texture obtained vary depending upon the way of carrying out the reaction operation.

    (a) The reaction liquid may be slowly added to the protein solution while stirring the solution vigorously.

    (b) The protein solution may be slowly added to the reaction liquid while stirring the liquid vigorously.

The protein coagulates differ depending upon which of the above two methods is used. The texture of the coagulate produced by method (a) is similar to that of animal meat which has been boiled and ground, whereas the tissue of the coagulate produced by method (b) is fibrous and is similar to corned beef.

(4) Neutralization: After the reaction (3) the excess of alkaline agent (usually calcium hydroxide) is neutralized with an acid, e.g., HCl, and then is removed in the subsequent water-washing step. Thereby, any strong alkaline taste due to the existence of an excess of alkaline agent is completely removed.

(5) Water Washing Steps: After completing the neutralization step in (4), the protein coagulate is filtered and pressed to reduce the water content to 40 to 70%, and thereafter, the protein coagulate is dispersed in water again, and the steps of filtering, pressing and dispersion are repeated.

Example: 200 grams of commercially available soybean protein powder was dispersed in hot water, and after adjusting the pH to 10.5 with 10% (by weight) K_3PO_4 solution, was heated to dissolve and the amount of liquid was adjusted to 2.0 kg. The composition of commercially available soybean powder was: 0.4% fat, 64.0% protein, 6.0% ash, 25.6% carbohydrate and 4.0% water. Next, a casein solution was prepared from 1.19 kg of lactocasein consisting of 2.5% fat, 84.0% protein, 1.5% ash and 12.0% water dispersed in 8 kg of hot water at 40° to 50°C, and 0.715 kg 10% by weight of K_3PO_4 solution was added, and after adjusting the amount of liquid to 10 kg, was heated under stirring to completely dissolve the casein and then heated to 90°C. The pH of the casein solution was 6.4 and the protein concentration 10%, by weight. 8 kg of the casein solution was mixed with the above soy protein solution, and the temperature adjusted to 60°C. Next, 0.285 kg of $Ca(OH)_2$ was dispersed in 3 kg water, 1.660 kg of 3 N HCl added, and, after adjusting the amount of liquid to 10 kg, heated to 85°C under stirring to make a reaction liquid.

The pH of the reaction liquid was 11.6 and it contained about 2.5 mmols of free calcium ions per 1 gram of protein. The protein solution was gradually added over 10 minutes to the reaction liquid under vigorous stirring, and, after the addition was finished, stirring was continued for about 10 minutes. After completion of the reaction, the pH was maintained at 11.6. After the above reaction was completed, 400 grams of 3 N hydrochloric acid was added thereto, and after stirring for about 30 minutes, the pH was adjusted to 6.8 by further adding 3 N hydrochloric acid.

The texture was a meat-like texture much like corned beef. A total of 3.50 kg (yield 94.2%) was obtained having the following composition: 0.34% fat, 25.80% protein, 1.15% carbohydrate, 3.21% ash, and 69.50% water. The protein coagulate can also be used as a binder or filler for sausage.

Soy Whey Protein-Polysaccharide Complex

E.E. Schmitt; U.S. Patent 3,792,175; February 12, 1974; assigned to American Cyanamid Company has produced a fibrillar soy whey protein complex by contacting the whey with a polysaccharide solution containing carrageenan.

According to the process, soy whey is contacted with a polysaccharide under various stringent conditions. The polysaccharide is preferably used as an aqueous solution and comprises any known polysaccharide in combination with at least 10%, by weight, based on the weight of the combination, of carrageenan. More specifically, from 0.001 to 5.0%, by weight, based on the weight of the protein in the soybean whey, of a polysaccharide containing at least 10%, by weight, based on the total weight of the polysaccharide, of carrageenan, is used. Carrageenan per se may be employed as such or in solution as long as the above concentrations are maintained. Examples of polysaccharides which may be used in the process in combination with carrageenan include Irish moss, gum karaya, sodium alginate, alginic acid, agar, gum tragacanth, sodium carboxymethylcellulose, gum arabic and the like.

The process is conducted by mixing the soy whey with the polysaccharide, at the pH of the whey, at 10° to 50°C, preferably at ambient temperature, and at atmospheric pressure. The polysaccharide and the soy whey are allowed to remain in contact for a time sufficient to dissolve all the ingredients. After the ingredients are dissolved, it is critical that the pH be adjusted to 3.85 to 4.35, preferably 3.90 to 4.25, with any suitable acid such as citric acid, acetic acid and the like.

Alternately, the polysaccharide solution can be adjusted to pH 4 and added, with stirring, to the whey solution pretreated with the appropriate acidulant such that the pH falls within the range of 3.85 to 4.35 and preferably from 3.90 to 4.25 when the two solutions are mixed.

It is also critical that the reaction be continually stirred while the pH is being adjusted and maintained within the above range. The stirring is necessary to reduce the pH of the entire reaction medium in a uniform manner and thereby prevent areas of the reaction medium from having lower or higher pHs and to afford maximum contact between the two solutions. In this manner, the proteinaceous material precipitates out of the solution in a fibrillar condition. The fibrillar soy whey protein complex can be recovered by direct filtration through any suitable filtering medium having a mesh of at least 20 to 40. Alternatively, the entire reaction medium may be suspended in a suitable dehydrating agent with which the protein complex may be in contact without a deleterious effect. The resultant fibrous complex floating on the top of the resultant medium may then be removed by decantation or filtration, etc.

The protein complexes of this process can be utilized as a food supplement (5 to 50%) in the production of frankfurters, whipped toppings, etc. The complexes are well below the limits required by food products in regard to their bacteria, fungus, etc. content and the method of isolating soy whey protein results in amino acid compositions which closely resemble whole whey protein as shown in the following table.

Comparison of Essential Amino Acid Compositions

Essential Amino Acids	Whole Whey	Complex of This Process
Isoleucine	4.6	4.2
Valine	5.3	5.1
Leucine	7.2	7.0
Cystine and methionine	4.6	3.8
Threonine	5.4	5.8
Phenylalanine and tyrosine	8.3	8.6
Lysine	8.4	7.3

Soy whey is obtained from soluble fractions resulting from isolate production or can be obtained in the following manner. 10.25 parts of defatted soy flake are suspended in 12.00 parts of water which is then adjusted to pH 4.6 with 0.3 N HCl. This mixture is vigorously stirred for 3 hours at room temperature before the undissolved portion is centrifuged off. The mother liquor is further clarified by filtration through a bed of filter aid.

Example: 100 parts of soy whey are adjusted to pH 3.8. A haze develops but centrifugation brings down a trace of solids which are discarded. The clear solution is stirred while 4 parts of a 0.5% aqueous carrageenan solution is slowly dropped in. After the polysaccharide is completely added, stirring is continued for 1 hour. Throughout this time period, the pH is measured and adjusted, when required, to match the original value. The product is filtered through paper. Two types of material can be observed, a fibrous product and a fine gelatinous dust. The mixture of dust and fiber is filtered through a 30 mesh screen. In this manner, the fibrous materials are isolated from the fine gelatinous dust which passes through the screen with the mother liquor.

The amino acid composition of the recovered fibrous complex is set forth in the above table, under the column headed Complex of This Process. Stereomicroscopic examination of the product shows that the predominant character is fibrous although some of the material exists as thin ribbons and sheets. Most of the ribbons curl or roll up at their edges thereby creating a more fibrous appearance. All the fibers are twisted and bent, analogous to randomly crimped textile fiber, and entangled with neighboring fibers in clumps. This clumping results in very loose aggregations of fluffy texture.

HEAT EXPANSION FIBER-FORMING PROCESSES

Textured Fibers by Pressurized Heating

R.A. Hoer; U.S. Patent 3,662,672; May 16, 1972; assigned to Ralston Purina Company has developed a pressurized heating process for producing a tender, bland fiber which can be directly incorporated into simulated meat or other products to provide a pleasing taste and mouthfeel.

The process involves heating a slurry of protein by conducting the slurry through a heat exchanger under high pressure. The slurry can have a solids content of 0.3 to 35% or more by weight, depending on the limitations of the pump; the solids should be protein at a fairly high purity. After the protein slurry passes through the heat exchanger, it may be cooled and pumped into a collecting zone. The textured protein is recovered from the collecting zone.

A variety of protein can be used to produce fibers by the process; vegetable protein such as soy or other oilseed protein concentrates, e.g., high purity soy isolates and soy meals or flakes; animal protein concentrates, such as albumen and casein; and microbial protein, from sources such as brewer's yeast, torula yeast, or petro protein can be used in the process. It is important that the protein used be of fairly high purity, preferably above 70% pure protein.

The exact reaction which produces the fibers is not known. However, it is believed that the protein reacts under the temperature and pressure conditions of the process to produce elongated multimolecular protein polymers. To produce elongated filaments which are most desirable, the protein starting material should be in a sufficiently reactive form. The protein should be in a hydrated or soluble state to be reactive. If the native protein is not in a reactive form, it may be hydrated or made more soluble in several ways: by finely grinding an aqueous slurry of the protein; by subjecting an aqueous slurry of the protein to a change in pH, either by raising the pH to a point well above the isoelectric point or by lowering the pH to a point below the isoelectric point; or by a combination of finely grinding and pH change treatment.

Care must be used in adjusting the pH so that the material is not hydrolyzed to a point that it will not react when processed by this method. A high pH between 8 and 12 has proven to be satisfactory. After the protein has been treated to make it sufficiently reactive, the pH of the slurry can be readjusted to the isoelectric point range, for example, between 4 and 6 for soy protein, and the slurry reacted by the described process.

If the proper reagent materials are used, the protein filaments can be produced over a much broader pH range. The salts and hydroxides of certain polyvalent metals act as linking agents which permit the reaction to be carried out to form protein filaments from a slurry having a pH range of 4 to 11. Calcium hydroxide, calcium chloride, aluminum sulfate and other salts and hydroxides of bivalent and trivalent metals such as magnesium and copper have proven to be suitable linking agents. It is believed that polyvalent metal ions promote or participate in the reaction which forms the protein structures.

The reaction is a function of time, temperature and pressure. Generally, temperatures as low as 165°F may be satisfactory to produce the structured material, depending on the protein used. Temperatures of between 240° and 315°F are preferred, especially for soy protein or egg albumen. The protein will degrade if heated at too high a temperature for too long. After the protein is heated it may be cooled, generally to a temperature of 160° to 210°F, depending on the protein used.

Pressures above 50 psig are satisfactory to produce the desired texture. Preferably, back pressures of between 50 and 5,000 psig are used. A restrictive orifice may be placed in the exit line from the process equipment. The orifice serves to help maintain a back pressure on the system and to control the shape of the product. In general, circular orifices between 0.015 and 0.030 inch in diameter have proven satisfactory. Rectangular orifices have also proven to be satisfactory for some applications. A rectangular orifice 0.375 times 0.0625 inch has been used to produce flattened filaments. The method of producing edible protein structures will be more apparent from the following examples.

Example: A dried soy protein isolate obtained from the Ralston Purina Company (Edi-Pro-A) was slurried with water, the pH adjusted to about 10 with NaOH solution, and the protein precipitated by reducing the pH of the slurry to 4.5 with 85% H_3PO_4 solution. The hydrated protein was separated and reslurried with water at a solids content of 25% by weight. The protein purity of the solids was about 95%. The pH of the slurry was then increased to 5.0 by adding 50% NaOH solution. The slurry was pumped at a pressure of 5,000 psig through a four coil heat exchanger made of 80' of 3/8" x 0.209" I.D. seamless stainless steel tubing in a 6" pipe.

The temperature of the heat exchanger was set at 300°F. The retention time in the heat exchanger at this pressure was about 5 minutes. The slurry passed through the heat exchanger, was expelled through a 0.0135" diameter circular nozzle, and was cooled by dropping 20' through ambient air to a collecting vessel. The fibers were recovered and the excess water was removed by centrifuging. The fibers were light and tender, were about 40 to 60 mm in length, about 1/64" in diameter and had a moisture content of 55 to 60% by weight.

The recovered fibers could be combined with other materials to produce a high grade meat extender. To make a quality meat extender 48 parts by weight of recovered fibers were mixed with 48 parts by weight water and 4 parts by weight of combined salt, albumen, whey and a tracer. The meat extender could be used to make up as much as 50% by weight of hamburger patties when combined with lean muscle meat. The extended patties had good meat-like flavor and texture. The extender has the additional advantage of retaining moisture and fat upon frying which prevents shrinkage and improves the nutritional value of the meat.

A slight modification of the process described in U.S. Patent 3,662,672 has also been disclosed by *C.W. Frederiksen and W. Heusdens; U.S. Patent 3,662,671; May 16, 1972; assigned to Ralston Purina Company.*

The starting material for this modified process need not be protein of high purity. It is possible to make a light, bland, flatulence free product directly from proteinaceous materials having a high content of off flavoring materials and only a moderate protein content. The resulting product has a high protein content and can be directly incorporated into simulated meat, meat extenders, or other protein products without extensive washing steps. The process can be carried out on simple, readily available equipment and does not require a large amount of reagent chemicals. The process is inexpensive and simple to operate.

In this modification, the addition of phosphate ion is beneficial, particularly when the starting material contains protein of moderate purity such as soy flakes, which contain about 50% protein. Trisodium phosphate is a convenient reagent to use to supply phosphate ion, since it can be used to perform two functions. The trisodium phosphate supplies the needed phosphate ions and also raises the pH to a point well above the isoelectric point to make the proteinaceous material in the slurry more reactive.

Example 1: 20 lb of solvent extracted soy flakes were slurried with 80 lb of water. $Ca(OH)_2$ was added to raise the pH of the slurry to 10.3. The slurry was ground on a Fitz mill, left standing for 30 minutes, the pH was adjusted to 4.5 by adding 50% H_2SO_4 solution, and the slurry was ground again. The solids content of the slurry was about 20% by weight and it had a protein purity of about 57%. The slurry was pumped at a pressure of from 200 to 1,000 psig through a four coil heat exchanger made of 80' of ⅜" x 0.209" I.D. seamless stainless steel tubing in a 6" pipe. The temperature of the heat exchanger was set at 290°F. The slurry was then cooled to 185°F by conducting it through a cooler. The cooled slurry was expelled through a 0.028" diameter nozzle into a collecting zone. The protein was recovered as small, discrete textured particles.

Example 2: 50 lb of high DPI soybean flakes containing 50% by weight protein was ground to a particle size so that 99% of the material would pass through a number 200 U.S. Standard Sieve. The ground material was then slurried with 72 lb of water and 2 lb of Na_3PO_4 for 30 minutes to obtain a slurry having 35% solids and a pH of 8.0. The pH of the slurry was reduced to 5.0 by adding 4,000 ml of a 25% citric acid solution. The slurry was pumped through the heat exchanger described in Example 1 at a pump pressure of 2,000 to 3,000 pounds per square inch gauge and a heat exchanger temperature of 290° to 300°F. The treated slurry was expelled through a 1/16" x ½" rectangular orifice onto a screen.

As soon as the treated slurry was observed to leave the orifice the slurry fed to the heat exchanger was cut off and a stream of cold water was pumped through the process. When the material leaving the orifice was observed to be clear the slurry was again fed to the process. The cycling of slurry and cold water was repeated throughout the run. Light, bland fibers were recovered from the screen. The recovered fibers contained about 50% protein on a wet basis. On a dry basis the stachyose (flatulence factor) was reduced from 6 to 1½%.

Wheat Gluten Fibers by Heat Expansion

Simulated meat products have been prepared by *H.C. Palmer; U.S. Patent 3,645,747; February 29, 1972; assigned to Kal Kan Foods, Inc.* from a heat coagulable vegetable protein material having viscoelastic properties similar to wheat gluten. The simulated meat may be prepared by combining this vegetable protein in an amount from 15 to 45% by weight, and preferably from 25 to 35% by weight, with 55 to 85% by weight of water and preferably to a water content of 65 to 75% by weight. The gluten and water are mixed to form a substantially uniform slurry. An excess of water forms a thin liquid slurry which would not thicken to a dough on standing and which would result, upon heating, in an overly soft product texturally, whereas an excess amount of heat coagulable protein in the slurry results in a stiff and resistant dough which is difficult to handle and which forms an exceedingly tough and rubbery product upon heating.

The heat coagulable vegetable protein addition may be partially replaced by either a vegetable or animal protein, providing such substitution does not result in loss of the required viscoelastic characteristics of vital wheat gluten found necessary for formation of this product. Typically, vegetable proteins of oilseed or legume origin such as soybean, cottonseed, peanut, safflower and rapeseed proteins or the like; and animal-derived proteins, such as meat and fish protein or milk-based proteins such as casein or the like, may replace part of the heat coagulable vegetable protein if desired. These proteins need not be highly purified. When incorporated with the gluten dough prior to heating, these materials may be included in amounts up to 15% by weight, and desirably in an amount from 6 to 12% by weight. All the additive materials may, however, alternatively be subsequently added by absorption or injection into the simulated meat matrix as produced by the process, in which case substantially higher levels of incorporation may be achieved.

A layer of slurry typically ¾ to 1 inch in depth is then deposited in a suitable container. The slurry is heat processed, typically in a hot air oven, at a temperature of 380° to 430°F for between 5 to 30 minutes, although higher or lower temperatures and times may be used, if desired. For example, a satisfactory product may be prepared using a steam oven at 212°F for a processing time up to 50 minutes, depending upon the layered thickness of the slurry used. Regardless of the type of heating employed, it is necessary and essential that a surface skin be developed on the slurry to entrap, in situ, gases and vapors generated, as during heat processing.

Whatever method of heating is employed, the internal temperature of the slurry will typically not exceed 250°F, although the skin temperature may be higher due to loss of moisture and its immediate proximity to the heating medium. The time and temperature for processing a ¾ inch deep slurry is approximately 20 to 30 minutes at 400°F for a dough initially at room temperature, and approximately 15 to 20 minutes at 400°F for a dough initially at 120° to 140°F prepared with boiling water.

During the heating, expansion of gas bubbles and generation of steam within the dough causes the dough to rise and, because of the viscoelastic properties of the wheat gluten, during the rising, fibers, filaments and expanded cellular structures are formed which extend substantially parallel to the direction in which the dough is rising. The further addition of heat coagulates the gluten in its stretched form so that elasticity is lost and the structure remains permanently stretched.

The prepared simulated meat food product is found to possess a certain amount of porosity and is capable of absorbing or being injected with a wide variety of ingredients such as additives as desired to produce a simulated meat flavor and structure.

The absorbed portion may be formulated to contain a substantial proportion of heat coagulable protein materials including, egg, blood and homogenized meat ingredients and after absorption the sponge-like product may be subjected to a second heating to coagulate those materials disposed in the liquid or semisolid phase.

The product typically comprises areas of coagulated protein of homogeneous appearance intermixed in a fibrous matrix giving the appearance of organ meats such as those of liver and spleen.

Example: A high protein simulated meat is prepared by combining the following materials.

Ingredients	Percent by Weight
Vital wheat gluten	29.04
Water (195°F)	48.39
Fresh beef blood (anticoagulant added)	21.78
Sodium nitrite solution (2% aqueous solution)	0.26
Titanium dioxide (food grade)	0.53

The hot water, fresh blood, sodium nitrite solution and titanium dioxide are slurried together with stirring for 1 to 2 minutes to form a substantially homogeneous mixture. The vital wheat gluten is then added with continued mixing and mixing is continued until a satisfactory discharge of the gluten is effected to form a homogeneous slurry. The slurry has a consistency like thin dough and may be characterized as being soft and pliable. A layer of dough about ¾" in depth is deposited onto a standard baking tray and transferred to a hot air oven preheated to 400°F. During the first few minutes of heating, a surface skin is formed on the slurry which gradually commences to rise as the heating is continued.

After 15 minutes of heating, the skin has risen over the whole surface of the tray to a height of 2½" to 6". Heating is continued for 5 minutes to coagulate the stretched protein filaments and membranes within the skin encased structure, after which time the product is removed from the oven and permitted to cool for 5 minutes. The partially cooled product may then be cut into ¾" cubes and allowed to cool to ambient temperatures. The yield of finished product is approximately 90% of the starting material and upon close examination, is found to resemble cooked muscle meat in color, general appearance and texture and includes adherent connective membranes with fibers and adjoining filaments.

Shredded Protein Texture

A process to produce a shredded texture has been developed by *I.I. Rusoff, W.J. Ohan and C.L. Long; U.S. Patent 3,047,395; July 31, 1962; assigned to General Foods Corporation.* It provides a meat-like texture and appearance in a product having the ability to be cooked by deep fat frying, roasting, boiling, etc. where high temperatures may be used without causing disintegration and loss of the structure and texture. The product can be dehydrated without loss of texture or appearance upon rehydration and cooking.

The process involves the rapid orientation and coagulation of protein material in a substantially undenatured, finely-divided, hydrated state under conditions which produce a shred-like structure. Orientation and coagulation of the protein must be related so that coagulation follows orientation of the molecules so that on coagulation, the molecules are set in a shred-like form.

In carrying out the process, a protein source such as meat, poultry or fish muscle, trimmings or scraps or vegetable protein flour, is comminuted to a homogeneous mixture. It is preferred that the protein content of the protein source be 30 to 100% of the solids on a dry basis and in the case of vegetable protein, as high as possible for reasons of control.

The protein source is mixed with sufficient water or other aqueous liquid to prepare a slurry or paste having a 50 to 80% moisture content. The paste or slurry is placed in a reaction vessel, typically an autoclave or continuous reactor such as the Votator and heated rapidly to 300° to 400°F, preferably 330° to 360°F. Where the slurry or paste is heated to less than 300°F, little or no fiber formation takes place whereas above 400°F the fibers start to decompose with a resultant loss of cohesiveness, chewiness and texture.

The heating of the paste from room temperature to maximum temperature should be rapid, generally less than 5 minutes, 1 to 2 minutes being most preferred. If the rate of heating is so slow that a temperature of 400°F is not reached in 5 minutes, the resultant product is soft, lacks the desired cohesiveness and shred character, and contains excessively degraded protein. During heating, agitation should be maintained to provide uniform heat transfer and prevent charring and also to aid in the orientation and stretching of the molecules so that they are placed in proper juxtaposition with respect to one another prior to coagulation. Preferably the agitating device should rotate 30 to 500 rpm to obtain the desired shreds and chunks of shreds. At agitation rates over 500 rpm, shreds and chunks of shreds tend to break down into smaller units which are not as desirable.

As soon as the slurry has been heated to the desired temperature, cooling is commenced and is carried out as rapidly as possible until a temperature of at least 200°F and preferably 100° to 150°F is reached. The resulting product will be obtained in a yield of 30 to 85% of the total starting solids in the form of shreds or masses of shreds, the remainder existing as a soupy mass which may contain small fragmentary shred particles. The soupy mass may be recycled in subsequent batches depending upon the raw material source to provide further shred formation.

The cooled shreds are separated from the supernatant liquid by any common means such as screening, decanting, etc. The shreds generally have a solids content 5 to 10% higher than the solids content of the starting material. The shreds may be washed, sliced, flavored and color added or they may be first bleached and the subsequently flavored and colored. The shreds may be used in food preparations which would employ meats such as beef, chicken, fish, and may be used in a variety of products calling for the use of such protein shreds or for nutritious nonmeat components.

Example: The starting material is hexane extracted soy flour substantially undenatured and having a moisture content of 10% by weight. About 8 lb of this soy flour was mixed with 10 lb of water and slurried in a Day Mixer to produce a thick paste. The paste was then introduced to an autoclave having a 5-gal capacity. The autoclave was equipped with a jacket piped for steam heating and water cooling, a thermowell extending into the center of the cavity to contain a thermocouple, a valved vent opening for the relief of gases from the head of the autoclave, a pressure indicating gauge, a separate opening piped to a rupture disc as a safety relief, and a flush valve in the base of the autoclave normally acting as a discharge port but adapted to serve as a steam injection port during the process.

The interior of the autoclave was also fitted with cooling coils and an anchor type agitator. The agitator blades were approximately 9¼" across and the width of the vessel 10", thus providing close clearance for prevention of material adhering and building up on the side walls. Variable agitator speeds were provided via the use of various sheaves. The coils piped for water cooling, described a cylinder above the agitator blades and proximate to the side walls.

Following introduction of the soy paste charge, the autoclave was closed and steam introduced through the steam-injection port to heat the paste. Initially a period of 5 to 10 seconds of injection of 400 psig steam was required in order to reduce the viscosity of the paste sufficiently to start agitation. Thereafter, steam was continuously introduced to the autoclave and the agitator was operated at 400 rpm. Agitation and heating under the above specified conditions for approximately 1 minute and 15 seconds provided a charge temperature of 360°F whereafter steam injection was terminated and cooling water circulated through the cooling coils and the jacket to cool the charge down to approximately 100°F. Upon opening the autoclave, it was found that the protein was coagulated to shred-like fibrous material. 40% by weight of the solids fed to the reactor were recovered in the desirable shred form.

The composition, in addition to its shred-like gross appearance, had a plastic-pliable somewhat elastic nature which permitted physical compaction of the fiber, if desired, to give various elongated continuous structures, meat-like in texture. The material was dehydrated and thereafter rehydrated readily upon water cooking. It was also possible using the product of this example to freeze the meat-like structure and to rethaw it without loss of texture in the manner of ordinary cuts of meat. Upon cooking either after recovery from the reaction vessel, freezing or dehydrating the material very substantially retains its original shape and texture. In the mouth, the composition could be generally characterized as non-

PROTEIN CURD FROZEN INTO FIBROUS SPONGE-LIKE MASS

A vegetable protein isolate is converted by a freezing step into a fibrous, sponge-like protein mass lacking the taste of soy or other starting material in a process described by *G.K. Okumura and J.E. Wilkinson; U.S. Patent 3,490,914; January 20, 1970.* The resultant mass can then be further processed into a meat-like solid food or into a milk, additional steps of finely comminuting the fibrous mass and adding water and other additives being utilized in producing the milk. In addition to defatted soybeans, sesame seeds or cottonseeds, converted to defatted flakes could serve as starting materials, as could alfalfa plant material.

Example: (1) Defatted soybean flake is soaked in water at room temperature a few minutes to soften it. Any hulls contained in the flake will be softened and loosened.

(2) The cleaned flake is then ground without loss of moisture to particles which vary in size from colloidal dimensions up to discrete particles. The ground flake is converted to a slurry by adding water to bring the total water content up to 8 times or more the dry flake content, by weight, the slurry being subjected to agitation to maintain the solids in suspension.

(3) The slurry is then strained or centrifuged and the pulp separated from the fine solids and dissolved proteins, and discarded.

(4) The dissolved proteins and fine solids are then precipitated with a precipitating agent such as calcium sulfate, calcium chloride, magnesium chloride, hydrochloric acid, citric acid or lactic acid, or lemon juice and the mixture gently stirred until the precipitate collects as a curd at the bottom of the vat.

The curd is washed with fresh water which dissolves the excess precipitating agent, and the wash water is largely removed by draining. More water is then removed by pressure extraction through a strainer bag under a gentle pressure of 1 to 4 oz/in^2, over a period of 15 to 30 minutes. Much of the soybean flavor is removed in the preceding steps.

(5) The residue curd, after washing, is ground (e.g., in a stone grinder) and is mixed with water in a freezing container to develop a fine milk-like suspension, the proportions, by weight, being approximately 15 parts curd and 85 parts water. This suspension is then frozen, at about 20°F over a period of about 6 hours.

As freezing progresses, the resulting expansion develops pressure which compacts and toughens the protein sponge structure so that it will retain its fibrous characteristic during subsequent processing. The freezing step has the additional function of eliminating such carbohydrate content of the precipitate as has not been eliminated in the preceding steps. Such carbohydrate, being largely in solution, will be trapped in the ice crystals and separated from the protein fiber which develops between the ice crystals.

(6) The frozen material is then melted, e.g., by immersion in hot water of about 100°F or by bathing the container in a jet of steam of about 180°F and the water and carbohydrates removed by draining the resulting melt. During this step the carbohydrates will remain dissolved in the water and separated from the protein material, which in the meantime has developed into solid fibers which do not readily absorb the water or the dissolved carbohydrate. The soy taste is further reduced during the freezing and this result is attributed to the elimination of residual carbohydrates which appear to be carriers of the soy flavor.

(7) The protein fiber residue is then placed in a centrifuge and continuously bathed in a copious stream of water poured into the centrifuge and centrifuged off so as to continuously bathe the protein fiber with fresh water to remove the residual carbohydrate remaining in the melted liquid clinging to the fibers.

(8) For soy milk production the washed fibers are then comminuted by subjecting the fibers to the action of a high-speed rotary cutter known as a food disintegrator. Grinding is not effective for this step.

(9) During the comminuting, the reconstitution of the protein into milk is begun by adding selected additives including water, oil (e.g., cottonseed, peanut, corn or other equivalent oil), sugar, salt, calcium gluconate and ferrous sulfate. The disintegrator also functions to start homogenization.

(10) After the comminuting and additive-mixing steps, the mixture is homogenized in a piston-restricted orifice type homogenizer to produce a stable emulsion that will not separate. The resulting product is quite similar to natural milk and lacks any unpleasant soy taste.

(11) As a final step, the milk is refrigerated and packed in containers.

Steps (1) through (11) are required if a soy milk is desired. Steps (1) through (7) are sufficient for the preparation of a fibrous food product (solid or semisolid) other than milk.

TEXTURED PROTEIN GELS AND EXPANDED PRODUCTS

These gels and textured protein products are used as extenders and supplements for meat products such as hamburger and chili. They can also be used in simulated meat products as binders or as meat-like particles.

EXPANDED TEXTURED PROTEINS

Expanded Hydratable Proteins

W.T. Atkinson; U.S. Patent 3,488,770; January 6, 1970; assigned to Archer Daniels Midland Company discloses a process preparing a protein product in dehydrated form which does not disintegrate on contact with boiling water, and which, therefore, can be hydrated and on hydration by steaming or boiling has the texture, appearance, and coherence of cooked meat.

This protein product is obtained by extrusion of moistened, protein material in the form of a plastic mass at elevated temperatures through an orifice into a medium of lower pressure to result in a porous, protein-containing product of plexilamellar structure characterized by its open cell structure in which a majority of cells have a length-to-width ratio of greater than one, the length being measured in the direction of extrusion and the width being measured in the transverse direction.

The protein extrudate is a tough, resilient, dry to semi-dry, open celled, funicular structure made up of interlaced, interconnected funiculi of varying width and thickness. The majority of the cells defined by this plexilamellar protein structure are irregular in shape but are characterized by greater length in the direction of extrusion than average width as measured in the transverse direction.

The process is not limited to any particular type of protein. Any type of edible protein of vegetable, fish, or animal origin can be employed. The term proteinaceous material or protein-containing material as employed herein is intended to define an edible material having a protein content of at least 30% by weight. The protein can be employed in substantially pure form, in water-soluble form, or, as is preferred, in the form of flakes or flour, generically referred to as meal.

The nature of the extrudate, for any given starting material, is principally governed by the concentration of water in the protein mix, the temperature to which the protein mix is heated during the extrusion, and the pressure developed in the extruder. The presence of water is essential for two reasons: it plasticizes the protein mix to form the necessary plastic mass and it causes the expansion of the extrudate. It therefore follows that an increase in the concentration of water will result in greater plasticization and a higher degree of expansion. However, proteins contain water which is not released at the temperatures employed for extrusion and thus is unavailable for plasticization. The concentration of such nonreleasable water is not constant and increases with decreasing extrusion temperatures as well as with increasing protein concentrations in the protein mix fed to the extruder. It will also vary with the nature of the proteinaceous material employed.

Hence, the minimum concentration of water necessary to obtain the formation of the plexilamellar extrudate will vary, but should be at least 10 to 15% above the nonreleasable water concentration.

Since the expansion of the extrudate normally occurs at atmospheric pressure, the minimum temperature to which the protein mix must be heated in order to cause expansion of any degree is the boiling point of water, 212°F, in order to cause steam expansion of the extrudate. However, to produce a product having good stability in the presence of boiling water, it is desirable for the extrudate to emerge from the extruder at a temperature of at least 250°F and preferably above 300°F. The application of a vacuum to the extrudate may, of course, allow the use of somewhat lower temperatures.

The formation of the plastic mass from the protein mix and its extrusion into plexilamellar protein also requires sufficient pressure to maintain the plasticizer, i.e., the water, dispersed in the protein mix and also sufficient pressure to shear the protein particles and cause the protein to become the continuous phase. The pressure is also employed to cause the unidirectional flow of the plastic mass, i.e., flow through the helical path formed by the extruder screw and barrel, in the plasticizing section of the extruder and out of the extruder orifice.

In order to produce a commercially acceptable extrudate which can be rehydrated in hot water or steam while retaining a meat-like structure, it is necessary that both temperature and pressure be maintained well above the minimum levels which would be barely sufficient to cause extrusion through the orifice and steam expansion of the extrudate. In general, temperatures above 250°F, preferably above 300°F, and pressure within the extruder above 250 psi, preferably above 500 psi, will yield a satisfactory product. A reduction of either the temperature or the pressure to a lower level will have serious adverse effects upon the structure of the extrudate and its resultant ability to retain its form upon rehydration in hot water. The process is further illustrated by the following example in which all parts are by weight unless otherwise indicated.

Example: The following components, listed in the order of their addition, were mixed in a ribbon blender at 120°F for about 20 minutes; 11,350 grams of extracted soybean flakes prepared according to U.S. Patent 3,100,709 by T.M. Paulsen, containing 50% soy protein and 6.5% moisture; 45 ml of 50% hydrogen peroxide for purposes of flavor and odor control diluted in 380 ml of water; 1,700 grams of imitation beef seasoning; 3,785 ml of water; 90 grams of 97% pure sodium hydroxide; and 340 grams of calcium chloride dissolved in 500 ml of water.

The resulting composition contained 30% moisture and 3% calcium chloride and was extruded in a Prodex 1¾ inch extruder equipped with a medium compression screw and an extrusion die containing eight three-sixteenth inch diameter orifices. The extruder was maintained at a temperature of 350°F at the extrusion die and the front end of the barrel. The screw was rotated at the rate of 176 rpm.

The product expanded rapidly on emerging from the die while releasing steam. Substantially dry plexilamellar protein strands were obtained which were cut into 0.5 inch lengths by a rotating knife. The resulting product was autoclaved at 15 psi steam for 60 minutes. The hydrated product resembled beef in appearance and had firm and chewy eating characteristics.

A modification of the above product is also produced by *W.T. Atkinson; U.S. Patent 3,812,267; May 21, 1974; assigned to Archer Daniel Midland Company* using a blend of vegetable protein (e.g., soy) and fish protein. It was found, however, that mixtures containing up to about 25% by weight of fish protein concentrate which has a protein concentration of about 75 to 80% and contains less than 0.5% fat and solvent extracted oilseed proteinaceous materials as the remainder of the protein containing component of the extrusion mix could be extruded by the process to give rise to a product having an expanded cellular structure and on rehydration the texture and chewiness of cooked lean meat.

Examples 1 through 4: The components listed in the following table below were mixed in a ribbon blender at 120°F for about 20 minutes.

INGREDIENT	EXAMPLE			
	1	2	3	4
Solvent extracted soy bean flakes	95	90	80	75
Fish protein concentrate (Viobin)	5	10	20	25
Water for extrusion	40	40	40	40

Each of the resulting compositions was extruded in a Prodex 1¾ inch extruder equipped with a medium compression screw and an extrusion die containing eight $3/16$ inch diameter orifices. The extruder was maintained at a temperature of 350°F at the extrusion die and the front end of the barrel. The screw was rotated at the rate of 146 rpm.

The products expanded rapidly on emerging from the die while releasing steam. Substantially dry plexilamellar protein strands were obtained which were cut into 0.5 inch lengths by a rotating knife. The resulting products were autoclaved at 15 psi steam for 60 minutes. The hydrated products had meat-like textures and had firm and chewy eating characteristics. Products 3 and 4 were adjudged to be somewhat dry and tough.

Use of Flow Inducing Salts for Expanded Proteins

A plastic flow inducing salt is also used by *F.E. Calvert and W.T. Atkinson; U.S. Patent 3,498,794; March 3, 1970; assigned to Archer Daniels Midland Company* to form a hydratable food product. The product has a dry, dense, mechanically strong structure which can be hydrated without disintegration into a firm, flexible, tender, spongy structure which retains the general outline and shape of the original dehydrated structure.

The process comprises mixing a vegetable protein, having a protein concentration of at least 70%, with releasable water such that the total water concentration does not exceed 50% of total composition, and with 0.1 to 10%, based on total composition, of a soluble salt of a weak acid having a dissociation constant below 1×10^{-5}, the pH of the mixture being adjusted to be within 5.0 to 8.0. The mixture is then subjected to plastic flow at 120 to 210°F and a pressure of at least 1,000 psi and a coherent, translucent, homogeneous protein product is recovered which can be hydrated without disintegration.

It is believed that subjecting dry powdered protein to high pressure and controlled amounts of moisture and heat, in the presence of a plastic flow-inducing agent, ruptures many naturally occurring intermolecular bonds between the protein chains. This permits the molecular chains to slip and slide into and around each other thereby forming the translucent, homogeneous, coherent protein product.

The presence of controlled quantities of moisture in the protein during the hot plastic flow phase is an essential factor of the process. The moisture serves not only to induce swelling and reduce protein interchain attraction and hydrogen bonding, and thereby to assist plastic flow at the elevated temperatures and pressures of the process, but it further produces innumerable criss-crossing microvoids or pathways in the protein product. It is believed that these microvoids allow the water to penetrate the protein product and permit ready hydration and swelling without loss of structure and shape.

The hydratable protein food product can be prepared from any vegetable protein concentrate, although it is generally preferred to use protein derived from defatted oilseed meals. The concentrate is prepared from the oilseed meal by known isoelectric point precipitation methods of protein solubilized from the meal.

Most protein concentrates are isolated at a pH in the range of 4 to 5. At these levels the protein is too sour for human consumption nor can such protein concentrates readily be subjected to plastic flow in standard equipment in view of the high pressures required. The pH is generally adjusted to a value between 5.0 and 8.0 and preferably to a value between 6.5 to 7.5 by the addition of an alkaline reagent such as preferably sodium hydroxide.

A variety of acids can be used in the form of soluble salts as plastic flow inducing agents in this process. From the standpoint of operability, the soluble salt of any weak acid that has the required dissociation constant can be used and, hence, the choice is dictated by factors such as toxicity, taste, price, and nutritional value. The ammonium and alkali metal hypophosphites are outstanding plastic flow inducing agents and are advantageous for edible products because they impart no flavor and are nontoxic in the concentrations used in the process.

Another group of highly effective plastic flow inducing agents, which have nutritional value, are the alkali metal or ammonium cysteine salts. Sulfites and sulfides of ammonia or the alkali metals are good flow inducing agents but have the disadvantage of imparting undesirable flavor. Although as much as 10%, based on the protein mix, of the plastic flow inducing agent can be used, it is generally not necessary to use more than one percent and, hence, the preferred concentration of the plastic flow inducing agent is from 0.25 to 1%, based on the protein mix.

Many mechanical devices or machines are suitable for the plastic flow step of the process. It is only necessary that they should be of sufficient ruggedness, power and proper design to apply pressures in the range of 1,000 to 10,000 psi and that they are capable of being heated to the necessary temperatures in the protein mix. Examples of suitable equipment are roll mills having differential or even speed, heatable revolving screw and hydraulic ram presses, injection molding machines, thermoplastic extrusion and compression molding machines, and many other machines which compress a plastic mass at the desired temperatures under sufficient pressure.

The mechanically strong, translucent protein product is obtained in the form of sheets, bars, rods or other shapes depending on the type of equipment employed. The larger shapes may be cut into smaller chunks or pieces and then air dried to remove excessive moisture.

The cut dried protein food product obtained by the process is readily packaged, handled and stored and does not require refrigeration. The cut pieces can be swollen, with hot water, into edible, flexible, tender, nutritious structures many times their original volume without disintegration. These pieces, furthermore, retain their original shape and structure on prolonged heating in water.

Example: Into a rotating change-can mixer is placed 6,000 grams of a soybean protein obtained by isoelectric point precipitation of protein solubilized from defatted soybean meal. To the protein is added 60 grams of sodium hypophosphite dissolved in 2,240 ml of water in fine streams under agitation. The temperature of the mixer is maintained below 120°F to prevent prereaction. After sufficient mixing to obtain a uniform distribution of the water, there is then added slowly 140 grams of sodium hydroxide in the form of a 50% aqueous solution to result in a pH of 6.7. The mixing is continued until the protein mix appears to be uniform. A total mixing time of 20 to 60 minutes is usually sufficient to obtain the desired degree of uniformity in the protein mix; which, at this point, has no transparency whatsoever, and appears to be a moist particulate mass.

The protein mix is then fed into the nip of a roll mill in which the rolls operate at a differential speed. The slow roll is maintained at a temperature of 180°F and the fast roll at a temperature of 200°F. The distance between the rolls is set at 1/8 inch. The feeding of the protein mix is adjusted to permit a milling time of about one minute although longer or shorter milling times can be employed. On forming the translucent appearance caused by plastic flow, the product is taken off the roll mill in the form of a moist sheet. The sheet is cut into pieces and dried in an air oven. The resulting pieces swell when placed into water at 120°F and pick up 3 to 5 times their weight of water without disintegration.

α-Cellulose in Expanded Protein Products

Plexilamellar products from oilseed proteins are modified in the process of *M.M. Hamdy; U.S. Patent 3,623,885; November 30, 1971; assigned to Archer Daniels Midland Company* by the incorporation of small amounts of α-cellulose. This permits the incorporation of up to 5% fat in the resulting meat-like fiber product. This process allows the addition of up to 5% of an edible fat component directly to an extrudate prepared from solvent extracted protein materials. The term fats as used here includes liquid and soluble materials comprising triglycerides of fatty acids which are edible. This method may also be applied to the extrusion of protein materials containing fats or oils which are difficult to extrude because of their oil content. As mentioned above, extrudates containing up to about 5% fat or oil can be satisfactorily extruded by the expedient of inclusion of a small, but effective amount of finely divided α-cellulose in the extrusion mix.

By the term small but effective amount is meant that amount of α-cellulose necessary to provide satisfactory extrusion of extrudates containing up to 5% fat. The amounts of α-cellulose found to be effective are from 0.1 to 10% by weight based on solids of the extrudate. Substantially more than about 5% by weight of fat cannot be satisfactorily incorporated by this process.

The process is not limited to any particular type of protein. Any edible protein of vegetable, fish, or animal origin can be used. The protein can be used in substantially pure form, in water-soluble form, or, as is preferred, in the form of flakes or flour. Preferred materials are obtained by solvent extraction of oilseeds such as peanuts, cottonseeds, sesame seeds, or soybeans.

The oilseed meals which have protein concentrations of 40 to 70% are preferred since they can be extruded into the plexilamellar product desired over a broad range of conditions. Finely divided protein flour is less preferred because of its higher lubricity and its lesser tendency to shear and orientate. The protein concentration of the material to be extruded should, however, be maintained above 30% since otherwise the nonproteinaceous ingredients will interfere in the formation of the continuous protein phase and its orientation in the masticating step.

Example: This example illustrates the preparation of an unflavored textured vegetable protein containing fat. The following ingredients were blended in a ribbon blender at 120°F for about 20 minutes.

Ingredient	Parts by Weight
Solvent extracted soy flakes	90
Avicel (α-cellulose)	5
Crisco (hydrogenated shortening)	5
Water added for extrusion	45

The resulting composition was extruded in a Prodex 1¾ inch extruder equipped with a medium compression screw and an extrusion die containing eight 3/16 inch diameter orifices. The extruder was maintained at a temperature of 350°F at the extrusion die and the front end of the barrel. The screw was rotated at the rate of 146 rpm. The product expanded rapidly on emerging from the die while releasing steam. Substantially dry plexilamellar protein strands were obtained which were cut into 0.5 inch lengths by a rotating knife. The product was canned in water and autoclaved at 15 psi steam for 60 minutes. The product had a firm fibrous structure and a fair, meat-like texture.

Expansion Using Microwave Energy

According to the process *R.A. Boyer, A.A. Schulz, E.V. Oborsh and A.V. Brown; U.S. Patent 3,662,673; May 16, 1972; assigned to Ralston Purina Company* an expanded textured protein product is produced by the steps of mixing together oleaginous seed material containing at least 35% by weight protein and an aqueous liquid, subjecting the mixture to temperatures high enough to form an expanded, substantially water-insoluble, irreversible cross-linked structure, and cooling the resulting product.

The product has been expanded and heat-set to retain the expanded shape or configuration. The product in that form exhibits excellent physical properties. For example, a dried product has been found to sorb about four times its weight in water and yet retain its desirable physical properties. There is no physical deterioration such as crumbling due to sorption of large amounts of moisture as is characteristic of bread-type products which tend to physically disintegrate upon exposure to excessive amounts of water. Furthermore, the product substantially maintains its integrity and desirable physical characteristics even when subjected to severe wet cooking conditions such as cooking under high temperatures and pressures.

The oleaginous seed materials used may be in a variety of forms. For example, full fat soy flours and defatted soy flours have been found to produce expanded products which offer the highly desirable characteristics discussed. Materials containing a higher percentage of protein, including isolated soy protein, may also be used to obtain the same desirable characteristics.

The most favorable results are obtained when the ratio of protein material to the aqueous liquid is from 1:0.2 to 1:4 and preferably 1:0.4 to 1:2 by weight. Other ingredients such as color, flavoring, etc. may be added to the mixture to obtain specific end products, also appropriate chemicals may be added to modify the protein properties.

After mixing the protein material and the aqueous liquid, it may be necessary to adjust the pH of the mixture for expansion. It has been found that the best results are obtained where the mixture has a final pH of from 4.5 to 7 and preferably from 5.5 to 6.5. Where the final pH is below 4.5 for most protein or 4.0 for partially hydrolyzed protein, it has been found that the product gels and discolors to form a crumbly product. Where the final pH is above about 7, the resulting product has poor color and undesirable physical characteristics.

The aqueous protein mixture is then subjected to elevated temperatures at atmospheric pressure to expand and heat the mixture. The elevated temperatures may be attained using radiant energy (e.g., infrared, microwave, induction ovens). It has been found that different heat sources produce expanded products having different physical characteristics.

For example, electronic wave sources produce products having a soft, flexible surface, whereas infrared radiant sources generally produce products having a less flexible, hard surface. It is preferred that electronic waves be used, since other heat sources tend to form an undesirable hard surface or crust on the expanded product caused by heat denaturation of the protein or surface dehydration.

The temperatures used to produce and retain the expanded structure of the product are dependent upon the heat source. For example, microwave treatment depends mainly upon the residence time of the product being exposed to the electronic waves and power input. That is to say, since the temperature of the product generally does not exceed 212°F (the boiling point of water), the residence time and power would determine the desired amount of expansion, final moisture content, etc. of the product.

Radiant heat sources such as infrared must be subject to closer control, since the temperature must be sufficient to obtain the desired expansion, yet not excessive as to cause the formation of an undesirable crust on the surface of the product. After the product has been heated and expanded, it may be cooled or dehydrated and subsequently processed into various food products.

Example: One hundred grams of isolated soy protein (95% protein, pH 4.5) was mixed with 175 ml of water in a Hobart food mixer for 5 minutes. The mixture had a pH of about 4.5 and was formed into a small loaf and placed into a microwave oven (8.8 kw, 115 to 200 volts, 2,450 mc) for a period of 3½ minutes. The resulting product was removed from the oven and cooled. The product was sliced and the internal structure was seen to be cellular and the product would not tear easily. A slice of the material was dried, weighed and subsequently placed in water and weighed again. It was found that the slice sorbed 4 times its original weight in water and would not tear easily even though containing the large amount of water.

In a comparative example, the above procedures were repeated, except that wheat flour (14% protein, pH 6.8) was substituted for the isolated soy protein. The resulting product had a cellular structure similar to that obtained from soy as described above; however, when placed in water, the wheat flour product lost its integrity and separated into individual particles.

Extruded Expanded Proteins Containing Mg or Ca Oxides

Expanded protein material having an open, cellular structure which is ultimately used in food products has been developed by *M. Glicksman, R.E. Klose and R.D. Kirkeby; U.S. Patent 3,684,521; August 15, 1972.*

This process is considered an improvement over the prior art in that small amounts of additive, namely magnesium oxide and/or calcium oxide, greatly increase the amount of expansion in the protein-containing material that do not contain expansion additives other than the necessary water. However, the most important advantage in this process is the improvement in flavor qualities in the product.

The protein material can be derived from various food materials and preferably contains protein at least 30% by weight of the dry material. This minimum level of protein is necessary to obtain the desired result of a greater expansion of the protein material. The protein material may be from vegetables, meat, fish, milk and egg, as well as mixtures of these materials. Vegetable source of protein examples are soybeans, peanuts, cottonseed and sesame seeds. These may be in any form such as ground flour or meal, pellets, bits or grits.

The amount of water that is added to the protein should be such that a moisture content of from 15 to 50% by weight of the mix is obtained. Use of less water does not allow the desired expansion. Excessive moisture similarly interferes with the expansion.

The percentage of added magnesium oxide and/or calcium oxide on a dry basis preferably should be less than 2%. The preferred percentage of the magnesium oxide and/or calcium oxide is between 0.2 and 0.6% (dry basis) for optimum expansion.

The temperature is raised during mechanical mixing above the boiling point of water and preferably above 220°F. Temperatures below the atmospheric boiling point of water may be used if vacuum conditions are employed at the extrudate outlet such as temperature as low as 190°F. A preferred range of temperature is about 220° to 400°F, and more preferably about 250° to 370°F.

The pressures employed during the mechanical mixing may be varied within rather wide ranges. The critical feature involving pressure is that a sudden release of pressure be obtained. Illustrative of pressures that may be employed are pressures in the range of 100 to 300 psi. However, much higher pressures may be employed to obtain the desired expansion of the proteinaceous material. The upper limit of pressure will be determined to some extent by the type of apparatus employed.

Example: A mix containing 99.8% by weight of soy grits and 0.2% magnesium oxide was blended until homogeneous. The mix was continuously metered into a Wenger Model X-25

Extruder at a rate of 300 lb/hr along with a stream of water. The water and the soy grits-magnesium oxide were thoroughly mixed in the extruder. The amount of water was 25%. Steam pressure in the extruder jacket was 50 psig. Two ⅜" diameter dies were used. Screw speed was 340 rpm. The temperature recorded at the die face during extrusion was 295°F and the pressure was 120 psig. The extruded product was dried to less than 10% moisture by an air stream at 250°F. The pH of a 2% water slurry of this material was 7.3. The density was 0.40 g/ml.

Exit Gate for Texturizing Apparatus

It has been discovered that finely-divided particulate protein material may be texturized by passing the material through an elongated cylinder or pipe and applying elevated pressure and temperature. Although this method provides highly satisfactory texturized protein, certain problems were confronted during sustained operation. For example, at times pieces of texturized protein were produced that plugged or blocked the apparatus thus necessitating shutting down operation and disassembling of the apparatus.

H.N. Dunning, P.K. Strommer and G.J. Van Hulle; U.S. Patent 3,707,380; December 26, 1972; assigned to General Mills, Inc. have developed an improved exit or outlet means for such texturizing apparatus. This can be illustrated by the following figure.

FIGURE 13.1: EXIT GATE FOR TEXTURIZING APPARATUS

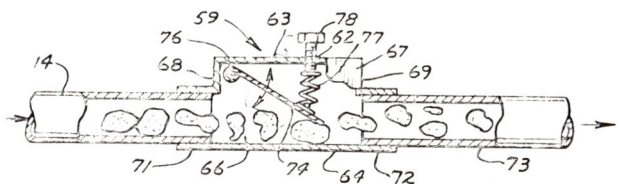

Source: H.N. Dunning, P.K. Strommer and G.J. Van Hulle; U.S. Patent 3,707,380; December 26, 1972

In this apparatus the nozzle **59** has a body portion **62** which may be rectangular in shape having an upper wall **63**, a lower wall **64**, a pair of side walls **66, 67** and a pair of end walls **68, 69**. The end walls are each provided with a tubular portion **71, 72**, respectively, for connecting the nozzle to the pipe **14** and, if desired, to a conveying pipe **73**. The nozzle further includes a flapper or gate **74** which is pivotably mounted in body portion **62** such as by pin **76**. The gate preferably sealingly engages the side walls. However, the gate must be free to move or pivot with respect to the side walls. The nozzle has a spring **77** which urges the gate towards the closed position and provides compressive force on the gate. The spring, for example, may be a coil spring, a leaf spring or a pneumatic spring. A screw **78** is shown for adjustment of the compressive force exerted by the spring and thus permits a certain amount of adjustment of the pressure in the pipe **14**.

Other forms of this outlet means may have dual gates with dual spring controls for control of pressure within the apparatus.

Example: Protein material was texturized using texturizing apparatus having a nozzle substantially like that shown in Figure 13.1. The protein material was a mixture of 70 parts soybean concentrate (Textrol), 30 parts soybean isolate (Promine R) and ½ part glycerol monostearate (Myvaplex 601). The mixture contained 20% moisture by weight. The feed rate of the mixture to the apparatus was 14 lb/min. The temperature of the steam fed to the apparatus was 450°F. The pressure in the texturizing chamber was about 78 psig. The product was texturized and had a shear press value of 900. The product had a water holding capacity of 2.0. Shear press values were determined by weighing out 75 grams (dry weight basis) of texturized protein material. The sample was placed in an excess of cold water and soaked at about 40°F for 1.5 hours. The sample was drained for five minutes and divided into 3 equal parts by weight. The three parts were wrapped in plastic and allowed to stand at room temperature for 20 minutes. Each of the parts was tested in the Allo-Krammer Shear Press Model 5-2H according to conventional techniques using a 2,500 pound ring and a ten-bladed head. The three values are added together. The term water holding capacity refers to the total amount of water the protein material is able to hold and is determined by soaking the protein in an excess of water for 20 minutes and then draining for five minutes. The water holding capacity is equal to the wet weight minus the dry weight and that value divided by the dry weight.

Textured Products from Defatted Cooked Soybean Flour

Defatted oilseed protein such as soybean protein is converted by a process disclosed by *J.K. McAnelly; U.S. Patent 3,142,571; July 28, 1964; assigned to Swift & Company* to a bland tasting, textured material. The process involves cooking a dough of defatted oilseed protein under conditions which insure that the dough will assume an elastic form which will recover from deformation caused by pressure or stress and return to its original shape. During this cooking the protein content of the dough is converted from the water-soluble to water-insoluble form while undesired flavor elements and precursors, as well as pigments remain in the water-soluble form. The product resulting from this cooking step is then contacted with an aqueous solvent and the water-soluble materials are extracted from the cooked protein product.

While the method has particular application for treatment of soybean flour, it is also applicable in the manufacture of protein supplements from other vegetable protein products such as peanuts, zein, etc., which have not been used for this purpose because of flavor problems.

The soybean flour, a desolventized, untoasted, oil-free soybean flour, is mixed with water to form a dough which can be shaped and formed for cooking. A dough containing 40 to 60% flour with the remainder being water is very satisfactory. Larger water-to-flour ratios provide a fluid product which is not easily handled, while the use of less water results in the production of a dough which is somewhat crumbly and more difficult to form.

The natural pH of the dough is around pH 6.3 and in the preferred form of the process pH is not altered. If the dough is made more acid by the addition of nontoxic acids, to adjust the pH to 4.5 to 5.5, the final product is harder and more brittle than that resulting from a dough held at the natural pH of 6.3. A more alkaline dough having a pH around 8.5 is quite soft and mushy and the final product is much more frangible and can be broken down with less force or pressure than required for the product made from dough with a natural pH.

Shaping of the product prior to cooking can be carried out by any of the conventional means such as slicing, extrusion, molding, pressing, etc., to produce filaments, strands, strips, particles, etc. It is preferred to place the dough in a form which exposes the greatest amount of surface area in the subsequent cooking step.

The dough in the shaped form is then subjected to live steam and cooked under pressure. The cooking step serves to coagulate the protein in the dough and convert water-soluble protein to the water-insoluble form. The live steam also assists in volatilizing certain undesirable flavor components and serves to impart a structure to the dough. The dough assumes a cellular or spongy structure as a result of the swelling when the pressure is re-released and water vapor is volatilized.

After removal of the cooked dough from the cooking vessel the strands, rods or pellets are further subdivided by slicing or grinding to produce pieces of a small size with a large number of freshly cut surfaces. The ground or sliced product is then contacted with water or other polar solvent to leach the water-soluble materials from the cooked soybean flour. The divided cooked material is placed in hot water or other polar solvent and steeped or is washed with hot water or other polar solvent to remove water-soluble materials. The steeping or percolation step serves to extract water-soluble flavor substances, flavor precursors, some water-soluble proteins, carbohydrates, and coloring materials.

Steeping can be carried out by placing the denatured flour in water and heating to a temperature of 32° to 212°F, sufficient to extract undesirable flavor substances, coloring materials, etc.

Polar solvents, as well as water, can be used in the leaching step to extract soluble proteins, bitter and beany flavor materials and pigments. Aqueous ethanol containing up to 95% ethanol has been employed as the leaching solvent with good success. Where organic solvents are used it is advisable to follow the leaching with a water rinse to remove traces of the organic solvent.

The steeped product is dried with hot air or by vacuum-drying or any other suitable drying technique to remove water and any other solvent. The temperature of drying preferably is 180° to 185°F and the length of drying is not greater than 4 to 6 hours. Higher temperatures and/or longer drying times yield a product with an undesirable flavor and a dark brown color. The dried material is then further subdivided, if desired, to a particle size suitable for inclusion into any of a number of food items.

Example: 550 grams of 60 mesh soybean flour was mixed for 3 minutes in a mixer with 450 grams of water. The resulting dough was passed through a food grinder equipped with a ¼ inch plate. The extruded strands were placed on a screen in an autoclave. The temperature was maintained for 5 minutes. At the end of this time the pressure within the autoclave was released within 1 to 1½ minutes. The cooked strands were chopped in a food chopper to obtain smaller pieces and 100 grams of the cooked, chopped strands were placed in a beaker containing 900 grams of water heated to 180°F.

The beaker contents were stirred occasionally for 15 minutes and the water then was poured off and replaced with an equal volume of fresh water at 180°F. The contents of the beaker again were stirred occasionally for 15 minutes at which time the water was poured off and replaced with an equal volume of fresh water at 180°F. This water was allowed to remain in contact with the cooked pieces of product for 3 minutes. At the end of this time the water was poured off and the particles were dried in a forced air oven at 180°F for 2 hours. The dried particles were light tan in color and when rehydrated and tested they were found to be spongy and elastic and possessed no flavor.

Mechanically Tempering Prior to Heating

A process is disclosed by *D.H. Waggle; U.S. Patent 3,810,764; May 14, 1974;* assigned to *Ralston Purina Company* for producing textured protein particles having surface orientation characteristics. The method comprises the steps of mixing together ingredients comprising secondary protein source material containing at least 35% by weight protein and an aqueous liquid, the mixture having a pH from 5 to 10, mechanically tempering the mixture, subjecting the mixture to elevated temperatures and a constant pressure, the temperatures being sufficient to cause expansion of the mixture to form an expanded, substantially water-insoluble, irreversible cross-linked structure and cooling the resulting product. A wide variety of protein or oleaginous materials can be used in the process, for example, high DPI oilseed flakes or meals, toasted flakes or meals (those materials which have been denatured to some degree by heat), protein isolates, and full fat oilseed flours. However, materials which have been severely heat treated (highly denatured or burned) are not satisfactory.

A variety of methods may be used to mechanically temper the protein mixture. The mixed material may be rolled, stretched, folded, or worked with mechanical beaters or even subjected to several combinations of mechanical tempering steps. The mechanical tempering aids in imparting a high degree of orientation to the protein structure which is later set in the isobaric heating-expansion step. The high degree of orientation of the structure improves the resilient, chewy, and meat-like properties of the material. It was found that a rolling step is particularly advantageous in imparting the desired meat-like characteristics in the final expanded product. The rolling operation imparts a degree of mechanical tempering which provides a product which has the ability to absorb water and aqueous liquids, is tough, resilient and chewy, and which will maintain its structure and physical properties when subjected to heat.

The mechanical tempering operation and the texture of the final product are greatly influenced by the presence of various humectant and preservative solvent materials in the aqueous protein mix. Typical preservative organic solvent and humectant materials are glycerol and 1,2-propanediol. Other reagents may also be added to the mix to influence the mechanical tempering operation or to impart other properties to the protein product.

The mechanically tempered material is then subjected to elevated temperatures at isobaric pressure to accomplish expansion and heat-setting. The elevated temperatures should be accomplished by means of a radiant energy source which will generate the heat uniformly throughout the mass. Radiant energy having a wave length which will penetrate the mass is effective to generate the heat uniformly throughout the mass. Devices of the microwave type which have a wave length tuned to resonate the water molecules are most satisfactory.

However, the magnetostriction or induction devices which will resonate the water or protein molecules may also be used. The temperatures employed to achieve and retain the expanded structure of the product are dependent upon the residence time of the product being exposed to the radiant energy and the boiling point of water at the pressure used in the process. The pressure and temperature conditions must be sufficient to heat denature the protein in an irreversible structure after the structure is formed by the expanding steam from the boiling water. A practical minimum temperature required to set soy or other oilseed materials is about 180°F. In addition, the temperature must not be so high as to destroy or degrade the protein structure when it is formed. The oilseed materials have a maximum temperature of about 400°F.

For an isobaric process operating at atmospheric pressure the temperature limit is about 212°F, the boiling point of water. A typical residence time for the product in a process operating at atmospheric pressure would be about 30 to 90 seconds. After the product has been heated and expanded in accordance with the above procedures, it may be cooled or dehydrated and subsequently processed into a suitable form to further processing into various food products.

Example: One hundred grams of solvent extracted soybean meal having a protein content of 50% by weight and a DPI of 70% was mixed with 175 ml of water in a Brabender Sigma blade food mixer for about five minutes. The mixed material was separated into discrete chunks of about 60 grams and the chunks were rolled into rods about 1" in diameter and 3" in length. The material was rolled on a pair of canvas belts for about 15 seconds to form the rods and to impart the desired surface characteristics to the material. The formed rods were then placed on a moving belt and passed at atmospheric pressure through a Varian microwave oven with a residence time of about 60 seconds. (1 kwh, 220 v, 2,450 mc). The product removed from the microwave oven was a puffed, expanded product which had a tough, resilient structure and would not tear easily.

The product was sliced and the internal structure was observed to be cellular and it had a definite orientation due to the shape of the cells and their connective membranes. The material had a chewy resistance and mouthfeel similar to that of meat. A slice of the material was dried and weighed. When the dried material was immersed in room temperature water and removed, it was observed that the slice had sorbed 120% of its original weight in water in twelve minutes and would not tear easily even though containing the large amount of water. The material exhibited a very bland flavor free of the typical soy or beany taste.

TEXTURED PROTEIN GELS

Chewy Protein Gel Binders for Fibers

M.L. Anson and M. Pader; U.S. Patent 2,813,025; November 12, 1957; assigned to Lever Brothers Company have found that protein filaments can be made into food products which closely resemble meat and meat products by using a chewy protein gel as a binder for the filaments. Such products are made by applying to the filaments a protein solution or dispersion (gel precursor) capable of forming a chewy protein gel on being heated and then applying sufficient heat to convert the protein solution or dispersion to a chewy gel.

It is advantageous if protein in the gel precursor is in a denatured, preferably heat-denatured form. This may be achieved by the application of heat before, during or after the precipitation of the protein from the extract, the system normally being heated to a temperature of at least 50°C, generally no less than 70°C and preferably, within the range of 85° to 100°C.

The protein concentration and the pH are the two main factors which determine whether a heat-irreversible gel will be formed from the gel precursor upon the application of heat. The two factors are largely interdependent; the lower the concentration, the higher the pH below which satisfactory gel formation does not take place. The protein concentration of a peanut protein gel precursor with which the filaments are to be treated is usually in a range between about 20 and 30% by weight. A concentration in the vicinity of 25% is preferred.

At the preferred range of protein concentration, the pH is generally adjusted to a value of about 6 or greater. There has been observed an increase in the tendency of certain samples of protein toward the development of undesirable color and flavor of the gel ultimately obtained, as the pH of the gel precursor is raised. Too high a pH may also lead to a certain undesirable rubberiness.

Example: Preparation of Gel Precursor — The commercial peanut meal was extracted with aqueous hydroxide solution and the insoluble residue removed by centrifuging. The extract was heated to 95°C and maintained at that temperature for 5 minutes before 10 N hydrochloric acid was added to reduce the pH to 5.

The precipitated protein was then separated by centrifuging. The protein suspension so obtained, which had a solids content of about 35%, was put through a mincer whose plate had holes of ½ inch in diameter and then through a 16 square per inch sieve. The resultant mass was mixed with sufficient water and sodium hydroxide in a mechanical mixer and mixing continued to form a smooth plastic mass whose solids content was about 25% and whose pH was 6.9.

Admixture of Ingredients — Eight parts by weight water, 40 parts by weight gel precursor and 25 parts by weight of a mixture of farina, skim milk powder, hydrogenated vegetable oil, diced pork fat and flavoring and coloring agents were thoroughly mixed into a slurry. Twenty-five parts by weight of a peanut protein of filament tow, chopped into about ½" lengths, were added and the mixture mixed in as short a time as possible to avoid undue disintegration of the filaments.

Final Processing — The mixture was packed into 8 ounce sanitary cans which were sealed immediately. The cans were then autoclaved for 2½ hours with steam at a pressure of 10 psi. The cans were then allowed to cool.

Chewy Gel from Denatured Proteins

M.L. Anson and M. Pader; U.S. Patent 2,830,902; April 15, 1958; assigned to Lever Brothers Company have disclosed the preparation of a product having the chewiness and texture of meat. The three essential steps are (a) adjusting the protein-water system, in particular the pH and the solids content, to a composition conducive to gel formation; (b) the shaping of the system and (c) the application of such heat as is necessary to produce a chewy gel. The shaping may take place before or after either of the other steps of the process.

Among the proteins preferred for the process are heat denaturable proteins, for instance oilseed proteins, such as those obtained from peanuts and soybeans, and fish proteins and also casein when processed under certain special conditions.

The nature of the gel precursor and the manner of its formation may vary widely, depending in part upon the type of product desired, the source of the protein, the composition of the outer additives and other factors.

One preferred procedure for preparing a suspension of denatured protein involves separating the protein from substances associated with it in the natural product to an extent sufficient to allow subsequent gel formation and precipitating the protein before, during or subsequent to the application of heat to the system. In this type of procedure the system is heated to a temperature above about 70°C preferably within the range of about 85° to 100°C.

A preferred form of this procedure, referred to as hot precipitation, involves forming a protein extract by suspending the ground meal of such parent materials as peanuts, soybeans and the like in water adjusted to a pH above the isoelectric point of the protein to dissolve the proteins and to leave undissolved the carbohydrates and certain of the coloring materials and off-flavors, and thereafter subjecting the suspension to filtration, centrifugation or the like to separate the extract, i.e., the dissolved protein from the undissolved material. The extract is then heated and maintained at an elevated temperature for a short interval of time, whereafter its pH is adjusted to a value at or near the isoelectric point of the protein. This results in the precipitation of the protein from the extract.

It is believed that the superior structure of the protein gel precursor made of protein precipitated by slow addition of acid to a hot extract may be due to the additional hydration which is imparted to the protein precipitate when it is heated on the alkaline side of the isoelectric point. This extra hydration is of advantage when the shaping is accomplished by extrusion.

Whatever the method of producing the gel precursor, it is usually heated for a time and at a temperature sufficient to result in the formation of a gel. Obviously, the larger the mass, the longer it will require to gel the entire mass. Within reasonable limits excessive heating does not have any deleterious effects. While heating is not necessarily essential to gel formation, it is preferred to heat the mass above the boiling point of water while avoiding dehydration of the mass. It is desirable to avoid temperatures higher than 300°F. A preferred method of forming the gel is that of autoclaving by subjecting it to steam at superatmospheric pressure. This, in addition to raising the temperature of the mass to a point at which gel formation takes place readily, has the additional advantage of avoiding dehydration before the chewy gel forms. The following example is illustrative of the process.

Example: Five kilograms of peanut meal prepared by solvent-extraction of lye-dipped peanuts, were suspended in 95 liters of water at 22°C and 240 cc of a 2.09 N aqueous sodium hydroxide solution added. This suspension, which had a pH of 7.2, was stirred for 60 minutes and then centrifuged to remove the insoluble fraction of the meal.

84 liters of the extract were heated to 95°C by the introduction of live steam and kept at that temperature for 5 minutes. While stirring the heated extract, 304 cc of 3.58 N hydrochloric acid were added and the protein precipitated. The suspension of protein, having a pH of 4.6, was kept at 95°C for an additional 5 minutes and then centrifuged at 1,800g to collect the protein. The protein suspension thus obtained had a solids content of 34%.

60 cc of 2.09 N sodium hydroxide, 155 cc of water and 12 grams of locust bean gum were added to 1,000 grams of the protein suspension. The materials were thoroughly mixed and

then broken down to a smooth paste by passing the mixture through a roller mill. The resulting smooth plastic mass (gel precursor) was at a pH of 7.1 and had a protein content of 28%.

The gel precursor was placed in the cylinder of a commercial macaroni extruder. The extruder had a die 2½" in diameter with several rows of holes 0.007" in diameter. The holes were staggered so that the extruded material, when deposited on a flat surface, was laid down as a single layer of parallel cylinders. A 100 mesh stainless steel screen was inserted into the cylinder of the extruder immediately over the die to prevent clogging of the die by large pieces of protein which might not have been broken down in the roller mill or by foreign material.

The gel precursor was extruded onto a reciprocating table, the table being moved at a rate just sufficient to collect the protein cylinders in substantial parallelism. As the cylinders were deposited on the table, they were dusted with a mixture of three parts wheat starch to one part skimmed milk powder and were gently compressed by Teflon-covered rollers. This procedure was continued until a mat was built up to a height of 3". The mat contained 15% by weight of the wheat starch-skimmed milk powder mixture.

The mat was then placed in a wire mesh cage having substantially the same interior dimensions as the mat. The mat, in its cage, was then placed in an autoclave and heated for 15 minutes at 15 psig steam pressure in an atmosphere of live steam. The steam pressure was then released and the product removed.

The autoclaved product resembled, in texture, a piece of meat such as pot roast and was capable of being roasted, broiled, fried or boiled and thereupon converted into an entirely edible product having the chewiness characteristic of meat. The product can be used in forming meat-like products, using binders and additives.

Conversion of Fibers to Chewy Gel in Coagulation Bath

This process of *M.L. Anson and M. Pader; U.S. Patent 2,833,651; May 6, 1958; assigned to Lever Brothers Company* involves converting the filaments (formed in the coagulating bath of the spinning process) at least partly to a chewy protein gel.

The conversion to a chewy protein gel involves the adjustment of the pH of the protein filaments, which may, but need not have been stretched, to a value conducive to gel formation on being heated and the subsequent application of sufficient heat to form a chewy protein gel. The term gel precursor will be used to refer to protein filaments in a condition in which at least part of the protein is capable of forming a chewy protein gel on being heated.

The process is particularly suitable for making food products which simulate not only the texture and chewiness of various kinds of cooked meat or meat products but also their moistness. The process allows products to be made which also simulate the color, general appearance and taste of a variety of kinds of cooked meat or meat products.

Usually, the product also contains substances other than protein filaments, such as dyes, flavoring materials, fat, starch or flour which may be incorporated in order to impart color, flavor or a desired texture to the product. Such additives may be incorporated at one or more stages of the process. Thus, in brief outline, a typical process may comprise the following steps: the preparation of the filaments; the treatment of the filaments with an alkaline agent to convert them to gel precursors; the admixture of additives, for example by passing the gel precursor through a bath containing the desired additives in the form of a liquid; the removal of excess liquid from the mixture of gel precursors and additives, for instance by squeezing; the shaping of the resultant mixture to the shape desired of the product, for example by packing into a can; and the application of heat to the mixture in an autoclave.

EXTRUDED GRANULES OF GLUTEN-SOY FLOUR BLENDS

R.V. MacAllister and T.P. Finucane; U.S. Patent 3,102,031; August 27, 1963; assigned to General Foods Corporation describe the food product made from inexpensive sources of vegetable protein, of which wheat gluten and soybean flour are preferred materials. The product is highly nutritious and is capable of simulating not only the flavor and texture of cooked meat but also its appearance.

The soy flour and gluten increase or enhance the value of each other so that the use of both together provides advantages over and above their individual advantages. From a physical standpoint, the gluten is gummy and difficult if not impossible to extrude, but by the addition of the soy, an extrudable product is formed. From a nutritional standpoint, the amino acid content or pattern of the soy complements that of the gluten, and vice versa.

Generally speaking, the product comprises irregularly shaped, substantially dry, granules capable of absorbing water and comprising essentially wheat gluten and soybean flour. Suitable flavoring materials are present to give the cooked product the taste of cooked meat. When cooked in the presence of water, the granules develop a flavor and chewy texture characteristic of cooked meat and they acquire the appearance of cooked chopped meat.

It is desirable to add minor proportions of egg albumin and starch. The albumin adds to the chewiness of the cooked granules, while the starch supplies binder-like properties. Additional ingredients comprise at least one meat flavor, salt and spices.

The gluten used is preferably wheat gluten because of its availability and low cost, although rye gluten may be used. Glutens high in lysine are particularly desirable. Vitalized or undenatured gluten is one that has not been subjected to high heat for any time such as would denature its protein content. The vitalized gluten lends chewiness to the granules and also holds together the other ingredients by virtue of its gummy or film-forming properties. The gluten may suitably comprise about 49.5% by weight of the granules and may range from 40 to 60%.

The soybean flour is any clean, bland material which has been solvent extracted to remove the undesirable bitter and beany flavors common in untreated soybean flours. A suitable amount of flour is 20 to 28% by weight of the product granules, although it may range from 10 to 35%.

The albumin is preferably obtained from eggs and to facilitate handling, is used in dehydrated form. The albumin may be used in an amount of 2 to 10% by weight of the product, preferably 3 to 5%, and a particular amount is about 4.8%. The preferred starch is ungelatinized wheat starch. It is also useful as a binder, in conjunction with the gluten, to hold together the various ingredients of the product.

Example: A 50 pound (dry basis) batch of protein granules was prepared. A mixture of the following dry ingredients was made up:

	Percent
Vitalized wheat gluten	49.5
Defatted soybean flour	27.9
Egg albumin	4.8
Wheat starch	4.2
Hydrolyzed fish protein	1.5
Onion powder	1.0
Salt flake	2.0
Garlic powder	0.02
Ground black pepper	0.04

All parts are by weight based on the dry protein granules. These ingredients were dry blended in an Abbey (Sigma blade) mixer for 10 minutes.

A beef extract emulsion was prepared by adding 12.25 pounds of water to a steam jacketed kettle, and then adding 3.33 pounds of beef extract. The mixture was heated with agitation to 190° to 200°F to dissolve the beef extract, after which 13.35 pounds of hydrogenated cottonseed oil (HCSO) was added and melted in the mixture.

Then 75.7 grams of sodium stearate was dissolved in the mixture and the latter was mixed in a Kolloidtek Puc Mill; then 281 grams of calcium chloride dissolved in 500 ml of water was added to the foregoing mixture and the batch was subjected to further mixing in the Puc Mill. This produced an emulsion batch weighing 30.82 pounds, of which 7.95 pounds was added to the above dry-blended dry ingredients, and the resulting mixture blended in the Abbey mixer for 10 minutes. Next a water-dye solution was prepared comprising 24 pounds of water plus conventional amounts of FD&C Red No. 1, FD&C Blue No. 1, Black Shade B and Jell-O Shade C. This water-dye solution was added to the material in the Abbey mixer and blended for 7 minutes, forming a colored dough ready for extrusion.

On the basis of the dry protein granules, the beef extract concentration was 1.7% by weight, the hydrogenated cottonseed oil was 6.9%, the sodium stearate was 0.09%, and the calcium chloride was 1.3%.

The dough was transferred to a Braibanti macaroni press having an auger which fed the dough to an extrusion head having a die with a multiplicity of 0.13 inch (approximately one-eighth inch) diameter openings. As strands of dough came through the die openings, a cutter adjacent to the die severed the strands into lengths one-sixteenth inch long.

The granules, having a moisture content of 36% by weight, were collected in containers and passed to a gas-fired tray dryer where the granules were disposed on 6 trays for approximately 20 minutes while air at 220°F was circulated by them. After 20 minutes, the dried granules were cooled by means of fans to a temperature below 100°F and were collected. They had a moisture content of about 7% by weight.

The dried granules were then screened in a Rotex screener using numbers 4 and 10 mesh screens, that is, those granules passing through the number 4 screen, but not the number 10 were collected. Stainless steel balls were present on the number 4 screen to break up oversized material.

CONSUMER PRODUCTS

SIMULATED MILK PRODUCTS

High Yield Process from Soybeans

In the past a common method for the production of soy milk has been to subject soybeans to prolonged soaking, grind the soaked beans into a slurry, centrifuge the slurry to extract the liquid, subject the extracted liquid to pressure cooking and formulating (additive) steps, homogenize the formulated soy milk, and then process the product into containers. In that process a considerable percentage of the finer edible content of the soybean was discarded along with the coarser fibers removed by centrifuging.

This process described by *C.P. Miles; U.S. Patent 3,288,614; November 29, 1966; assigned to Loma Linda, Foods* departs from the older methods in a number of respects. The presoaking of the soybeans and the grinding of the water-softened beans are eliminated. Hulls are removed at the outset from the unmoistened beans by cracking and winnowing, whereas in the older process, reliance was placed upon the softening and loosening of the hulls in the soaking process followed by a series of washing steps. Instead of the grinding of water-softened beans, this process reduces the cracked, dehulled beans to fine flake which is converted directly to slurry by addition of water. Instead of precentrifuging the coarse slurry developed from the ground soaked beans prior to homogenizing to protect the homogenizer valves from damage from solid particles, this process uses a high pressure homogenizing step in which the slurry is homogenized before it is passed through a centrifugal clarifier.

In the older methods, the time elapsed between the soaking of the beans and the treatment of the extract in the cooker was long enough for a substantial bacterial growth to occur at the temperature of the intervening processing steps, and the extraction process was too slow to match the speed of vacuum pans of substantial capacity that were then available. Avoidance of bacterial contamination was a substantial problem. In this process, bacterial growth and enzymatic action are arrested almost at the beginning, i.e., immediately after mixing the soybean flakes with water. This is accomplished by cooking the slurry and then promptly subjecting it to high pressure homogenization.

By developing an instant slurry from flaked unmoistened beans mixed with water immediately before cooking and homogenization, this process gives a savings of more than 15% reduction in the quantity of beans required for the production of a given quantity of milk. For example, one batch utilizing 350 pounds of bean flakes represents a savings of about 70 to 80 pounds of beans over the previous method of soaking and grinding the whole beans of a comparable batch (producing the same volume of end product).

This process is carried out in the manner described below.

Example: A satisfactory variety of clean soybeans having a moisture content in the range of 10 to 14% of water by weight are used. They are cracked in their dry condition. The hulls are loosened by the cracking. Dehulling is accomplished by winnowing, e.g., by passing a stream of air through the cracked dry beans during free-fall from one portion to another of a conveyor system by which the cracked beans are conducted to flaking rolls. The hulls are blown away from the cracked beans.

The dehulled, dry, cracked beans are passed through flaking rolls and rolled under heavy pressure into very thin flakes (0.003 to 0.008 inch). This flaking step is performed promptly following the cracking and dehulling steps so that the cracked beans may have no opportunity to dry further. Water (either hot or cold) is now added to the soy flakes at a ratio between 1 pound of beans to 1 gallon of water to 50 pounds of beans to 32 gallons of water. The slurry is agitated from a minimum extent where blending will be barely attained up to a period as long as one-half hour.

A stabilizer of the phosphate type, or a sequestering agent such as EDTA is then added to the slurry. The slurry is then pressure cooked at 220° to 250°F for a period of time from a mere flash to a maximum of 10 minutes. The cooked slurry, without being subjected to any separation step, is directly subjected to high pressure homogenization, (e.g., by using a Manton Gaulin homogenizer) at a pressure of 5,000 to 8,000 psi. The slurry made from the flaked beans, after pressure cooking, will be in a sufficiently finely divided state to avoid damage to the homogenizer valves, which would occur from the forcing of particles of substantial size through the valves.

The slurry is then clarified by pumping it through a centrifugal separator such as a De Laval or Westphalia, with a flush cycle of 2 to 5 minutes. More specifically, the slurry is pumped into the centrifuge for 2 to 5 minutes, the pumping is then arrested, and after a short interval sufficient to empty the centrifuge, it is back-flushed to remove the solid residue accumulated during the centrifuging cycle. The liquid extracted is then pumped into a formulation tank where oil, carbohydrate (including sugar), vitamins and minerals are added in accordance with the requirements for a whole milk, a skim milk, or for mother's milk where the product is prepared for infants.

The formulated milk is then subjected to the conventional milk homogenizing step at a pressure of 1,500 to 3,500 psi. Any homogenizer may be used for this step, which can be performed either before or after condensing. The homogenized milk is then evaporated in a vacuum pan to produce a proper density for direct use, or to produce a condensed milk. The condensed milk is then passed through a heat exchanger and on to a cold storage tank for standardization. From the cold storage tank the milk is packaged into containers to be sold as fresh milk.

Simulated Cow's Milk from Mixed Proteins

A simulated milk powder having minimum enzyme destruction and a naturally occurring antioxidant material is produced by *G.W. Johnson; U.S. Patent 3,386,833; June 4, 1968.* The composition of the milk comprises three principal ingredients: soybean flour, sesame seed flour and coconut meal. The three components are ground to extremely fine powder with the soybean being provided preferably in 4 parts by weight to 3 parts by weight of the sesame seed and 2 parts by weight of the coconut meal.

Somewhat varying amounts of the three components may be employed generally, the soy flour should be present within the range of 2 to 6 parts by weight, the sesame seed flour within the range of 2 to 6 parts by weight and the coconut meal within the range of 1 to 2 parts by weight. The soy and sesame seed flour are best employed for a superior milk respectively in the range of 3 to 5 parts by weight per 1 to 2 parts by weight of the coconut meal. The coconut meal is prepared from dried coconut meat (copra). Copra typically contains 2 to 5% water by weight. A small amount of lecithin, usually 0.5 to 2%

by weight of the three principal ingredients, is incorporated as a water dispersing agent for the three ingredients. The powdered mixture may be placed in liquid suspension by stirring in water, preferably using a Waring type blender. The product will remain in suspension for a long period of time without the aid of an emulsifying or suspending agent. However, a superior product is had by incorporating a small stabilizing amount of an edible emulsifying agent such as extracts of Irish moss, *(Chondrus crispus)* or other edible alginate base material. The composition may be enriched by adding vitamins or other desired ingredients.

In the preparation of the liquid simulated milk, 6 to 9 teaspoons (approximately 4.9 milliliters per teaspoon) of the powdered composition are added to one quart of water and thoroughly mixed in a blender. The precise amount of powdered composition used may be varied somewhat to suit the taste of the user. In the plant manufacture of the composition of this process, it is advantageous to first mix the finely ground powder composition with water to make the simulated cow's milk and then spray dry to obtain a product which may be placed by the consumer more readily into a liquid form.

Cow's milk and conventional soybean milk have high protein contents of about 3.4% compared with human breast milk of about 1.4% protein. This higher protein content of cow's and soybean milks is frequently not acceptable to infants and it becomes necessary to dilute the milk where it is being substituted for human milk and there is thus a resultant lowering of other desirable nutritional values.

This simulated milk contains a significantly less protein content of approximately 1.5% which more nearly approximates that of human milk than either cow's milk or soy milk. The lower protein content is attributable to the significantly less protein found in sesame seed flour and coconut meal than in undiluted soybean flour. By use of this mixture of vegetable proteins, a milk product is produced having an improved methionine content and a higher calcium content than if soy protein were used alone.

Soy Milk from Sprouted Beans

A method of preparing soy milk of improved taste and flavor is described by *G.K. Okumura and J.E. Wilkinson; U.S. Patent 3,399,997; September 3, 1968.* In general, the process is one where selected soybeans are first subjected either to a sprouting step or to a prolonged soaking step in which the hulls are loosened and the beans are softened. The softened beans are subsequently reduced, by grinding, to a slurry which is then diluted by water, and wherein the milk is then extracted from the diluted slurry and finally treated by mixing with additives. The preferred method is given in the following example.

Example: Cleaned, selected soybeans are moistened and sprouted at a temperature of 90° to 100°F for approximately 3 days to produce sprouts ½ to 1½ inches in length. The sprouting is carried out by spreading the beans on a screen or sheet of porous fabric in a layer of several beans depth. This layer is sprinkled from above by a water spray at intervals of between 2 and 3 hours to maintain the beans adequately moist. In the spray, use of water much colder than the sprouting temperature, so as not to arrest the sprouting action is avoided.

The sprouting step improves the vitamin content of the beans and also results in the hulls being loosened to a considerable extent. A substantial percentage of the hulls will drop off the beans during the sprouting and the ensuing transfer to the next processing area. The sprouting also modifies the taste of the beans to eliminate the disagreeable flavor. Subsequent to the sprouting step, the beans, while still in a softened, moistened condition, are transferred to a stone grinder and ground into a slurry or mash. They are reduced to particles varying in size from colloidal dimension up to discrete particles. After the stone grinding, water is added to bring the total water content up to approximately 8 times the dry bean content, by weight. Water is added to the slurry until it attains a substantially liquid state, with all of the bean solids in suspension, the slurry being subjected to agitation to maintain the suspension. The diluted slurry is then cooked in a pressure cooker or in an open vat, using live steam injected directly into the slurry, in both cases. In the pressure cooker, cooking is continued from 5 to 7 minutes. In the open vat cooking, it is continued for about 20 minutes. Cooking temperature in the pressure cooker is approximately 250°F. In the open vat cooker, it is approximately 215°F. Following the cooking, the slurry is treated to separate the liquid and suspended fine solids from the pulp (remaining solids of the ground beans). The suspension liquor is extracted from the cooked diluted slurry by a suitable extraction process, (e.g., by straining it through a fabric bag strainer with the application of pressure to abstract as much of the liquid as possible) or by centrifuging.

To the extracted suspension liquor is then added a mixture of calcium sulfate and magnesium chloride as a precipitant and the milk is stirred very gently until the solids are precipitated as a curd. The precipitated curd is then washed in the precipitating container by first removing as much as possible of the excess liquid standing above the curd without disturbing the curd and then filling the container with fresh water which dissolves and washes away the excess precipitant that has not been chemically combined with the protein. The percentage of the chemicals that is actually absorbed is far below the percentage that could be harmful to the person eating the product.

The rinse water is stirred gently for a few minutes. A wire mesh basket is then inserted into the container on top of the curd to contain it while the excess water is removed, (e.g., by pumping, siphoning or dipping). More water is then removed by pressure extraction in a mold lined with a fine fabric cloth liner in full-bag form, having outlet openings in the bottom for release of water and having as a cover or top, a movable pressure plate or piston which is moved slowly downwardly exerting a gentle pressure of about 8 ounces per square inch, (e.g., 100 pounds for a pressure plate of 200 square inches area) against the bag, the pressure being applied by a dead weight and over a period of about 15 to 30 minutes until the drainage of excess liquid substantially stops.

The residual bean curd is then comminuted to a very fine state. This can be done by grinding, though preferably by treatment in an emulsifying machine or colloid mill in which extremely finely divided state is attained. The milk is then formulated by the addition of ingredients for adjusting the milk to resemble as nearly as possible natural cow's milk or other natural milk which it may be desirable to simulate. The emulsified product is then pasteurized to eliminate bacterial contamination acquired during the preceding processing steps.

Following the formulation step, the milk is homogenized at a pressure of 3,000 pounds per square inch to preserve the suspension of the oil in the milk and prevent separation. A conventional milk homogenizer can be used in carrying out this step. The homogenized milk is then cooled to 40°F and is then packaged. Where the milk is canned, the pasteurization and cooling steps are eliminated and the homogenized milk is directly packaged into sealed cans, the cans pressure cooked for 30 minutes in lieu of pasteurization.

Deodorizing Soybean and Peanut Milks

Improved enzymatic and/or microbiological methods have been developed by *K. Fujita, E. Sato and T. Moroe; U.S. Patent 3,460,950; August 12, 1969; assigned to Takasago Perfumery Company Limited, Japan* to deodorize soybean milk and/or peanut milk. It was found that the conidiospores separated from the koji starters were more effective in deodorization of bean milk while the enzymatic activities were very low compared with those of the koji. Therefore, when the conidiospores were employed instead of the koji in the deodorization of bean milk, characteristic disagreeable flavors and odors were removed more effectively without development of bitter taste, brown discoloration, and rancid flavors, etc.

The conidiospores which were found effective in deodorization of bean milks were those of *Aspergillus oryzae, Aspergillus niger, Aspergillus glaucus, Aspergillus ochraceus* and *Aspergillus versicolor*. Typical data on *Aspergillus oryzae* are summarized in the table on the following page in which the enzymatic activities and the deodorizing effects of the koji starter, the mycelium and the conidiospores separated from the koji starter of *Aspergillus oryzae* are given.

	Protease activity (u./g.) [2]	Amylase activity (u./g.) [3]	Lipase activity (u./g.) [4]	Phosphatase activity (u./g.)	Deodorizing effect
Koji or culture on solid medium [1]	5,350	127,800	92.7	a,[5] 1,019 b,[6] 0 c,[7] 54,000	++
Mycelium [1]	7,250	156,000	103.4	a, 294 b, 0 c, 2,620	+
Conidiospores	0	9,960	84.6	a, 956 b, 99 c, 34,260	+++

[1] Containing culture medium.
[2] Assayed with the modified Anson method (Ref. B. Hagiwara; Ann. Rep. Fac. Sci. Osaka Univ. 2, 35–79 (1954)).
[3] Assayed with the 3,5-dinitrosalicylic acid method (Ref. E. Borel et al.; Helv. Chim. Acta 35, 115 (1952)).
[4] Assayed with the modified Nord method (Ref. K. Yamada, H. Machida; Nippon Nogei Kagaku Kaisha 36, 860–864 (1962)).
[5] Phosphomonoesterase
[6] **Phosphodiesterase** (Ref. Omori; Enzymol. 4, 217 (1937)).
[7] Pyrophosphatase (ref. S. Akamatsu; J. Biochem. 39 203 (1952)).

The preparation of conidiospores is as follows: The conidiospores of the above Aspergilli were transferred from potato-dextrose agar slants to 500 ml Fernbach flasks containing 50 grams of solid media of the following compositions: wheat bran, 37.5%; defatted soybean flakes, 12.5% and water, 50%. The inoculated flasks were incubated at 28°C for 72 hours with occasional shaking to prevent clumps. The cultures were dried, filtered with 100 mesh sieve and the conidiospores were collected. In the cases of large scale preparations, specially designed trays or boxes were employed instead of Fernbach flasks.

The preparation of soybean milk is as follows: One kilogram of soybeans or defatted soybeans were macerated, ground with addition of water and heated to 85°C to inactivate soybean enzymes and to enhance the solubility of soybean constituents. The product was then filtered. The final volume of the filtrate was adjusted to 5 liters. The solids content of this soybean milk was 12 to 15%. 500 mg of any of the above conidiospores was added to this soybean milk, singly or in the mixed state, which were previously suspended in water containing a small amount of surface active agents.

The inoculated milk was incubated at 45°C for about 2 hours, then heated to kill the conidiospores and homogenized. The odorless milk thus obtained may be dried and pulverized if necessary. The detailed procedures are illustrated in the following examples.

Example 1: One kilogram of soybeans was soaked in water until the beans contained about 1.5 kilograms of water (usually for 12 to 16 hours) and drained and the macerated beans were ground with the addition of 4 kilograms of water. The slurry thus obtained was heated to 85°C for 15 minutes, filtered with a cotton filter, cooled, pH adjusted to about 6.0 and inoculated with 500 mg of the conidiospores of *Aspergillus oryzae*. The inoculated soybean milk was incubated at 45°C for 2 hours with gentle agitation, then heated to 85°C to kill the conidiospores, and was homogenized. By this procedure, 5 liters of odorless soybean milk were obtained.

Example 2: One kilogram of shell-free dried peanuts was treated in the same manner as Example 1 and inoculated with 1 gram of the mixture (1:1 by weight) of the conidiospores of *Aspergillus oryzae* and *Aspergillus glaucus* and incubated in the same manner as in Example 1. About 5 liters of odorless peanut milk were obtained. This may be dried, with addition of the proper amount of antioxidant, by means of a spray dryer or other drying equipment. This deodorizing process as described is much simpler in operation, economically advantageous as compared with processes developed up to the present time, and much shorter in processing time as compared with the enzymatic or microbiological processes reported to date, and therefore free from putrefaction or coagulation of bean milk.

Wet Milling of Vegetable Protein

Products which form a suspension having a "mouthfeel" or smoothness similar to cow's milk have been produced by *G.C. Mustakas, W.J. Albrecht and G.N. Bookwalter; U.S.*

Patent 3,639,129; February 1, 1972; assigned to the U.S. Secretary of Agriculture. Generally, this process by which vegetable protein flours are converted into a dried vegetable protein powder that forms a stable emulsion when mixed with water comprises the following steps: dispersing soybean protein flour in water to make a slurry containing about 20% or less solids; reducing the particle size of the flour to 5 to 40 microns in diameter by wet milling; homogenizing and emulsifying the wet-milled slurry and spray drying the resulting emulsion. A highly nutritious and palatable dried protein beverage base is also prepared by the same steps as above but adding vegetable fats and emulsifiers to the slurry before wet milling to give a protein and fat content similar to cow's milk and to give a slurry containing 11 to 20% solids. Also, flavorings and sweeteners in sufficient quantity to give the final product a flavor similar to milk were added. This can be done at any step of the process.

The preferred starting material used in the process is full-fat soy flour. Defatted soy flour gives an equivalent product with respect to dispersibility and mouthfeel but requires the addition of greater amounts of vegetable fat in the formulation of milk substitutes. Any vegetable protein flour or grits from oilseeds or other difficultly dispersible protein product can be processed by this method to produce a highly water-dispersible powder similar to that produced from soy flour. Enough water is added to give a 10 to 20% by weight flour in water slurry. This ratio of flour to water is specified because thicker slurries, greater than 20% flour, clogged the colloid mill and homogenizer used in the example. However, another kind of mill could be used which might operate with a thicker slurry.

The flour-water slurry was run through a mill which can be any one of the many different types used industrially; hammer mills, roller mills, single- and double-disc mills, or disintegrators such as the Rietz type. The clearance of the mechanism through which the slurry will pass should be 0.001 inch or less to insure a material ground fine enough for the next step. The above dimension is the minimum clearance for the equipment used in the example. One of the major criteria of the process is the small particle size of the final powder. The smaller the particle size, the greater the dispersibility and the better the mouthfeel of the final product.

Homogenizing and emulsifying can be done simultaneously in apparatus such as ultrasonic vibrators, impulse rotary hydromills, high pressure piston type, or gear type homogenizers. The equipment used in the example was a Manton-Gaulin two-stage homogenizer which was run at 3,500 psi. The maximum psi of apparatus is about 5,000 psi which is also operative for the process. At lower psi, about 2,500, the emulsifying action is much less efficient.

Spray drying was effected in a Bowen-type apparatus at an inlet temperature of 250° to 275°F. At 275°F brown specks appeared in the product which make it unusable for a beverage base. The lower limit of spray drying temperature depends on air flow. A greater air velocity permits the use of lower temperatures. The preferred temperature is between 250° and 265°F.

Example: Six pounds of full-fat soy flour were slurried with 48 pounds of water and the slurry of about 11% by weight flour in water processed through a colloid mill having a clearance of 0.001 inch. The Manton-Gaulin two-stage homogenizer emulsified the milled slurry at 3,500 psi, after which the emulsion was spray dried in a Bowen-type spray dryer at 250° to 265°F inlet air temperature and 150° to 160°F outlet air temperature. The soy protein powdered product had a NSI of 57.2 and a particle size of 5 to 40 microns as measured by scanning electron microscopy. The product showed good stability on storing at elevated temperatures for up to 182 days.

Dispersible Soy Protein for Milks

W.E. Koski, D.E. Smith and A.R. Touba; U.S. Patent 3,653,912; April 4, 1972; assigned to General Mills, Inc. have produced a soy material of improved solubility, dispersibility and taste for use in beverages such as soy milk. The modified soy material of the process is produced by treating soy material such as soy milk or other high protein soy material.

The soy material is dispersed in water and solubilized by the addition of an alkaline substance, typically sodium hydroxide. The solution may be raised to a pH of at least 12. An acidic material is then added to lower the pH to neutral or near neutral. The resulting solution is then spray dried.

Processes are known in which soy protein is purified by solubilizing the soy protein with alkali and then acidifying the solution to precipitate the soy protein. Such prior processes differ markedly from this process in that the pH in the prior processes is lowered to the isoelectric point where precipitation of the protein takes place. The pH at the isoelectric point is generally in the range of pH 4.2 to 5.2. In this process the pH is lowered only to about neutral, typically 7.0 to 8.0 and thus above the isoelectric point. The proteins, therefore, remain in solution.

Starting material may be a soy milk or other bland soy material preferably containing between 30 and 70% protein, usually 40 to 60% protein. Soy materials having a larger protein content, however, may be used with less preferred results. Such materials may be obtained by removing at least a part of the nonprotein constituents of defatted soy flour, meal or flakes by any of various means.

In carrying out this process, any water-soluble alkaline materials can be used to raise the pH of the starting soy protein material. Preferred alkaline materials are the inorganic bases and salts such as sodium hydroxide, ammonia, ammonium hydroxide and trisodium phosphate. The amount used will depend somewhat upon the particular alkaline material. However, when an aqueous dispersion containing about 5 to 20% by weight soy protein is treated with sodium hydroxide, it is preferable that the weight ratio of soy protein to sodium hydroxide be between 8:1 and 20:1.

Any of a variety of acids may be used in lowering the pH to neutral or near neutral. Typical acids would include citric acid and phosphoric acid. However, other acids such as hydrochloric and lactic acid may be used. During the addition of the acid, it is preferable that the soy protein solution be stirred in order to disperse any beads of precipitated protein that might form as the acid is being added. The pH of the solution should be reduced to about 7.0 to 8.0. The modified soy protein, at or near the neutral point, has a substantially improved solubility and therefore beverages of very high protein content can be produced.

Example: A beverage was prepared using Textrol which contains about 58.9% soy protein, 5.7% ash and 7.1% moisture. Prior to modification, about 38% of the protein was soluble. The Textrol was modified to improve flavor and dispersibility. Approximately 150 pounds of the Textrol was placed in a 200 gallon tank containing 1,134 pounds of water and mixed for 15 minutes using two Lightnin Mixers. Eighteen pounds of a 50% aqueous sodium hydroxide solution was added raising the pH to 12 and mixing was continued for about 10 minutes. Twenty-six pounds of a 50% aqueous citric acid solution was added at the rate of approximately one-half pound per minute. The mixing was continued during the addition of the acid and the final pH was 8.0.

The product was then stored in a room having an ambient temperature of 0°F for about 6½ hours and then in a room having an ambient temperature of 40°F for about 8 hours. The product at this point may be used as a beverage. The product, however, in this instance was spray dried using a De Laval Package Spray Dryer. The spray drying conditions were inlet temperature, 480°F; outlet temperature, 200°F; feed pressure, 5,000 pounds per square inch gauge. The spray dried product contained 50.5% protein, 85% of which was soluble. The product further contained 7.8% moisture and 11.5% ash. A tasty beverage was prepared by mixing 12 grams of the dry modified soy material, 20 grams of Hershey's Instant Chocolate and 8 ounces of cow's milk. The resulting beverage had 6% protein content which is substantially above that of the original cow's milk.

Soy-Milk Combinations

The characteristic beany flavor of soy milk is eliminated in the milk substitute described by *A.C.-Y. Peng; U.S. Patent 3,798,339; March 19, 1974; assigned to Swift & Company*. It was found that soybeans can be processed in a solution of certain milk materials or by-products, particularly cheese wheys, without any further adjustment of alkalinity so as to overcome the off-flavor and bitterness. As a result, it is possible to produce an economical and improved soybean beverage base and also to profitably use a wasted milk by-product such as whey left from cheese manufacture.

Generally, the method comprises forming a mixture of a full-fat soybean material in an aqueous solution of milk material or milk by-products including whole milk, skim milk, neutralized buttermilk (the liquid remaining following butter churning of sour cream previously neutralized to about pH 6.6), lactose, and particularly certain cheese wheys. Cheese whey usually is slightly acid due to the ripening process in cheese manufacture. The degree of acidity is dependent upon the extent ripening is continued, but it is believed that most fresh cheese wheys, except cottage cheese whey, immediately upon removal from the processing vat, are pH 6.0 or higher. (Cottage cheese whey is usually pH 5.0 or lower.) However, the whey will continue to become more acid, if bacterial action is not halted, until pH declines to 4.5, whereupon most bacterial action will cease. Pasteurization of whey will also halt bacterial action and fix the pH value. Similarly, spray drying whey to form a powder will arrest the pH value, and dried food grade wheys are usually found to be about pH 6.3.

It is necessary that the mixture of milk material and soybean material be no more than slightly acid, displaying about pH 6 or higher. Subsequently, the mixture is heated to boiling (which is about 214°F for cheese whey) and held for a period of about 0 to 40 minutes and preferably for 10 to 30 minutes. This cooked product may be used without further processing as a nutritional and relatively flavorless and odorless milk substitute type product. However, the mixture may be dried to remove moisture and produce a powdered milk substitute type beverage base product. This powdered product can easily be rehydrated, either immediately or after storage in dry form, into a bland, nutritive beverage of the milk substitute type. The product may also be used as a milk solid substitute in the baking industry and the like.

Example 1: A 1:1 ratio (solids basis) of clean, dehulled cracked soybeans (7 pounds) and fresh Italian cheese whey (100 pounds with 7% whey solids) of pH 6.3 were soaked for about 8 hours. Then the mixture of pH 6.5 was heated to boiling in a 45 gallon round processor and held at boiling for 15 minutes. The whole mixture was then ground and filtered through a wire screen. The resulting slurry was then jet dried. The dry product was a slightly yellowish-white powder, and it had a sweet taste and a bland flavor. The dried product was readily rehydrated into a milk substitute type liquid product of neutral pH around 6.8.

Example 2: The following demonstrates that various forms of soybean material may be processed according to this method. In each sample a 1:1 ratio (solids basis) of soybean to whey was prepared by mixing 105 grams soybean material with 105 grams whey powder and 1,500 grams tap water. Each mixture was ground and then heated and boiled for 15 minutes in a steam jacketed vessel. Thereafter each mixture was evaporated at 27 inches vacuum for 15 minutes, homogenized in standard equipment at 2,500 psi and freeze dried. Samples of each dried product were reconstituted on the basis of 10 grams dry powder to 100 grams tap water, and the reconstituted products evaluated by two taste panels, each comprising 8 trained members, according to the scale 1 to 6, representing a range of no bean flavor to very much bean flavor, respectively.

Soybean material	Average taste score
Full-fat flour	2.855
Full-fat toasted flour	3.285
Full-fat untoasted cracked bean	3.855
Defatted toasted flour	2.645
Defatted untoasted flour	2.355

Dual Inoculation for Flavor Improvement of Soy Milk

T. Kanno and H. Kobayashi; U.S. Patent 3,718,479; February 27, 1973; assigned to Showa Sangyo KK, Japan have an economical method for producing soy products free of bitter taste and off-flavor. In brief, the method comprises submerging whole soybeans in water containing 0.1 to 0.5% of sulfurous acid and mother starter of lactic acid bacteria, crushing, skinning, and steaming the soybeans, and subjecting them (water content about 60%) to fermentation by inoculating a culture of microorganisms possessing proteolytic and soybean cotyledon cell macerating activity and then drying and pulverizing the soybeans. One of the main characteristics of the method is that immersion of soybeans in sulfurous acid solution and lactic acid fermentation are carried out simultaneously.

By blowing sulfurous acid gas into the water in which soybeans are immersed, the pH of the water comes to nearly the isoelectric point of soybean protein, and reduction of the yield of the product by the soybean protein dissolving in the water can be avoided. The fermentation by inoculation of microorganisms can be carried out in the pasty medium, water content of which is about 60%. The drying of the product can then be carried out more economically compared with usual soybean milk (water content of which is about 90%) production.

The characteristics of this method are as follows. The object of immersing the soybeans in sulfurous acid solution is to soften and collapse the bonds of the protein body and spherosome granule in the cotyledon cells of the soybeans and to remove the substances such as weak acids, soluble sugars, nonprotein nitrogen compounds, off-flavor substances and most of the water-soluble physiological inhibitors. By inoculating the mother starter of sulfurous acid resistive lactic acid bacteria, for example, *Lactobacillus delbrueckii*, in the immersion solution, lactic acid is produced. The cotyledon cells, which absorb moisture and swell, are subjected to softening and collapsing action of the sulfurous acid and lactic acid.

Heating by steam increases the effect of the immersion in sulfurous acid solution and inactivates the physiological inhibitors such as urease and trypsin inhibitor and enzymes such as lipoxidase. The soybeans are pulverized after the inactivation of lipoxidase, thereby avoiding the formation of off-flavor.

Fermentation is characterized by adding a culture of microorganisms, having the proteolytic activities and macerating activities of cotyledon cells, to the soybeans treated by the immersion and steam heating. Microorganisms having the foregoing activities are *Neurospora sitophila, Aspergillus niger, Rhizopus niveus,* and *Bacillus subtilis,* which are commonly available. By this process, a product of good flavor which entirely differs from usual soybean milk and processed soybean products can be obtained.

Example: One hundred parts of selected soybeans are washed with water and are soaked for 24 hours at 55°C in water to which 0.5% of sulfurous acid and mother starter of lactic acid bacteria, for example, *Lactobacillus delbrueckii,* have been added. After dehydration, the soybeans are coarsely crushed, washed with water, skinned, and steamed at 100°C for 5 minutes. Soybeans thus treated are ground into paste by a chopper.

To this paste 5 parts of culture of *Neurospora sitophila* are inoculated and the mixture is fermented for 4 hours at 30°C. After this fermentation, the fermented material is neutralized, dried under a vacuum of 10 mm Hg at 40°C, pulverized, and 80 parts of powdered soybean milk product (moisture content 8%) of less than 40 mesh is obtained. This product retains a mild and delicious flavor without bitter taste and off-flavor.

Lipoxygenase Inactivated Full-Fat Soy Flour for Milk

G.C. Mustakas; U.S. Patent 3,809,771; May 7, 1974; assigned to the U.S. Secretary of Agriculture; describes a method of producing full-fat oilseed-protein beverages comprising the steps set forth on the following page.

(1) suspending full-fat oilseed flour in from 8 to 10 times its own weight of water, which has been adjusted to a temperature of from 150° to 212°F and a pH of from 3.5 to 4.5, for a time sufficient to inactivate lipoxygenase enzyme present in the full-fat oilseed flour;

(2) cooling the suspension to and holding at a temperature of from 50° to 75°F for a time sufficient to precipitate lipid-protein present in the full-fat oilseed flour;

(3) separating the precipitated lipid-protein from the resulting liquid whey;

(4) resuspending the precipitated lipid-protein in water at a pH of about 9, the amount of the water equaling from 7 to 10 times the weight of the oilseed flour of step (1);

(5) heating the suspension of step (4) at from 205° to 212°F for from 1 to 5 minutes;

(6) cooling the suspension of step (5) to 70° to 75°F and adjusting the pH to about 7; and

(7) clarifying the cooled and neutralized suspension to form a full-fat oilseed-protein beverage.

A method was also developed for producing defatted protein isolates in high yields comprising essentially the steps (1) through (7) described above, and the additional steps of removing the water and extracting the fat with a suitable solvent from the neutralized suspension described in step (7) above.

To obtain a soybean product having optimum nutritional value and consumer acceptability, it is preferred that the starting material be flour prepared in the following manner: crack and dehull whole raw soybeans; lightly cook the dehulled cracked beans in dry heat at about 150°F for about 5 minutes; flake and pin mill to a mesh size such that 95 to 100% of the product will pass a 100 mesh screen.

The object of step (1) of the process is to impart to the final product a superior flavor and shelf stability by deactivating the lipoxygenase enzymes that catalyze oxidation of fats, thereby causing rancidity. Enzymes were inactivated in less than 1 minute when soy flours were slurried in boiling water at a pH of 7. Complete deactivation also occurs within about 10 minutes when the starting materials were slurried in water which had been previously adjusted to pH 3.5 to 4.5 and temperatures of 150° to 212°F. Details of the process are illustrated in the following example.

Example: Preparation of Raw Preheated Full-Fat Soy Flour — Field dried soybeans were subjected to a single break (one pass) cracking on corrugated rolls (10 corrugations per inch, 0.040 inch clearance). The cracked meats were then passed over a two-deck screen shaker (one-eighth inch round hole top screen and a No. 14 bottom screen) with dehulling aspiration on the meats as they entered the top screen and on the No. 14 screen tailings. The dehulled meats were then dry heat treated by passing through a screw paddle conveyor maintained at 150°F by jacketed steam and having an overall retention time of about 5 minutes. The hot meats were passed through flaking rolls, cooled, and ground (99% through 100 mesh screen) in a wide chamber air swept pin or stud mill. The resulting flour had an NSI value of 97.

Preparation of Soy Milk — Five hundred grams of soy flour prepared as described above (39.5% protein, 22.0% fat, 4.8% moisture, all by weight) were slurried in 4,000 cc of tap water that had previously been adjusted to pH 3.5 with 10% sulfuric acid, heated with sparge steam at 206° to 210°F, held at that temperature and stirred for 10 minutes, cooled to room temperature (70° to 75°F) and stirred for 15 minutes. The slurry was maintained at pH 3.5 with additions of 10% sulfuric acid. The slurry was then fed over a 5 to 10 minute period to the bowl of a perforated basket centrifuge running at 3,800 revolutions per minute (1,500 x gravity). The perforated basket was covered with a cotton duck cloth.

The resulting cake (lipid-protein curd) was washed twice with 500 ml portions of tap water, pH 7, spun dry for 5 minutes. The liquid whey was recycled through the centrifuge and the recovered curd combined with the cake. The wet curd, (i.e., cake) was reslurried in 4,000 cc of tap water and the suspension adjusted to pH 9 with 10% NH_4OH solution. The alkaline suspension was quickly (2 to 3 minutes) heated to about 205°F using open sparge steam, held at 205°F for 1 minute, and quickly (2 to 3 minutes) cooled to room temperature (70° to 75°F). The alkaline-cooked suspension was fed to a Manton-Gaulin colloid mill having a plate clearance of 0.001 inch, then to a Manton-Gaulin single-stage homogenizer at 8,000 to 10,000 psig. The milk was adjusted to pH 7.2 with 10% H_2SO_4 and centrifuged. The small amounts of cake obtained at this stage can be recycled to the soy flour starting material at least 5 times and with no detrimental effect on the finished product.

The final beverage products of this process are suitable for further processing such as spray drying to obtain dried reconstitutable soy milk powders as food supplements. Additives such as fats, vitamins, minerals and flavorings can also be used to make beverages suitable to any taste and nutritional need. When extra fat is added, it is preferred that it be added to the suspension as it is being fed to the colloid mill.

OTHER DAIRY-TYPE PRODUCTS

Soy Yogurt

A process for the manufacture of a synthetic yogurt from soybeans is described by *H. Ariyama; U.S. Patent 3,096,177; July 2, 1963.* In this process, washed soybeans are steamed for 25 to 30 minutes and soaked for 1 hour in either 0.1 or 0.2% caustic soda solution. Then the skin of the soybeans is removed and the soybeans are soaked again in the caustic soda solution which was used in the preceding step for 4 to 5 hours.

The thus treated soybeans are then ground with the caustic soda solution and filtered. The resultant filtrate is neutralized with hydrochloric acid or citric acid, adjusting the pH of the solution at 6.8 to 7.0 and then about 75 grams of sugar (as cane sugar) is added to 500 ml of the adjusted filtrate. Then the solution is boiled, agitating the same continuously for about 5 minutes and after it is cooled, the cooled solution is inoculated with *Lactobacillus bulgaricus*. The whole solution is incubated at a temperature of 37° to 43°C for 4 to 6 hours until the protein which has been extracted from soybean and dissolved in the solution coagulates.

Example: Well selected soybeans were steamed for 25 to 30 minutes and soaked in 24 liters of 0.1% caustic soda solution for 1 hour. The beans were taken out and skinned using a mixer equipped with rotary blades. The skinned beans were again soaked for 4 to 5 hours in the caustic soda solution which was used in the preceding step and ground with the caustic soda solution by means of a suitable grinder.

The whole solution was filtered through a cloth bag under pressure, yielding 18 liters of concentrated protein solution. To the resultant solution hydrochloric acid or citric acid was added, adjusting the pH to 7.0. Seventy-five grams of cane sugar was added to 500 milliliters of this neutralized solution which was then boiled for 5 minutes with continuous agitation. After cooling, 10 ml of ordinary milk yogurt or a suitable amount of *Lactobacillus bulgaricus* was added as a starter and then 20 to 40 drops of lemon or orange essence were added. The whole solution was agitated slowly, homogenized, then transferred into sterilized containers and kept at a temperature of 37° to 43°C for 4 to 6 hours, yielding a final product.

The constituents of the product are shown in the table on the following page. It may be seen that protein, mineral, and lecithin contents of this synthetic soy yogurt are 3 times, 6 times, and 4 times those of ordinary milk yogurt, respectively.

	Percentages	
	Milk Yogurt	Soybean Yogurt
Water	94.2	76.9
Crude protein	3.4	9.8
Fat	1.1	0.4
Carbohydrate	10.6*	8.6**
Minerals	0.7	4.3
Phosphatide (lecithin)	0.0062	0.024

*Lactose and sucrose
**Mainly sucrose

Heat Stable Cheese Curd from Soy Milk

Until the process described by *E. Lundstedt and F.Y.-Y. Lo; U.S. Patent 3,743,516; July 3, 1973; assigned to The Hong Kong Soya Bean Products Co., Ltd., Japan* soybean curd, upon heating, has become rubbery and noncoherent. It also has been impossible to alter its flow characteristics to a point where it can be pumped, homogenized and packed hot into containers.

It has been found that by addition of fat, the basic characteristics of the soybean milk curd are changed and the curd becomes heat stable, meltable and suitable for homogenization. It is believed that the fat added to the curd acts as a lubricant and makes it suitable for homogenization. The amount of fat added may be varied upward from a minimum of 1%, but practical considerations indicate that the amount of fat added should be between 1 and 10% when the soybean milk has about a 6% solid content. The fat may be any edible vegetable or animal fat. Another feature of this process is the fact that superior results are achieved by the incorporation of skim milk or skim milk solids in conjunction with the fat in soybean milk.

The method used for the preparation of the heat stable, melting curd differs from previous methods. After addition of a fat in the portion of 1 to 10%, the milk is heated, preferably to 200°F with live steam for 10 minutes and homogenized at 2,500 psi pressure.

It is then cooled to 75°F and treated with a coagulating agent, for instance 0.3% anhydrous calcium chloride in a 10% aqueous solution. The calcium chloride is added under strong agitation, the mixture is heated slowly to 160°F and the coagulated milk allowed to stand until the curd particles gather to form particles of sufficiently large size. The mix is then cooled and for the drainage, nylon bags are advantageously used at a temperature between 40° and 60°F. Drainage is slower at low temperature, but low temperature gives a superior and smooth, silky curd. Although many variations may be made in the temperature, the type of bags, and the time involved, it is advantageous to drain the bags for about 2 hours in a cheese vat and then transfer them to a room kept at 50°F or lower where the bags are hung on racks. A moisture content of about 75% is achieved after draining for 14 to 18 hours.

The curd prepared as described above is not entirely satisfactory because if it stands in the heating kettle too long, it undergoes hydration. In the preferred example of this process, a small amount of skim milk or skim milk solids is added to the soybean milk containing fat. The amount of skim milk solids may vary between 0.5 to 5% and is preferably 1.5% when the fat content is 2%. The beneficial effect of the skim milk solids is probably due to the fact that the casein and the proteins present in the soybean milk precipitate together in a finely dispersed suspension which ultimately after drainage produce a meltable curd. In accordance with this example, the soybean milk containing, for instance, 6% solids, 2% fat and 1.5% skim milk solids, is heated to about 200°F for 10 minutes, homogenized, cooled to about 75°F and treated with the coagulating agent.

Many coagulating agents may be used in this process. In addition to calcium chloride, organic or inorganic acids may be used. Lactic acid forming bacteria are to be preferred.

A butter culture consisting of 15% *Streptococcus diacetilactis*, 8% *Leuconostoc citrovorum* and the balance *Streptococcus cremoris* is satisfactory. This culture may be added in the amount of about 0.5% to the soybean milk fortified with fat and skim milk as described above. At a temperature between 70° and 80°F for 15 to 18 hours, a solid curd is formed at a pH of 4.0 to 4.5, usually 4.3 to 4.4. The curd is then heated under vigorous agitation to a temperature of 160°F or to a point where the curd becomes drainable. By this process, the curd is obtained in a very finely divided form. After cooling to below 85°F, the mix is standardized to a pH of 4.8 with about 0.1% sodium bicarbonate. The curd is placed in bags and separated from the liquid. When calcium chloride is used as the coagulating agent, the pH of the curd is about 5.3.

The curd so obtained is smooth, of fine texture, heat stable and suitable for the preparation of cheese by the hot pack process. The curd is suitable for the incorporation of other flavors, for instance, other cheeses, fruit, or condiments. For the purpose of better illustrating the process, the following example is described in detail.

Example: One hundred pounds of soybeans were washed and soaked in 1,000 pounds of water at 72° to 75°F for 4 hours, ground and then filtered or centrifuged to give soybean milk. The total recovery of solids from the beans, which contained 89% solids, after cooling and centrifuging, was 65%. The total amount of soybean milk was 965 pounds, containing 57.9 pounds of soybean solids.

Butterfat in the amount of 19.3 pounds and skim milk solids in the amount of 14.5 pounds were added and the material heated to 200°F for 10 minutes. The material was then cooled to 75°F, inoculated with a starter consisting of 15% *Streptococcus diacetilactis*, 8% *Leuconostoc citrovorum*, the balance being *Streptococcus cremoris* in the amount of 0.5%. On standing overnight at 75°F, the pH dropped to 4.3 to 4.4 and a solid, highly aromatic curd was formed. The milk was heated under agitation to 160°F, cooled to 85°F, and 0.1% sodium bicarbonate added to obtain a pH of 4.8. The curd was obtained in a finely dispersed form and was drained by means of nylon bags which were left in a cooler overnight. This curd was suitable for the preparation of cheese by the hot pack process and contained 75% moisture.

After formation of the curd, it is found that the recovery of the soybean solids is 37.6 pounds, corresponding to 65% of the soybean milk solids. The recovery of the fat used is 18.9 pounds, corresponding to 98% of the 19.3 pounds of fat used. The recovery of nonfat milk solids is 38% of the 14.5 pounds used, that is 5.4 pounds. Thus the total amount of cheese solids is 61.9 pounds. In view of the fact that the cheese contains 75% moisture, this corresponds to 248 pounds of cheese.

Blue Cheese from Soy Milk

A blue cheese is also prepared by *E. Lundstedt and F. Y.-Y. Lo; U.S. Patent 3,743,515; July 3, 1973; assigned to The Hong Kong Soya Bean Products Co., Ltd., Japan* using a process which takes only about 15 days. This cheese with the same flavor, taste, and blue veins as Roquefort cheese is made from a soybean curd which is prepared from soybean milk fortified with butterfat and nonfat milk solids by inoculation with *Penicillium roqueforti* and also preferably *Streptococcus diacetilactis*. Mold ripening requires only two weeks for full flavor development.

The starting material for this process is the soybean curd produced by the method of U.S. Patent 3,743,516. For the purpose of preparing a cheese with the flavor, odor and characteristic blue veins of Roquefort cheese, the fat added to soybean milk prior to curd formation is butterfat because the flavor and taste are the result of the action of the mold on the butterfat. The viability of the culture of *Penicillium roqueforti* is also essential because a mold powder of inferior quality considerably affects the flavor, taste and appearance of the cheese. Spores isolated from a high grade Roquefort cheese and inoculated in bread or in a liquid substrate are satisfactory.

According to a preferred method, a citrated whey culture of *Streptococcus diacetilactis* is sprayed over the soybean curd for the purpose of enhancing the growth of the mold and increasing the ability of the mold to break down the fat. After the curd is inoculated with the mold and sprayed with the citrated whey culture, the trays are placed in Pliofilm bags which are allowed to hang down from the trays about 4 inches. By this arrangement, air contamination is prevented and the water is allowed to evaporate until the moisture content is about 95%.

The trays are then placed on pipe racks and left there for 5 days until the curd is covered with green mold. The curd is then turned over, the salt content adjusted to about 4% and the trays left on the rack for an additional 2 days. The temperature during this first curing stage is 50° to 55°F. Finally the curd is placed in Pliofilm bags and transferred to the cooler room where it is allowed to stand at a temperature below 40°F, preferably 26°F, for an additional 7 days. The curing room is advantageously cooled by recirculation of air and the room is preferably kept dry.

The cheese obtained after a two-week curing time has the appearance, texture and spreadability of the traditional blue cheese. The product can be stored at 10°F for about 1 year. The product consists of about 75% water, about 15.16% soybean solids, about 7.6% butterfat and about 2.18% nonfat milk solids. The ripened cheese can either be cold packed or hot packed or it can be hot packed in a blend with the soybean curd.

Cheese from Peanut (Groundnut) Protein

A. Hirsch, B.M. Gibbs and B.D. Hemmings; U.S. Patent 2,919,192; December 29, 1959; assigned to Lever Brothers Company provides a process for the preparation of a cheese-like product from groundnut protein. Fat is added to an aqueous suspension or solution of groundnut protein, the suspension or solution is homogenized and the suspension or solution is heated either before or after the addition of fat or homogenization, a curd is precipitated from the suspension or solution by the action of acid, the curd is cut and the cut curd is drained, pressed and ripened to a cheese-like product.

It is preferable to use an aqueous suspension or solution of groundnut protein which contains a proportion of soluble carbohydrate, preferably a mono- or disaccharide which will support lactic fermentation. Such suspension or solution can be prepared by extracting the substantially oil-free protein-containing material with mild alkali such as, for example, 0.01 N sodium hydroxide solution. The alkali solution may then be separated from the residue, for instance, by centrifuging.

The concentration of protein used in the aqueous suspension or solution will generally be limited by economic considerations. As little as 1.5 to 2% of protein by weight may give rise to a suitable curd but generally about 2.5 to 3% will be used although higher concentrations, for instance, about 4.5 to 5% can be used if desired.

It is necessary to add fat to the solution or suspension, preferably before heating is carried out. The greater the amount of protein, the greater the amount of fat required; approximately equal concentrations of fat and protein are usually taken. Generally, a refined, deodorized vegetable fat is used, for instance, partly hardened groundnut oil and partly hardened palm oil.

The suspension or solution is heated, preferably to at least 80°C, preferably with agitation. The temperature to which the suspension or solution is heated is a major factor in determining the properties of the curd which is obtained, higher temperatures tending to give a firmer curd. The pH of the suspension or solution is preferably in the range of 7.0 to 8.0, most preferably about 7.5. Use of a suspension or solution having a pH above 8 leads to increased firmness of the curd but may also lead to undesirable effects such as brown color.

When soft water is used (hardness of not greater than 5 parts of $CaCO_3$ per 100,000 parts of water), it is generally sufficient to heat the suspension or solution to 90°C. When using hard water, on the other hand, it may be advisable to use a calcium sequestering agent such as sodium pyrophosphate or citrate and to heat at higher temperatures, about 95° to 100°C often being required. The suspension or solution must be homogenized after the addition of fat, either before or after heating.

A curd is precipitated from the suspension or solution after the heating by the action of acid. Normally, the hot suspension or solution is cooled prior to precipitation, preferably to a temperature of not more than 40°C. In the case where a starter organism is used, the suspension or solution must be cooled before inoculation with the starter organism to a temperature suitable for the growth of the organism.

Example: Groundnut meal, freed from oil by low temperature solvent extraction, was stirred with dilute aqueous sodium hydroxide, 1 part meal to 9 parts 0.03% NaOH, for 30 minutes. The water contained 40 parts per 100,000 hardness. The suspension was centrifuged to remove insoluble matter and the separated solids washed with a further 6 parts of water and recentrifuged. The liquors from both centrifugings were combined and given a further centrifuging treatment to remove fine particles.

The liquor contained nearly 1% sucrose and about 2½% protein. 2½% w/v of a refined, deodorized mixture of partly hardened groundnut oil and partly hardened palm oil was added, in addition to 0.1% sodium citrate, and the liquor was heated with constant stirring to 75° to 80°C. At this temperature the liquor was homogenized at 2,000 psi and rapidly heated in a plate type heat exchanger to 97°C and cooled to 32°C. The liquor was then inoculated with 2½% w/v of a skim milk culture of a sucrose-fermenting nisin-producing strain of *Streptococcus lactis,* 1% of mixed strains of *Lactobacillus casei* and 0.0075% of a commercial rennet preparation.

The inoculated liquor was held at 32°C and the lactic acid produced by the culture caused the pH to fall, the protein and fat gradually forming a curd. After about 4 hours, the curd was firm enough to cut with cheese knives, the pH value then being 6.0. The curd was then heated to 35°C over a period of 10 minutes, and the whey separated by placing a weighted perforated tray on the surface of the curd and pumping off the whey as it collected above the tray. When no further whey separated, the tray was removed and the curd was broken up, salted with 2% NaCl and packed into molds. The molds were pressed at gradually increasing pressures reaching a maximum of 3.6 kilograms per square meter.

After 24 hours at this pressure, the curd was removed from the mold, bandaged, and transferred to a ripening storeroom where it was allowed to ripen for 4 to 6 months at 16°C and about 80% RH with daily turning.

PASTE SPREADS

Paste from Moisturized Soybeans

A soy nut butter or spread without graininess and without a beany or grassy flavor is prepared by *M.J. Pichel and T.J. Weiss; U.S. Patent 3,346,390; October 10, 1967; assigned to Swift & Company.* The process comprises the steps of moisturizing dehulled soybeans with water to increase the moisture content by 5 to 15%, and not in excess of 50%, followed by quickly heating the moisturized beans to volatilize water and flavoring materials. The debittered beans are then promptly cooled and oil is added as required to provide 35 to 60% fat based on the weight of the beans.

The debittered bean-fat mixture is then ground to reduce the particle size of the beans and form a smooth, uniform paste composed of finely divided soybean solids, homogeneously dispersed in edible oil. The soybean butter can be used as such as a spread for bread or as an ingredient in candy or cake as well as with other confections. In addition, the soybean spread or butter can be blended in all proportions with nut butter such as peanut butter to form a uniform mixture which exhibits the most desirable features of both the peanut butter and soybean butter.

In carrying out the moisturizing step, the dehulled beans naturally containing 13 to 26% fat and generally around 18 to 22% fat are sliced or split to increase the available surface and the split beans are dipped in water, steamed or sprayed with water to add water to the beans. The moisture content of the beans is thereby increased by 8 to 15%, and not in excess of 50%, based on the weight of the beans.

Debittering of the moisturized beans is carried out by submerging the beans or frying the moisturized soybeans in edible oil. This frying or cooking is carried out at a temperature and for a time sufficient to remove a substantial proportion of the moisture from the bean as evidenced by the evolution of steam. Frying of the beans in a deep fat frying operation is carried out by immersing the moisturized product in vegetable oil or animal fat heated to 140° to 180°C for 1 to 10 minutes and preferably for about 2 to 5 minutes.

The debittered soybean after removal from the cooker is promptly passed into an oil cooling bath. This cooling bath contains an edible glyceride oil such as soybean oil, safflower seed oil or other oils which are liquid at room temperature and also at temperatures down to 0°C. The temperature of the cooling bath is maintained in the range of 0° to 50°C and preferably at 20° to 30°C. The cooling bath acts as a quenching medium, terminating the cooking of the hot soybeans, and also provides a source of oil to be absorbed by the hot soybeans.

After removal from the cooling bath, the oil content of the beans is increased by adding enough bland, edible oil to increase the oil content to 35 to 60% based on the weight of the soybeans. The adjustment of the oil content of the soybean-oil mixture prior to grinding is important since it is difficult to obtain the fine subdivision desired with beans containing less than about 35% oil. The soybean is next passed through an Urschel Model MG cutting mill or equivalent comminuting apparatus to form a smooth, homogeneous product of butter-like consistency.

Example: Twenty pounds of whole soybeans were split and dehulled and water was sprayed as a fine mist over the beans until the beans contained 8% additional moisture. The moisturized beans were then held at room temperature (20°C) for 15 minutes to permit the moisture to penetrate into the beans. The moisturized product was then placed in a container of hydrogenated soybean oil heated to 180°C. Frying of the beans in the hot oil was completed in 3 minutes, and the fried product was immediately immersed in a quenching bath of soybean oil held at 15°C. After 5 minutes in the cooling bath, the fried product was mixed with additional soybean oil to form a mixture which contained 50% oil.

This mixture was passed through the Urschel Model MG grinder equipped with a shaving head and the beans were passed through the comminuting chamber of the grinder where the pieces are cut and extruded to a very fine particle size so that 98% of the solids remaining after extraction of the product with diethyl ether passes through a 200 mesh screen. The ground product is smooth, homogeneous, and has a consistency approximating that of peanut butter. Although it has a pleasing nut-like taste, there is no noticeable beany or grassy flavor.

It is possible to enhance the soybean spread by incorporating sugar, salt, and other flavoring agents including artificial flavors as desired. The addition of 1 to 4% hardened cottonseed oil or other hard fat or hardened monoglycerides makes the product less sticky and also prevents separation in the product.

Attrition Milled Soy Protein and Oil

Researchers have attempted unsuccessfully for years to produce a protein type spread from purified soy protein. Failures have been due to the lack of a smooth mouthfeel, excessive graininess and loss of or dimution of flavors when added to the soy protein system. To eliminate the graininess of products produced from soy protein, water has been employed to solubilize or soften the protein. The use of water, however, contains many disadvantages. Water forms a suitable environment for microbiological growths so that the product cannot be stored without sterilization and/or refrigeration.

R.L. Hawley; U.S. Patent 3,469,991; September 30, 1969; assigned to Libby, McNeil & Libby has found that the graininess of the normal protein oil system is due to the encapsulation of the protein particle by the oil with little or no penetration of the oil into the protein particle. Unless the oil can enter the protein particle and/or the protein molecule, nothing can be done to eliminate the graininess unless water is added to the system. By introducing oil into the protein particle, the objectionable graininess is eliminated.

High flavor levels can also be produced using oil-soluble flavoring materials. Suitable flavoring materials include citric acid, which is solubilized in propylene glycol to form an oil-miscible flavoring mixture. For example, 0.1 to 5% by weight of the total composition of citric acid may be dissolved in 0.1 to 20% by weight of the final composition of propylene glycol. This oil-soluble flavoring is then added to the oil and protein composition.

To produce a soy protein product similar in texture to peanut butter, the protein must be modified by the introduction of oil into the protein particle. This is accomplished by use of an attrition mill which introduces energy into the system and changes the nature of the product. The mill produces a thick, nonviscous to free-flowing composition which, when it cools to room temperature, sets up into a paste resembling peanut butter. The oils which may be used include corn oil, soybean oil, animal fats, cottonseed oil, although preference is given to corn oil for its flavor. The oils are normally liquid at ambient temperatures.

The purified soy protein used in this process must contain an emulsifying agent such as lecithin or other specially prepared agents capable of being incorporated into the protein. The soy protein is used in amounts of 30 to 60% by weight. The oil is used in amounts of 30 to 70% by weight. The mixture of oil and the soy protein is subjected to the action of an attrition mill at room temperatures, however, it is preferable to carry out the process at sufficiently high temperatures to create a free-flowing liquid composition. Sweeteners, plasticizers and coloring agents may also be added to the final product.

Example: The modified protein, sugar and salt are blended and placed in a mixer. The oil or fat is placed in a steam jacketed kettle and monoglycerides are added as needed. Heat is applied until solution is achieved. After removal of the heat, a solution of citric acid in propylene glycol is added and the mixture agitated. Following this, the oil-soluble flavor is added and the mixture is then added to the protein blend and mixed until well blended. Glycerol or other color solutions are added during the blending period. The blended mass is then comminuted in any of a variety of variable speed devices.

TOFU, MISO AND TEMPA PRODUCTS

High Frequency Sterilization of Tofu

In general, the packaged soybean curds or tofu have been produced by packing cold soybean milk containing a coagulating agent into a container made of a synthetic resin and heating the package in hot water at 80° to 90°C to effect its coagulation and sterilization. Mere addition of a coagulating agent to cooled soybean milk hardly causes coagulation and it is necessary to heat it. In this process, the coagulating and sterilizing steps are of utmost significance. The coagulating temperature of tofu is about 60° to 70°C, although lower temperatures are applicable. On the other hand, the higher the sterilizing temperature, the more marked the sterilizing effect and the shorter the sterilizing time. Since the quality of tofu deteriorates when exposed to high temperatures for a long time, lower sterilizing temperatures and shorter sterilization times are preferred. The deterioration of soybean curds shows itself mainly in the releasing of water, (referred to as water release), and roughening which indicates that the smooth surface and the inside of the soybean curds become roughened and nonuniform. These deteriorating phenomena are detrimental to the taste, and in addition, consumers do not like water-released soybean curds even when they are not putrefied. Accordingly the water release markedly reduces the commercial value.

R. Ueno; U.S. Patent 3,712,823; January 23, 1973; assigned to Ueno Seiyaku Oyo Kenkyujo, KK, Japan has found that by heating packaged soybean curds with high frequency irradiation, the packaged soybean curds can be heated to a high temperature within a short time and a marked sterilizing effect can be achieved without substantially coagulating soybean milk. It has also been found that when soybean milk is subjected to high frequency irradiation and then allowed to stand at above 30°C to coagulate it, soybean curds substantially free of water release and roughening can be obtained.

In this process high frequency in the range of 500 to 10,000 mc/sec can be used. When the frequency is below 500 mc/sec, the efficiency of absorption of the electrical wave energy by foodstuffs is reduced. With frequencies above 10,000 mc/sec, it is difficult to produce tubes capable of generating electrical wave energy of large capacity with stability and good efficiency, and therefore these wavelengths are not commercially feasible. The use of high frequencies of wavelengths 2,450 ± 50 mc/sec is especially preferred in this process because of the absorbing efficiency of the electric wave energy on foodstuffs.

In the general practice, soybean milk containing a coagulating agent and packed in a container made of polyethylene is subjected to high frequency irradiation for 50 to 200 seconds, preferably 60 to 150 seconds to raise the temperature in the center of the soybean milk to 55° to 85°C, preferably 60° to 80°C. Alternatively, the carryover heating method may be employed in which soybean milk is subjected to high frequency irradiation for 50 to 100 seconds, preferably 50 to 70 seconds. Then the irradiation is stopped for 1 to 5 minutes, preferably 3 to 5 minutes, and the milk is again subjected to the irradiation for 50 to 100 seconds, preferably 50 to 70 seconds.

The packaged soybean milk subjected to the high frequency irradiation is substantially liquid, but is fully sterilized. Such soybean milk is thereafter coagulated at a temperature not lower than 30°C, preferably 40° to 80°C for a period of 10 minutes to 2 hours, preferably 20 to 40 minutes.

Example: Four containers made of high density polyethylene and having a size of 47 mm in length, 80 mm in width and 80 mm in height were each packed with 300 ml of soybean milk at 25°C containing 0.3% glucono-delta-lactone as a coagulating agent or 0.4% of calcium sulfate and heated by high frequency irradiation for 100 seconds. The temperature in the center of the soybean milk was measured at certain time intervals. As the high frequency irradiation apparatus, electronic range NE-500 (high frequency output 500 w, frequency 2,450 ± 50 mc/sec, product of Matsushita Electric Appliances Industry, Co., Ltd.) was used.

The containers were allowed to stand for 5 minutes with the stoppage of the high frequency heating and then again subjected to the high frequency irradiation for 50 seconds. The soybean curds obtained were allowed to stand under the conditions indicated in Table 1 on the following page. The coagulated condition and quality of the products were examined. In comparative runs Nos. 3 and 4, soybean curds were produced by the conventional method in which soybean milk packed in a container was immersed for 30 minutes in hot water maintained at 90°C.

It is seen from the results shown in this table that the temperature for coagulating soybean milk containing a coagulating agent should be at least 30°C, and at temperatures below this the coagulation of soybean milk is insufficient, and the resulting curds are too soft.

Although not shown in this table, the results obtained when the soybean milk is left to stand for 4 hours at 20°C are identical with those obtained when it is left to stand for 2 hours at this temperature. This indicates that at 20°C, there is no appreciable effect obtained by allowing the soybean milk to stand for a longer time.

TABLE 1

Run No.	Coagulating agent	Temperature of the central part, °C.	Heating time and temperature	Coagulated condition	Quality of soybean-curds	
					Water release	Roughening
1	Gluconodeltalactone	76.0	2 hours, 20° C	Exceedingly soft	Not observed	Not observed
		75.5	2 hours, 30° C	Good	do	Do.
		75.0	2 hours, 50° C	do	do	Do.
		75.5	40 minutes, 60° C	do	do	Do.
2	Calcium sulfate	66.0	2 hours, 20° C	Somewhat soft	do	Do.
		65.0	2 hours, 30° C	Good	do	Do.
		66.0	2 hours, 50° C	do	do	Do.
		66.0	40 minutes, 60° C	do	do	Do.
3	Gluconodeltalactone	81.0		do	Clearly observed	Clearly observed.
4	Calcium sulfate	68.0		do	do	Do.

The products obtained by this process suffer substantially no water release and roughening irrespective of the heating temperature and exhibit good quality. In contrast, the products obtained by the conventional method of heating in hot water undergo water release and roughening.

Example 2: Three hundred milliliters of soybean milk at 25°C containing 0.3% of glucono-delta-lactone as a coagulating agent were packed into each of high density polyethylene containers of the same size as in Example 1. Each of the soybean milk-containing containers was subjected to high frequency irradiation, and the soybean milk was coagulated under the conditions shown in the following Table 2. The quality and preservability of the soybean curds so produced were examined. The results are also shown in the following table.

In this experiment, the high frequency irradiation apparatus used was DR-121 (product of Hitachi Seisakusho, frequency 2,400 to 2,500 mc, output 1.2 kw). The irradiating conditions shown in the following table were attained by varying the plate current of the magnetron which is the high frequency generator of this apparatus. The putrefaction of the soybean curds was determined by the evolution of gas, release of water, surface roughening, and also the smell upon opening the container.

TABLE 2

Run Number	High frequency irradiation time and the temperature in the center of soybean milk		Heating and coagulating step		Appearance	Preservability	
	Seconds	°C.	Time, min.	Temp., °C.		Temperature, °C.	Days
5	55	50	20	70	No water release nor roughening; good	30	7
6	200	90	20	70	Water release and roughening observed	30	30
7	45	55	20	70	Because of very abrupt heating, the temperature rise is not uniform; a part of the soybean milk boiled and water release and roughening observed; sometimes the container ruptured.	30	20
8	220	85	20	70	Slight water release observed; in the latter half of irradiation, a part of the soybean milk is in a coagulated condition.	30	28
9	150	75	20	70	Good	30	27
10	150	75	20	70	No water release nor roughening	30	25
11	200	85	30	60	do	30	30
12	100	70	70	85	Water release and roughening observed because of long time exposure to high temperature.	30	30
13	100	70	5	35	Sufficient coagulation unobtainable		
14	100	70	5	85	do		
15	220	85	30	60	Slight water release; roughening observed	30	30
16 (control)	90	40			Water release and roughening observed	30	15
17 (control)	85	35			No water release nor roughening	30	10

Soybean Grits in Miso Preparation

Miso, a fermented soybean-rice preparation has been used for a long time as a valuable and important food in Japan. This food has been prepared by soaking and cooking rice which is then spread into trays and inoculated with viable spores of suitable strains of *Aspergillus oryzae*. The rice is allowed to be covered with mold mycelium. The mold rice becomes a source of enzymes for the breakdown of proteins, fats and carbohydrates during the soybean fermentation. At the same time, soybeans are washed, soaked in water for a number of hours and cooked with steam. The mold rice and cooked soybeans are put together with the addition of salt and inoculated with miso from a previous fermentation. The whole mass is then ground together and then placed in barrels, pressed down tightly and allowed to incubate.

No aeration is used during the fermentation which lasts for three months or longer. When the fermentation is completed the fermented material is allowed to age at room temperature for several months. At the time of harvest, the aged and fermented product is ground into a thick paste which is then sold as a finished food product.

It has been firmly believed that the desired fermentation of miso requires inoculation from a miso fermentation containing both yeasts and lactic bacteria. *A.K. Smith, C.W. Hesseltine, and K. Shibasaki; U.S. Patent 2,967,108; January 3, 1961; assigned to the U.S. Secretary of Agriculture* have now discovered that the latter organisms are completely unnecessary for the preparation of miso and that the yeasts alone are responsible for the development of the specifically demanded flavor. Accordingly, pure cultures of the specific Saccharomyces organisms, which are defined below, may be used in place of an inoculum from an active miso fermentation.

It was also found that removal of the soybean hulls prior to the use of the beans greatly accelerates the subsequent fermentation, apparently by the removal of substances which inhibit the activity of the organisms or of their products. In this process the soybeans are cleaned and then cracked by a roller-type crusher or other suitable apparatus. This effectively breaks the hulls from the embryo and cotyledons. The hulls can then be readily removed or separated from the remainder of the soybeans. The part of the soybean free of the hulls is referred to as grits. In this process the grits represent approximately 2 to 10 pieces of the seed minus the hulls. If the grits are crushed into fine particles they do not ferment as well as if they are divided only into a few pieces. The coarse grits are then soaked in water.

According to this process, grits after soaking for a short time in water, are cooked rapidly and uniformly. For example, when grits were soaked for 1.5 hours and cooked with steam for 40 to 60 minutes at 7 lb pressure, they were uniformly soft, rather elastic and uniformly cooked and possessed a highly desirable light color. Besides preparing the beans for a rapid fermentation, the soaking and cooking effectively destroy microorganisms associated with the soybeans.

About 26.7 and 41.6 parts of the cooked soybean grits are then cooled and mixed with 12.3 to 16.7 parts of *Aspergillus oryzae* mold rice, 1.7 to 3 parts of water, and 4.3 to 8 parts of salt, and are inoculated with osmophilic microorganisms adapted for growth and fermentation in this substrate. One part of inoculum is taken from a previously good miso fermentation containing *Saccharomyces rouxii* (NRRL Y-2547 and Y-2548) and mixed thoroughly with the soybean grits, mold rice, water and salt. This well mixed material is then placed in suitable fermentation vessels.

This process uses microorganisms which can tolerate high salt concentrations and which grow under nearly anaerobic conditions at temperatures up to 40°C. These osmophilic microorganisms belong to species of the yeast genus Saccharomyces. These microorganisms ferment the rice and soybean grits and the products formed by the enzymatic action of the mold rice to form a variety of fermentation products such as glutamic acid, lactic acid, succinic acid, ethyl alcohol, higher alcohols and their esters. Other products formed during the fermentation are amino acids and fatty acids. The fermentation of soybean grits is carried out in the presence of suitable sources of water, carbohydrates, nitrogen and salt concentrations up to approximately 14%.

Example: In this fermentation soybeans of the variety Hawkeye were used. The beans were crushed by a roller type crusher into grits. The size of sieve was $^{11}/_{64}$ inch round hole and it had a 4 x 22 inch slot. Grits prepared from this variety weighed 4,994 grams. The grits were soaked in 17,500 ml of tap water for 1 hour and 40 minutes and the excess water drained leaving 11,849 grams of wet grits. The grits were then cooked with steam at 7 lb for 1 hour. The weight of the grits were then 10,917 grams and had a moisture content of 59.3%. They were soft, uniform in texture, yellow colored and good in appearance. 1,000 grams of the above described grits were then mixed with 370 grams of mold rice, 200 grams of NaCl and 70 ml of tap water, and inoculated with 30 grams of good miso from a previous fermentation.

The mixture was then packed tightly into a suitable fermentator. The fermentation was allowed to go for 7 days at 28°C and then kept for 2 months at 35°C. No aeration was supplied. The fermentation was completed at this time and the miso allowed to age for 2 weeks at room temperature. The miso was then ground into a thick paste and this product had excellent appearance, excellent flavor and taste, and had an excellent odor.

By using soybean grits in place of whole beans, the final product after fermentation and aging is more uniform in color, and consistency. The color is a pleasant light color and has a pleasant odor and flavor. Since soybean grits were used and the more fibrous portion of the seed removed, the product is higher in protein. At the same time bad tastes and odors often associated with typical miso fermentation have been eliminated.

Steam Treatment for Preparing Soy and Miso Pastes

Kikkoman Shoyu workers have developed two methods for the preliminary steam treatment of soybeans for improved production of soy and miso. In the first process by *T. Yokotsuka, T. Aonuma, K. Mogi, D. Fukushima, A. Yasuda, H. Watanabe, N. Tsukada, and A. Arai; U.S. Patent 3,647,484; March 7, 1972; assigned to Kikkoman Shoyu Co., Ltd., Japan* the starting material is soybeans obtained by steaming soybeans of a moisture content of 30 to 70% by weight with saturated steam of a gauge pressure not less than 1.8 kg/cm^2 for time not exceeding 10 minutes, followed by rapid cooling to a temperature not exceeding 100°C.

The soybeans to be steam treated by this process may be of any form such as whole soybean, soybean flour, defatted soybean flake, or soybean meal, there being no particular restrictions so long as they contain soybean protein which can become the source of nitrogen. In the case of whole soybeans, those whose moisture content is 30 to 65%, and particularly 55 to 60% by weight, are preferred. In other forms of soybeans, those of a moisture content of 45 to 70%, and particularly 55 to 65% by weight are preferred.

For bringing the moisture content up to these levels, the beans are either soaked in water or the water added by sprinkling, spraying or other suitable means. If the moisture content is less than that indicated by the above ranges, the proportion of the protein remaining native in the soybeans after steaming becomes lost. For example, the dried whole soybean usually available has a moisture content of 10% by weight. Now, if such soybean is given a steaming treatment without increasing its moisture content, even though it has been steamed for 3 hours, the soy product forms precipitates upon dilution, and heating, which is caused by the native protein in the soybean. On the other hand, the use of moisture in excess of the foregoing range makes it difficult to carry out either the steam treatment operation itself or the subsequent Koji-making steps.

The steam treatment is carried out with saturated steam of a gauge pressure not less than 1.8 kg/cm^2. On the other hand, the steaming time employed (which is the period of time after the prescribed pressure of not less than 1.8 kg/cm^2 has been reached) is short, not exceeding 10 minutes. In general, when the gauge pressure is high, a relatively short period of time is employed, whereas when the gauge pressure is close to the aforesaid 1.8 kg/cm², a time preferably clost to 10 minutes is used. A pressure ranging from 1.8 to 15 kg/cm² gauge is usually chosen, but especially recommended is a pressure 4 to 7 kg/cm² gauge.

As soon as the steaming treatment has been completed, the soybeans must be cooled as quickly as possible to a temperature not exceeding 100°C. In general, it is preferred that the reduction to a temperature not exceeding 100°C be carried out at shorter periods of time in the case of higher gauge pressures. For example, in the case where the gauge pressure does not exceed 3 kg/cm² (143°C) the temperature of the treated soybeans should be reduced to below 100°C within 15 minutes, and preferably within 5 minutes. On the other hand, when the pressure exceeds 3 kg/cm², the reduction of the temperature to below 100°C should be accomplished within at most 5 minutes, and preferably within 1 minute. Thereafter, the temperature preferably should be cooled to the neighborhood of 40°C as promptly as possible.

The cooling can be accomplished by withdrawing the soybeans from the steaming zone, spreading them and allowing them to cool by exposure to air, at which time air can be positively blown against the soybeans. Alternatively, quick cooling can be carried out by use of reduced pressure such as a jet condenser or a vacuum pump and evaporating the moisture content of the treatment-completed soybeans, thus utilizing the latent heat of the moisture which is evaporated to accomplish the cooling.

An illustration is presented in the table below, showing the different rates at which the proteins are digested by means of the enzymes produced by the fungi which are typically used in the brewing of soy in the several cases of the soybeans treated by this process, and soybeans treated in accordance with the prior art conditions.

Experiment	Steaming treatment			Rate of digestion (percent)	Turbidity grade[*]
	Gauge pressure, kg./cm.² (temperature °C.)	Treatment time	Time required to cool to 100° C.		
Conventional method	0.9 (117)	45 minutes	20 minutes	86.13	0
Comparison 1	1.2 (123)	10 minutes	10 minutes	81.05	2
Present method	1.8 (131)	8 minutes	3 minutes	91.40	0
Comparison 2	1.8 (131)	15 minutes	...do...	80.23	0
Present method	2.0 (133)	5 minutes	...do...	91.60	0
Comparison 3	2.0 (133)	...do...	20 minutes	83.50	0
Present method	3.0 (143)	3 minutes	3 minutes	92.99	0
Do	4.0 (152)	2 minutes	1 minute	93.74	0
Do	5.0 (159)	1 minute	40 seconds	94.50	0
Do	6.0 (165)	30 seconds	...do...	94.90	0
Do	7.0 (170)	15 seconds	...do...	95.10	0

NOTE.—Soybean used was defatted soybean flake of a moisture content of 60% by weight.
*Grade: 1, clouds formed; 2, precipitates formed; 3, no discernible formation of clouds.

From the results given in the table above, it is possible to choose the gauge pressure, steaming time and the time to cool to a temperature not exceeding 100°C so that the rate of digestion becomes at least 90%, and preferably at least 92%, and the turbidity grade becomes 0. A rate of digestion and turbidity grade which both satisfy the foregoing requirements at the same time cannot be achieved with the hitherto proposed conditions.

In the process, excepting that soybeans which have been steam treated as fully described are used as the starting material, the preparation of soy is otherwise carried out in accordance with the brewing technique well known in the art. That is to say, parched and crushed wheat is added to the soybeans which have received the steaming treatment of this process, and the two are mixed together. The mixture is then inoculated with aspergilli and by carrying out the solid culture at 23° to 35°C for 40 to 90 hours Koji is obtained.

Next, a soy mash (moromi) is prepared from this solid Koji by mixing it with a solution of common salt. This mash is stored for 4 to 18 months in storage tanks until fully matured. During this period the mash is fermented by the activity of the naturally propagating and/or artificially inoculated microorganisms, principally soy lactic acid bacteria and soy yeasts. The matured mash is separated into filtrate (unpasteurized soy) and cake by pressing. Refined soy is obtained by heating this filtrate to pasteurize it.

T. Aonuma, A. Yasuda, T. Yuasa, A. Arai, K. Mogi and T. Yokotsuka; U.S. Patent 3,764,708; October 9, 1973; assigned to Kikkoman Shoyu Co., Ltd., Japan have also found that a further improved method of producing soy can be realized by using superheated steam instead of saturated steam, and the use of superheated steam also leads to the improvement of miso paste production. It has also been found that this method is applicable to the heat treatment of carbohydrates with excellent improvement.

The superheated steam used in this process should have a gauge pressure ranging from 4 to 8 kg/cm^2 and a temperature ranging from 200° to 280°C. The treating time should not exceed 15 seconds, and the preferred treating period is 3 to 10 seconds. Therefore, a procedure in which the material is conducted countercurrently with superheated steam in a vertical-type steam treating zone is not desirable because the material tends to be locally denatured excessively depending upon the shape, particle size, and specific gravity of the material. With both a vertical type or transverse-type treating zone, it is preferred that the material should be contacted with superheated steam concurrently in the tubular treating zone, and the treatment should be performed while the particles of the material are suspended and are carried by the steam current.

The starting soybeans and/or carbohydrates which have been treated with the superheated steam are then exhausted quickly into the atmosphere at normal atmospheric pressure or at pressures around atmospheric. By this rapid exhaustion, the treated soybeans and/or carbohydrates are rapidly cooled and puffed and thus excessive modification can be conveniently avoided.

The soybean and/or carbohydrates treated with superheated steam by this process are obtained in a dried state represented by a moisture content of 4 to 10% by weight. Hence, they have good preservability, and can be used after storage; there is no need to use them immediately after steam treatment as in the conventional method. Since this steamed, puffed material is in the dried state, there is no fear of putrefaction, and starch present is maintained in the alpha-form. Therefore, even when treated material is used after storage, it need not be subjected to the treatment, (denaturation of proteins and conversion of starch into the alpha-form) again by such means as steaming or parching, but it can be used as Koji-producing material or material for preparation of soy and miso paste usually with addition of a suitable amount of water. If desired, it can be used without addition of water.

Example: 30 kg of rice was dispersed and floated in a current of superheated steam having a gauge pressure of 6.0 kg/cm^2 and a temperature of 250°C, and while being transported continuously in this condition, were heated for 6 seconds. The heated rice was abruptly discharged into the atmosphere at atmospheric pressure. 17 liters of water was sprinkled onto the resulting steamed and puffed rice, and seed Koji was added to prepare Koji in a usual manner. It was subjected to the rapid brewing method together with soybeans (raw material soybeans 26 kg) soaked in water and steamed by a conventional method and 8 kg of common salt, and the mixture was matured for one month at 30°C to form miso paste (so-called rice miso) of good flavor.

Tempa Production in Perforated Bags

C.W. Hesseltine and A. Martinelli, Jr.; U.S. Patent 3,228,773; January 11, 1966; assigned to the U.S. Secretary of Agriculture have developed a method of producing uniform, high quality tempa in commercial quantities. Tempa, also known as tempeh, a fermented soybean preparation, has been used for centuries as a valuable food product in Indonesia. It contains proteins, carbohydrates, fats, vitamins, and other nutrients. Tempa is prepared in Indonesia in small factories by soaking whole soybeans in water, or in streams of water overnight. The soaked wet beans are then dehulled by hand and then boiled for some time in water which softens the beans and destroys contaminating microorganisms. The cooked soybeans are then spread out in thin layers in order to allow the water to drain and evaporate from the surface of the soybeans. The air-cooled soybeans are next mixed with the starter or inoculum consisting of molded soybean material from a previous tempa fermentation. The moist inoculated soybeans are then wrapped tightly in banana leaves and the material is allowed to ferment at room temperature until the soybeans are completely molded, the thusly molded soybeans being the product known as tempa. The consumer slices the tempa into thin slices, dips the slices in a salt solution and fries them in a vegetable oil.

The fermentation of the soybeans destroys the bad odor and flavor of the soybeans. The fermentation apparently allows the microorganism to produce enzymes which act on the proteins, carbohydrates and the oil to make the tempa palatable and nutritious and to give a desirable flavor.

The prior soybean fermentation for the manufacture of tempa has various objectionable steps. Since the required organisms are aerobic, the mass of soybeans which could be fermented has been restricted to a small mass of molded soybeans and requires much hand labor. The mass of soybeans being fermented must be covered by some film. The use of plant leaves such as banana leaves is extremely primitive, and the molded cake is irregular in shape and size. The progress of the fresh fermentation is uncertain and the tempa produced is highly variable as a consequence of using an inoculum consisting of a piece of old tempa which is contaminated by all sorts of molds, bacteria and yeasts. Since the inoculum is highly variable as to the viability of the mold even when the inoculum is relatively pure, fermentation time requirements for the formation of tempa are also extremely variable, as is the product itself.

One improvement in the process is replacing the old tempa as a source of fermenting microorganisms with certain phycomycetous fungi of the order Mucorales. This order is described by Bessey, *Morphology and Taxonomy of Fungi,* pages 150 to 172, Blakiston Company, Philadelphia, 1950. The order includes a number of genera including Mucor, Rhizopus, Absidia, Phycomyces, and Thamnidium. Among the several genera in this order, the genus Rhizopus has been found to be the most useful and preferred in this process. Species of this genus which are operative in the process include, for example, respectively *Rhizopus oligosporus,* NRRL 2710, *Rhizopus arrhius,* NRRL 1556, *Rhizopus achlamydosporus,* NRRL A-6997, *Rhizopus formosaensis,* NRRL A-10,180, *Rhizopus stolonifer,* NRRL 2233, and *Rhizopus oryzae,* NRRL A-6865.

In order to prepare spore inoculum for inoculating soybeans as above mentioned, a preferred species is grown on a medium containing potato extract, magnesium sulfate, calcium carbonate, glucose, tap water, and agar. This medium, called potato dextrose agar is described by Haynes et al, *Applied Microbiology* 3 361 (1955). The medium is dispensed in test tubes, sterilized with heat, cooled, and inoculated with a pure culture of one of the species of Rhizopus. When the Rhizopus has grown for 7 days at $25°$ to $280°C$, large numbers of sporangia are produced, and the sporangiospores may then be washed off the potato dextrose agar slant mold colony and the spore suspension used to inoculate the soybeans.

Soybeans that have been dehulled and cracked (10 to 15 pieces per bean) is preferred substrate for preparing tempa according to this process. The soybean must be softened and moistened by soaking in water and then boiled to sterilize and further soften the substrate. After cooking, the material is drained of excess water, cooled below $40°C$, and inoculated with a spore suspension of the Rhizopus spores prepared as described above.

The temperature at which tempa may be fermented in this process can vary from $25°$ to $37°C$. The sterilized full-fat grits or whole dehulled soybeans after inoculation with sporangiospores from a pure culture may be fermented in conventional nontoxic plastic bags such as those used for retailing of food but modified by the presence of 0.02 inch diameter perforations located not over 0.5 inch apart. When these flexible plastic containers are perforated as described, an ideal fermentation container for the fermentation of soybeans to tempa is formed. The sterilized inoculated soybean material is packed in the plastic containers, the end of the plastic bag heat sealed, and the soybeans are fermented in the package in which it is to be sold. Because of the very low bacterial count and substantially sterile condition of the plastic sack, one can successfully ferment soybeans to tempa without sterilization of the plastic container as it comes from the manufacturer. A further modification of the process is the fermentation in perforated flexible plastic tubing. The diameter of the plastic tubing is limited only by the requirement that sufficient air reaches the center for mold growth. The following example is illustrative of the process.

Example: In this fermentation Hawkeye variety soybeans were used. Dehulled whole bean halves and grits respectively were used. The beans, in the latter case, were crushed between rollers into full-fat grits, the particle size of which represented about $1/10$ to $1/15$ of the dehulled soybean. Both forms of soybeans were soaked in tap water at 25°C. The soybean halves were soaked overnight while the grits were soaked for 3 to 4 hours. About 3,000 milliliters water was used for every 1,000 gram of soybean material. The soaking water was discarded, and then the soybeans were boiled without pressure in excess water for ½ hour. The water was drained, and the soybean halves or grits placed on a sterile absorbent surface to drain further and to cool. The drained bean materials were now swollen and soft.

After cooling, each bean material was inoculated with spores of the mold, *Rhizopus oligosporus* NRRL 2710 in the following manner: two ml of sterile water was added to a sporulated culture of the organism on a potato dextrose agar slant and the wash water then poured onto 100 grams of soybean substrate. For larger amounts of soybean substrate a proportionate number of tubes was employed. After the beans or grits had been inoculated and mixed, 500 gram portions of the inoculated soybean material was placed in each of a number of perforated plastic bags (each 22 x 13 cm). The plastic bags had 0.02 inch perforations spaced 0.5 inch apart. The suitably filled bags were manually closed and heat-sealed and placed in an incubator at 31°C. After 22 to 23 hours of incubation the fermentation was finished. The tempa had an excellent color, excellent flavor, and excellent odor.

Tempa from Cereal Grains

C.W. Hesseltine and M.L. Smith; U.S. Patent 3,243,301; March 29, 1966; assigned to the U.S. Secretary of Agriculture have found the methods for producing tempa from cereal grains or mixtures of grain and soybeans. Formerly, tempa had been produced exclusively from soybeans. It is now found that varieties of tempa can be obtained from a pretreated cereal grain or mixed soybean-cereal grain substrate when the molding organism is limited to a fungus that produces large amounts of both proteolytic and lipolytic enzymes and but little if any amylolytic enzyme. Any appreciable amount of the latter class of enzyme would break down the cereal starch to simpler sugars that would then be fermented to disagreeably tasting organic acids and highly colored materials.

It was found that *Rhizopus oligosporous,* NRRL 2710, is completely operative for this purpose, whereas the closely related fungus *Rhizopus oryzae* and the mold *Aspergillus oryzae* on a cereal grain produce darkly colored, ill-smelling, and utterly unacceptable products. Although it would not be apparent that any pretreatment of the cereal grain could be important where the mold does not contain or produce amylolytic enzymes, it was found that only a very limited number of the mycelium developing from the germinating spores succeed in penetrating an unruptured seed coat and that precracking of the cereal grain kernels and a fully hydrated state are required if the mold is to grow quickly enough to avoid the development of adverse flavors. The principal steps of this process may be summarized as follows:

(1) Boil prewashed cracked cereal grain or soybean grits for at least 12 minutes to hydrate and sterilize at 100°C.
(2) After removal of excess water, inoculate the medium with an aqueous

(continued)

suspension of spores of *Rhizopus oligosporus*, NRRL 2710.
(3) Incubate at 31°C under aerobic conditions for 18 to 24 hours.
(4) Kill spores and enzymes with steam or by frying.

Example 1: Two 1,000-gram lots of Hard Red Winter wheat, differing only in that the kernels of one lot had been lightly cracked by mechanical means, were treated in an identical manner. Each lot was covered with water and soaked at 25°C for 20 hours, drained, washed in three changes of water, recovered with water, boiled at atmospheric pressure for 10 minutes in the case of the cracked wheat and for 30 minutes in the case of the uncracked (whole) wheat, cooled, inoculated with 30 ml of a sterile aqueous suspension of *Rhizopus oligosporous,* NRRL 2710, spores obtained from a 5-day potato dextrose agar slant, transferred to petri dishes and incubated at 31°C for 25 hours.

The dishes containing the uncracked wheat showed a very limited and incomplete growth that was not significantly increased by extending the incubation to 36 hours, at which time the slightly molded product was still very friable and could not be properly sliced for frying. The cracked wheat tempa at 25 hours consisted of a solid cake in which the cracked wheat was held firmly together by the extensive white mycelia of the mold, thus permitting slicing into wafers having a thickness of less than 5 mm. The fresh wafers having a pleasant yeasty odor were dipped in salt water for several minutes and then deep fried in vegetable oil to provide a crisp chip item having a flavor resembling that of popcorn.

The prefermentation boiling period should be adjusted for the particular legume or cereal so as to avoid gelatinization that would leave inadequate space for the mycelium to develop between the particles of substrate. Table 1 shows boiling times that are found operative.

TABLE 1

Substrate	Form of Substrate	Amount of Boiling at 100°C, minutes
Soybeans	Dehulled, coarse grits	25
Wheat (Hard Red Spring)	Cracked	12
White wheat	Cracked	12
Barley	Dehulled and cracked	12
Oats	Dehulled and cracked	8 - 10
Rye	Cleaned and cracked	12
Corn	Cracked	25
Sorghum	Cleaned and cracked	25
Peanuts	Roasted, dehulled and sliced into small pieces	25
Rice	Polished and cracked	10

Example 2: Other cereal grains, namely barley, oats, rye, corn, sorghum and rice that were lightly cracked and boiled for the time shown in Table 1 were otherwise treated precisely as the cracked wheat of Example 1 excepting that the fermentation was terminated at 22 hours instead of 25 hours. The results are shown in Table 2.

TABLE 2

| | ------Fresh Product------ | | |
Substrate	Fermentation— Development of Rhizopus	Odor	Ability of Substrate to Hold Together and Be Sliced
Soybeans	Excellent	Pleasant	Firm, solid and sliced well
Wheat	Good	Yeast-like	Firm, sliced satisfactorily

(continued)

Substrate	Fresh Product Fermentation—Development of Rhizopus	Odor	Ability of Substrate to Hold Together and Be Sliced
White wheat	Good	Yeast-like	Firm, sliced satisfactorily
Barley	Good	Yeast-like	Like wheat
Oats	Good	Yeast-like	Like wheat
Rye	Good	Faint	Like wheat
Corn	Poor	Faint to none	Poor—slices break
Sorghum	Poor	Yeast-like	Poor—slices break
Peanuts	Poor	Rancid peanuts	Poor
Rice	Good	Sweet, fragrant	Slices well
Rice (3 parts) and soybean grits (1 part)	Excellent	Sweet, fragrant	Slices very well
Wheat (3 parts) and soybean grits (1 part)	Good	Slightly yeasty	Slices very well

Powdered Tempa

A process for making tempa economically and in large quantities from comminuted soybeans has been developed by *P.P. Noznick and A.J. Luksas; U.S. Patent 3,489,570; January 13, 1970; assigned to Beatrice Food Co.* According to a preferred method of this process, powdered tempa is made as follows: whole soybeans are either ground to a powder and rehydrated with water or are soaked in water, e.g., overnight, and the wet cooked beans are ground or macerated without removing the hulls. The product is homogenized as a slurry in water at a solids content which is preferably about 10%. This slurry is then run through a heat exchanger to sterilize it. The preferred sterilization temperature is about 275°F for about 15 seconds, although lower temperatures at correspondingly longer periods of time may be employed.

The sterilized slurry is placed in a fermenter and inoculated with a strain of the tempa mold, such as *Rhizopus oligosporum* or *Rhizopus niger*, for 12 to 96 hours with constant agitation, e.g., by stirring and constant aeration. The fermented tempa slurry was then sterilized again by pasteurization to destroy the active fungi (the only organism present) and dried to form powdered tempa.

Although the tempa mold is conveniently available in previously fermented tempa, there are other sources. [See Hesseltine, Smith, Bradle and Djien, "Investigations of Tempeh, an Indonesian Food Flour, *Developments in Industrial Microbiology*, 275–288 (1963) for a list of the various strains of the tempa mold and their sources.] These authors report that one strain of the tempa mold has been isolated from Texas soil.

The pasteurized tempa slurry may be dried in any convenient manner. For example, it may be spray dried, roller dried, tray dried, freeze dried, etc. The resultant product is powdered tempa. The powdered tempa has all of the essential characteristics, e.g., taste and edibility, of the prior art tempa. Substantial savings in labor and time are effected by the method of this process because the hulls of the soybeans are no longer required to be removed at all, much less removed by hand. Since the inoculated soybeans are fermented in a slurry, this method permits large quantities of beans to be economically processed at the same time.

Example 1: Soybeans were soaked overnight in water. Whole and wet soybeans were macerated with their hulls on. The soybeans, after maceration, approached a solution. The solids content of the resultant slurry was adjusted to 10%. This slurry was sterilized by running through a heat exchanger at 275°F for 15 seconds. (Lower temperatures may be used with longer contact time.) A thousand gallons of the sterilized slurry were placed in a sterile fermentor under sterile conditions and inoculated with *Rhizopus oligosporum*. The slurry, after inoculation, is stirred with good agitation established. While the slurry is being stirred it is aerated with about 2.5 volumes of air per volume of slurry. This fermentation under the above conditions was continued for 24 hours. The fermented tempa was then spray dried to give powdered tempa.

The aeration can be accomplished with 0.1 to 5 volumes of air per volume of slurry. If a very high level of inoculum, as in natural tempa fermentation, is used, the fermentation time can be reduced to 12 hours.

Example 2: Example 1 was repeated except that the *Rhizopus niger* was used in place of the *Rhizopus oligosporum*, and that the sterilized slurry was incubated for 72 hours. Powdered tempa was again obtained.

FLAVORING MATERIALS

Alcoholic Extraction of Soybeans in Making Soy Sauce

It has been usual practice in brewing soy sauce that soybeans have to be thoroughly autoclaved on addition of water before introducing a mold of Aspergillus species. Imperfect steaming of the material will result in products of inferior quality, e.g., when diluted or heated, it produces a large amount of flocculent precipitation and can hardly be used for cooking.

Another form of soy sauce is a chemical product prepared with hydrochloric acid. The chemical soy sauce does not contain the potential turbid substance and shows a fairly good yield. However, its flavor is not as good as the brewed product.

D. Fukushima; U.S. Patent 3,170,802; February 23, 1965; assigned to Zaidan Hojin Noda Sangyo KK, Japan has now developed a method of extracting or heating the soybean with lower alcohols to give a product which produced improved soy sauce. In the process an amount of ordinary soybeans or defatted soybeans is heated for 30 to 60 minutes in the presence of lower alcohols of a certain concentration. The heating temperature is dependent on varieties and concentration of alcohols used for this purpose. The treatment will make the soybean proteins readily hydrolyzed by proteases. The material thus treated, though it would appear raw, can be made into koji by adding water and inoculating the mold, without steaming as needed in case of ordinary soy sauce brewing. The soy sauce produced by this method has been proved to have an excellent flavor and contain no potential turbid substance. In this method the yield is much higher than in the ordinary steaming method. The table below shows the comparison between the two.

	Total Nitrogen (TN), %	Amino Nitrogen (AN), %	NaCl, %	Direct Reducing Sugar, %	Acidity	AN/TN, %	Yield, %	Potential Turbid Substance
Present method	1.803	1.075	17.00	2.76	0.95	59.6	89.3	None
Ordinary method	1.620	0.895	17.05	2.25	0.72	55.3	80.0	None

The various methods of treating the soybeans are shown in the examples.

Example 1: Defatted or undefatted soybean flakes were put into a vessel which has a refluxed condenser and 84 volume percent ethanol solution was added in a sufficient amount to soak the mass of material. After having been boiled for 40 minutes under the ordinary atmospheric pressure, the treated material was separated from the ethanol solution and dried.

Example 2: Defatted or undefatted soybean flakes were put on a porous plate equipped in a closed vessel. A small quantity of 90 volume percent ethanol was put under the plate of the vessel. After air in the vessel was removed by a suitable means, the vessel was heated and the vapor in it was kept for 60 minutes at 95°C under the raised vapor pressure. Then the treated material was dried.

Example 3: Defatted soybean flakes were put into a vessel and subjected to contact with the superheated vapor mixture of ethanol and water (50:50) for 60 minutes at 140°C under the ordinary atmospheric pressure. The ethanol remaining in the treated flakes was then removed. The treated soybeans can then be converted to soy sauce, miso or the like.

Flavor Concentrates from Steffen Filtrate

In the manufacture of beet sugar, water-soluble constituents of the sugar beets are leached out and the extract is concentrated to recover the sugar. In the method known as the Steffen process the extract is diluted with water and a slurry of hydrated lime is added whereby the sucrose is precipitated as calcium saccharates which are removed by filtration. The filtrate, known as dilute Steffen filtrate, may be carbonated with CO_2 to remove residual lime and concentrated to about 60% solids content. This material is known as Concentrated Steffen Filtrate (abbreviated to CSF) and serves as a convenient raw material for the recovery of glutamic acid. The filtrate containing the glutamic acid mother substances is hydrolyzed and the glutamic acid is recovered from the hydrolyzate by crystallization at its isoelectric point.

Since both the hydrolyzed Steffen filtrate and the mother liquor following crude glutamic acid crystallization contain, in addition to glutamic acid, valuable materials such as amino acids, including aspartic acid, alanine, valine, methionine and leucine; betaine; and other organic acids, including lactic acid, Steffen filtrates and glutamic acid mother liquors obtained therefrom constitute an attractive nutritional source. Because of the flavor and odor characteristics, however, these materials have not found an outlet as a nutritional supplement for human consumption.

E.V. Heegaard and M.F. Hobbs; U.S. Patent 3,245,804; April 12, 1966; assigned to International Minerals & Chemical Corporation have developed the method of obtaining a flavor concentrate from hydrolyzed Steffen filtrate or glutamic acid mother liquor obtained therefrom which may be employed to enhance the existing flavor of food products.

The hydrolyzed Steffen filtrate or glutamic acid mother liquor obtained therefrom is treated with a solid adsorbent to remove generally from about 2% to about 40% of the dry substance of the filtrate and at least about 20% of the color materials to provide after removal of inorganics a brown-colored flavor concentrate for human consumption characterized by a desirable odor and flavor.

The concentrates of this process are characterized by a broth-like taste and a pleasing caramel-like color. The flavor concentrates have many characteristics of highly concentrated vegetable broths, are nutritionally valuable, and generally are characterized by a flavor which is quite reminiscent of an extract of beef. Consequently, in addition to increasing the existing flavor appeal of foods, the flavor concentrates increase substantially the flavor and color of such foods as meat sauces, gravies, soups, ground meat casseroles, and the like.

Example: Flavor concentrates were prepared from mother liquors from crude glutamic acid crystallization, viz., the liquors produced by hydrolyzing glutamic acid mother substances in Steffen filtrate and from which a portion of glutamic acid has been removed, according to the following procedure.

The mother liquor was polished by diluting 1:1 with water and allowing to stand for at least two hours to permit colloidal impurities to flocculate. A diatomaceous silica filter aid was added to the liquid and the solids were removed by passing the liquor through a filter. The liquor then was directly used for feed. A 10 cm diameter column 137 cm in length was employed as the adsorbent column.

The column was operated with gravity flow at a feed rate of 80 to 90 ml/min. The effluent was collected and adjusted to pH 6 to 7 with sodium hydroxide, and subsequently evaporated under reduced pressure to a 71 to 72% dry solids content in the liquid phase. The concentrated material was cooled and the precipitate removed by filtration. The resulting filtrate was diluted with water to a dry solids content of 70%.

Granular Food Seasonings from Protein Hydrolysates

A process for the preparation of a granular seasoning is disclosed by *T. Asogawa, E. Satani, S. Wada and Y. Funakoshi; U.S. Patent 3,711,301; January 16, 1973; assigned to Takeda Chemical Industries, Ltd., Japan.* This is an industrially feasible process for preparing a granular seasoning which consists mainly of the hydrolysate or extract of plant or animal protein and which has excellent properties for practical use.

The powdery hydrolysates or extracts of plant or animal protein such as hydrolyzed plant protein, hydrolyzed animal protein, yeast extract, meat extract and the like have been widely used as the flavor-enhancing condiments. However, these natural condiments are very hygroscopic and rapidly become sticky absorbing water in the atmosphere to form relatively large masses under ordinary conditions. Therefore, it is advantageous to use these natural condiments in a granular form coated with a nonhygroscopic material. But, these condiments cannot be granulated as they are owing to their high hygroscopicity, and, therefore, all of the known processes comprise diluting the condiments with suitable unhygroscopic vehicles such as lactose, starch, crystalline cellulose to decrease their hygroscopicity, granulating thus obtained mixture and coating the granules with nonhygroscopic coating material.

As in these known processes the nonhygroscopic vehicles would be added in such a high concentration, e.g., 70 to 80% by weight, it is impossible to prepare granular seasoning which comprises a high concentration of the natural condements.

This process provides that when these powdery and hygroscopic natural condiments are, directly or after the addition of a suitable amount of water and/or propylene glycol, heated at a temperature specified, they become molten and the molten condiments can be extruded into a desired shape and that extruded solids can be easily granulated in a desired size.

The hygroscopic powdery condiment composition should contain 2 to 8% by weight of water or a mixture of water and propylene glycol. In the latter case the content of propylene glycol is not higher than 5% by weight. The most preferable content of water or a mixture of water and propylene glycol ranges from 2 to 4% by weight. As such, powdery condiment composition contains about 2 to 8% by weight of water under ordinary moisture conditions. Owing to their high hygroscopicity, it is generally unnecessary to add a further amount of water or propylene glycol. However, water or propylene glycol may be added, as occasion demands to the composition in sufficiently small amounts not to make their total concentration in the composition higher than 8% by weight relative to the whole weight of the composition.

In the first step of the process, the hygroscopic powdery condiment composition is extruded in a molten state through a forming die of an extruder. This extrusion may be carried out with a conventional screw extruder which is popularly employed for the extrusion of plastics, e.g., polyvinyl chloride resin. The hygroscopic powdery condiment composition is fed into the extruder and is molten after or until it reaches the forming die. The melting is carried out by heating the composition up to a temperature not higher than 130°C. If the composition is heated up to a temperature above 130°C, the composition will be decomposed to cause undesirable coloration as well as scorching.

The molten composition is subsequently extruded through the forming die. It is advantageous to use a forming die having the interstices at least one dimension of which ranges from 1 to 30 millimeters, most advantageously from 2 to 10 millimeters. The molten composition can be extruded into a desired shape, e.g., linear or sheet-shaped, by choosing the shape of the interstices of the forming die.

In the second step, the extruded shaped solids are subjected to granulation. The extruded product is cut into a suitable length, e.g., 20 to 70 millimeters preferably after cooling to some extent, e.g., 60° to 80°C. The cut solids are further cooled to a temperature lower than 40°C and then is crushed to granules with a suitable crusher, e.g., Fitz mill. These granules are then coated with an edible and nonhygroscopic coating material (the third step). Examples of the coating materials include starch phosphate, sodium glycolate, gum arabic, guar gum, methylcellulose, sodium polyacrylate, sodium hydroxyethylcellulose, sodium alginate, sodium glutamate, lactose, calcium lactate, calcium carbonate and the like. It is preferable to use the coating agents in an amount from 1 to 50%, most desirably from 5 to 30%, by weight relative to the weight of the granules to be coated.

Example: Powdery hydrolysate of wheat gluten (moisture content of 3%), which is prepared by heating a mixture of 10 kg of wheat gluten and 15 liters of 24% (w/v) hydrochloric acid at its boiling point for 20 hours in the per se established manner, is fed into an extruder of twin screw type at a feed rate of about 10 kg per hour and is extruded through the forming die having interstices of 3 mm in diameter into linear shape after being kneaded at 80°C for 5 minutes. The linear product is cooled to 70°C, cut into about 30 mm in length, further cooled to 30°C and is then crushed into granules with Fitz mill. The granules are sieved to give those ranging from 10 to 35 mesh.

The granules are uniformly coated by sprinkling powdery monosodium glutamate in an amount of 30% by weight relative to the weight of the granules under an increased moisture condition in a blowing granulator. The coated granules are subjected to fluidized bed drying until their moisture content decrease to about 2%. This product is remarkably stable against moisture and keeps a high fluidity. For example, it shows 32° of angle of repose even after kept standing at 20°C and at a relative humidity of 75% for a week.

MISCELLANEOUS PRODUCTS

Surface Active Protein Products

It is known that surface-active protein derivatives may be obtained by splitting protein substances and reacting the peptide-like decomposition products thus obtained with acid chlorides of higher saturated or unsaturated aliphatic carboxylic or sulfonic acids. A serious hindrance to the large scale use of this method is the cost of preparing the acid chlorides.

S.B. Luce and H.H. Young; U.S. Patent 3,394,119; July 23, 1968; assigned to Swift & Company have disclosed an improved process for producing such materials from inexpensive reagents. The modified proteins used in this process are prepared by the treatment of an alkaline aqueous solution of protein with modifying agents, specific halogenated esters and ethers of carboxylic acids. Such a reaction using for example a carboxylic ester of glycerol monochlorohydrin and an aqueous alkaline solution of protein is as follows:

$$R-COOCH_2\underset{H}{\overset{OH}{C}}CH_2Cl + H_2NZCOONa \longrightarrow RCOOCH_2\overset{OH}{C}H\overset{H}{C}H_2NZCOOH + NaCl + HOH$$

where $H_2NZCOONa$ is the protein.

The modified protein still having a hydrogen attached to the nitrogen would be expected to be reactive with excess glycerol chlorohydrin ester. Also, the amide groups of peptides have available hydrogen which would also be reactive at this position, as would the active hydrogens normally present in protein, as part of hydroxyl or mercapto groups. Consequently, it is not desired to limit the process to a substitution theory involving attachment of the modifying agents at particular reactive hydrogen sites as it has been observed that under the conditions set forth there is a definite improvement in the surface activity of the modified proteins without regard to the particular mechanism involved in the reaction.

Protein materials with which the halogenated modifying agents couple include all those which can be placed in an alkaline aqueous dispersion. The proteins can be of either animal or vegetable origin and include simple and conjugated proteins. Included among the vegetable proteins are soybean, zein, cottonseed, peanut and casein of the alkaline soluble group, as well as their seed meals and hydrolysis products.

Example 1: 200 grams of lauric acid (1 mol) was mixed with 93 grams epichlorohydrin (1 mol) and refluxed in the presence of 5 grams of triethylamine until reflux had almost ceased, but the temperature within the liquid was 130°C. Reduced pressure was applied and any excess epichlorohydrin and amine was distilled over. The residue was the lauric ester of glycerol monochlorohydrin containing less than 1% free lauric acid and showing a chlorine content of 12.05%.

Example 2: 200 grams of extracted soybean flakes were agitated in 600 grams water containing 20 grams caustic soda. The solution was heated under pressure for 2 hours at 150°C. After flashing off steam there remained a 40% solid solution of hydrolyzed soy protein along with some carbohydrate material. The product was filtered to remove insoluble fiber and then reacted with 120 grams of a 1:1 mixture of the fatty esters of glycerol monochlorohydrin as prepared in Example 1. After 2 hours at 80°C the heavy product was blended with a mixture of detergent phosphate salts and dried to a washing powder having excellent detergent properties.

Whipping Agents from Soy Protein

N.L. Betz and N.J. Stepaniuk; U.S. Patent 3,674,501; July 4, 1972 disclose the preparation of new and useful compositions derived from soybean derivatives and more particularly to the preparation of the derivatives which are useful as egg white extenders and whipping agents for confectionary and bakery products.

According to this process ground soybeans from which substantially all the oil has previously been extracted are leached at a temperature of 12° to 25°C with an acid-water solution having a pH of approximately 4 to 6 and containing a trace of a proteolytic enzyme such as pepsin for a time sufficient to extract the water-soluble and water-dispersible soybean constituents. The residual solids are then separated from the aqueous extract by decantation or the like. A preferred method is continuous centrifugal separation, i.e., with continuous discharge of the solids and the supernatant extract. The extract is then concentrated and dehydrated to obtain dried soybean derivatives comprising approximately 30 to 60% water-dispersible, unhydrolyzed soy protein, the balance consisting essentially of water-soluble soybean carbohydrates.

In a preferred embodiment of the process the aqueous extract is first partially dewatered by reverse osmosis and then spray dried. In addition to removing water, reverse osmosis also removes objectionable impurities, whereby the color, odor, flavor and other properties of the product are substantially improved.

The resulting combination of soy carbohydrates and unhydrolyzed protein is useful per se as a replacement or extender for egg whites. It may also be combined with varying proportions of hydrolyzed soybean protein to yield a series of derivatives useful as whipping agents in a variety of food products such as candies, prepared cake mixes and the like.

The soybean residue from the leaching treatment also represents an improved source of soybean protein for enzymatic hydrolysis, since the aforementioned leach serves to solubilize and at least partially remove an enzyme inhibitor naturally present in soybeans which is not affected by any of the prehydrolysis treatments previously employed.

Example: The stability of a cold frappe made with the improved soy whipping agent was compared with the stability of frappes prepared with pure egg albumin and two commercially available hydrolyzed soy protein whipping agents identified in the following table as A and B. The improved soy whipping agent comprised the spray dried concentrate from the enzymatic hydrolysis extract which had been refined by continuous centrifugal separation and reverse osmosis. The hydrolyzed extract itself was obtained from soy grits that had previously been subjected to the acid-water leach also previously described. The composition of such an agent is typically as follows:

Protein, percent	62.5
Ash, percent	4.5
Moisture, percent	5.1
Carbohydrates, percent	27.9
pH of a 3% solution	6.6

The frappes were prepared as follows: First, 50 grams of an aqueous dispersion containing 6.9 grams of the whipping agent was beaten in an electric mixer at high speed for 5 minutes using a wire whip. Then, 75 grams of sugar were added over a period of 30 seconds at low speed, after which the mixture was again beaten at high speed for 15 seconds. Finally, 75 grams of corn syrup were added at low speed and the mixture again beaten at high speed for 30 seconds.

Stability was measured by placing the frappe in a 10 cm funnel suspended over a graduated cylinder. The amount of material collected in the cylinder was noted after 24 hours and 48 hours.

Whipping Agent	Cold Flow (ml) 24 Hours	48 Hours
Pure egg albumin	22	70
Commercial agent A	4*	48*
Commercial agent B	13*	43*
Improved agent	0	5

*Separation of the corn syrup from the sugared whipping agent also occurred.

Production of Dry Neutralized Proteinate

Heretofore, it has been the conventional practice to treat hydrophilic proteins, such as casein, soybean protein, peanut meal, and the like, with alkali to increase the solubility in water. One reason for this is to make glues and adhesives. Another is to prepare edible protein products. This is usually accomplished by suspending the protein in water, sometimes at its isoelectric point, then adding the alkali until the desired elevation of the pH takes place. Because of the amino acid content, the pH may be elevated toward, to, and past the neutral pH of 7, there being no precise end-point in neutralizing.

The drying of proteins as above described involves evaporating large amounts of water relative to recovered protein, in addition to the difficulties of handling thick and viscous solutions. At 20% concentration in water, alkali metal caseinate is so viscous that it is very difficult to dry.

L. Sair; U.S. Patent 3,440,054; April 22, 1969; assigned to The Griffith Laboratories, Inc. discloses a process to neutralize hydrophilic protein while in a solid state in the presence of a limited amount of water and with an hydroxide of alkali metal or alkaline earth metal. It is also possible by this process to alter the characteristics of neutralized protein by subjecting a hydrated solid form to mechanical pressure during or after neutralization.

Although the process is applicable generally to hydrophilic proteins, it is covered here by reference to processing food proteins, in particular, casein and soybean. These are used as food supplements, and as emulsifying agents, especially sodium and calcium caseinates for producing meat products.

The process is based on the discovery that the presence of free water in the protein casein, being neutralized is necessary to effect prompt and complete reaction. This may be done by first moistening the casein, as with water or a neutral nonreactive aqueous solution, preferably before adding the alkali, or by using water containing the hydroxide, such as that of sodium or calcium. The neutralized mass will contain total water derived from the moisture of the commercial casein, the moistening water, and the water providing the alkali. To produce an easy drying semidry crumbly mass of neutralized casein, the mass should have at least 60% solids and preferably 80% solids. Except for the water content of the commercial protein, the remaining water content may be variously distributed. The water content may be effected in several ways, including first moistening the protein, then adding dry soluble hydroxide or dissolved hydroxide; or by adding dry soluble hydroxide to the protein and then adding water; or by using the desired amount of water carrying the hydroxide.

The water in amounts shown in Table 1 was first added and allowed to hydrate the casein uniformly, to give a free-flowing mass. Then the fixed amount of 50% NaOH solution was added. Then after 10 minutes a specimen was extracted and titrated for free alkali, expressed in terms of titre, being ml 0.1 N sulfuric acid for a 10-gram sample. The results in terms of odor after 24 hours are given below:

TABLE 1

Example	H_2O, ml	NaOH Solution, ml	Titre	Odor
1	0	3.5	2.1	Bad, ammonia
2	3	3.5	5.1	Bad, ammonia
3	5	3.5	1.6	Bad, ammonia
4	7	3.5	1.3	Slight, ammonia
5	10	3.5	0.4	Good
6	18	3.5	0	Good

It has been discovered further that when the crumbly neutralized mass is mechanically compressed with high pressure, its physical form can be changed from opaque to glassy as the total water content is increased. It has also been found that the glassy forms have better emulsifying qualities than the nonglassy forms having the same proportions of reacted alkali and protein.

The following Table 2 shows the results of pressure on equally neutralized protein samples varying only in total water content. To 100 grams of 30-mesh commercial casein (from New Zealand) is added the amounts of water shown in column 2. Then to each sample was added 4 ml of 50% NaOH solution. Column 3 shows the water content as determined by oven drying at 100°C overnight with loss of all water. Column 4 shows the visual character of the products after forming pellets at 15,000 lb mechanical pressure for 5 minutes in a laboratory Carver press.

TABLE 2

1 Example	2 Ml. H_2O	3 Percent H_2O	4 Pellets
11	0	13.3	White, opaque.
12	3	15.3	Do.
13	6	17.4	Do.
14	12	21.3	White, opaque, slight translucency in spots.
15	24	27.9	Do.
16	30	31.8	About 60% white-opaque with remainder going translucent.
17	35	33.8	10% opaque, 90% translucent.
18	40	37.0	99% translucent, 1% opaque.
19	50	40.0	100% translucent, very hard pellet.
20	60	44.0	100% translucent. The pellet is much softer than Preparation No. 19.

Example: A Mixed Protein —

	Parts by Weight
Commercial casein	100
Wet soy protein*	100
Spray dried soy protein (70% protein content)	170
Calcium hydroxide (dry)	3

*A water suspension of 30% solids at pH 4.

The water of the wet soy protein serves to hydrate the mixed proteins before the dry calcium hydroxide is added. The product is pressed into pellets and dried at a denaturing temperature of 180°F, yielding an insoluble protein food additive. By further processing, an edible article may be produced, such as puffed flakes or chips.

Malt Treatment of Soy for Use in Cereals

A high protein cold cereal is produced by *W.T. Bedenk and E.R. Purves; U.S. Patent 3,682,647; August 8, 1972; assigned to The Procter & Gamble Company* using a malt treated soy protein. The soy-malt mixture is preferably combined with a cereal grain and further processed into a breakfast food having a protein content of greater than 20% and having a pleasing taste.

The soy protein used in this process is defatted and refined. Soy flour currently commercially available contains 40 to 60% protein on a dry weight basis and is to be contrasted with soy protein concentrate and soy protein isolate which contain on a dry weight basis 70% up to 90% and 90% up to 100% protein, respectively. As used here, soy flour contains on a dry weight basis 40% up to 70% protein. Depending on which protein source is used it is possible to add varying amounts of a cereal grain or other additives so that the final cereal will still possess a protein content of greater than 20%.

In accordance with the process the soy protein source is made more palatable, i.e., its crispness retention, tenderness, and taste are all improved by initially treating it with barley malt and water prior to processing it into a breakfast food. In the absence of such a treatment, a cold cereal containing a soy protein is very unacceptable. For a reason not precisely known, the barley malt and water treatment of soy protein under the critical conditions disclosed here has a very pronounced effect on the eating properties of the final product. The characteristic beany, bitter taste associated with the soybean is reduced to the extent that it is no longer objectionable. The taste improvement is even more pronounced with puffed products.

In the production of an all soy-malt cold cereal, the soy protein, barley malt and water are first blended in any suitable blending apparatus, either batch type or continuous. In a batch type operation the amount of water based on the total mixture is 45% to 75%. Greater amounts can be used but are avoided because of the necessity after the treatment to partially remove the moisture to facilitate handling in the subsequent process steps. Lesser amounts of water are avoided because of the difficulty in handling the resultant viscous mixture. Most preferably 50 to 60% water is added based on the weight of the total mixture. Barley malt added in the range of 2 to 10% based on the dry weight of the soy protein source results in an improved product. Most preferably 4 to 6% barley malt based on the dry weight of soy protein is added.

The mixture is first subjected to a temperature of 80° to 200°F for 1 minute to 120 minutes. Most preferably the mixture is held at 110° to 150°F for 15 minutes to 60 minutes to obtain a treated soy protein that when made into a cold breakfast cereal results in a superior product. After the soy protein malt mixture has been treated at the above temperatures the resultant dough is processed by any of several known methods of producing a ready-to-eat cereal product depending on form, type, or condition of the final product desired. In another embodiment of this process a soy sauce, barley malt and water mixture is heated to the elevated temperatures, and then combined with a cooked cereal grain to form a dough that is readily processed into a form suitable for human consumption. The cereal grain addition allows the use of higher malt percentages based on the soy protein source because the malt level based on the total product, i.e., soy protein source and cereal grain, is still less than that level over which the malt imparts a strong taste to the product.

The proportions of the treated soy protein and cereal grain combined depends on the particular protein source, the particular cereal grain and the desired protein content in the final product. Most preferably, the base cereal particles, i.e., the cereal particle consisting of soy protein source and cereal grain contains on a dry weight basis the following percentages of soy source with the balance being the listed cereal grain:

	Soy Flour	Soy Concentrate	Soy Isolate
Corn	18 - 68%	14 - 34%	12 - 26%
Rice	20 - 70%	16 - 37%	14 - 28%
Oats	10 - 62%	8 - 29%	7 - 21%
Wheat	14 - 64%	10 - 31%	9 - 23%

Example:

Formulation	Grams
Soy flour	100
Malt	5
Sucrose	10
Salt	2

The dry ingredients are blended with 165 grams water to form a mixture of about 58% water and is held at 120°F for 25 minutes. The mixture is then passed through an extruder. In the extruder the mixture is subjected to 500 psig and 150°F for 2 minutes. The extrudant is in the form of strands having a 3/16 inch diameter. These strands are tempered for 10 minutes and then sliced into pellets 3/16 inch long. These pellets are next passed through a two roll mill to produce flakes having thicknesses of about 0.010 inch. These flakes are next puffed by being contacted for 5 seconds with hot salt having a temperature of 350°F. After exposure to milk, the flakes are tender and have a pleasant taste. Control flakes made by the same process but with no malt are less tender after exposure to milk than are the flakes of this example.

Soy Protein-Soy Lecithin Mixtures

Materials useful as food emulsifiers have been developed by *S.J. Circle and E.W. Meyer; U.S. Patent 3,268,335; August 23, 1966; assigned to Central Soya Company, Inc.* These are soy protein-lecithin blends which may be used in foods in place of or in conjunction with egg yolk. The following outline indicates the various forms of soy protein (curd or concentrate) which may be used in the blend as well as the various stages of soy lecithin or phosphatides that may be used in the blend. The flow diagram in Figure 14.1 outlines the process.

When a modified form of lecithin is desired as the mixing reactant, the lecithin emulsion is dried to provide so-called dry natural lecithin. The dry natural lecithin has about two-thirds phosphatides and one-third oil, which may be separated by acetone treatment. The lecithin emulsion, for example, may be about 50% water, 33% phosphatides, and 17% oil. In the instance when dry lecithin is employed, it is preferred to reconstitute with water to provide the mixing reactant.

Following acetone treatment, the separated soy phosphatides may be used as such as the lipid reactant in the mixing step, or the so-called 100% phosphatides may be subjected to an alcohol treatment to separate the phosphatides into the alcohol-soluble phosphatides portion and the alcohol-insoluble phosphatides portion. Depending upon the concentration and treating conditions, the proportions of the soluble and insoluble fractions may be varied as desired. In any event, either fraction can be advantageously combined (preferably as an emulsion) with soy protein to provide a soy protein-lipid useful as an emulsifier for food preparation.

FIGURE 14.1: SOY PROTEIN-SOY LECITHIN MIXTURES

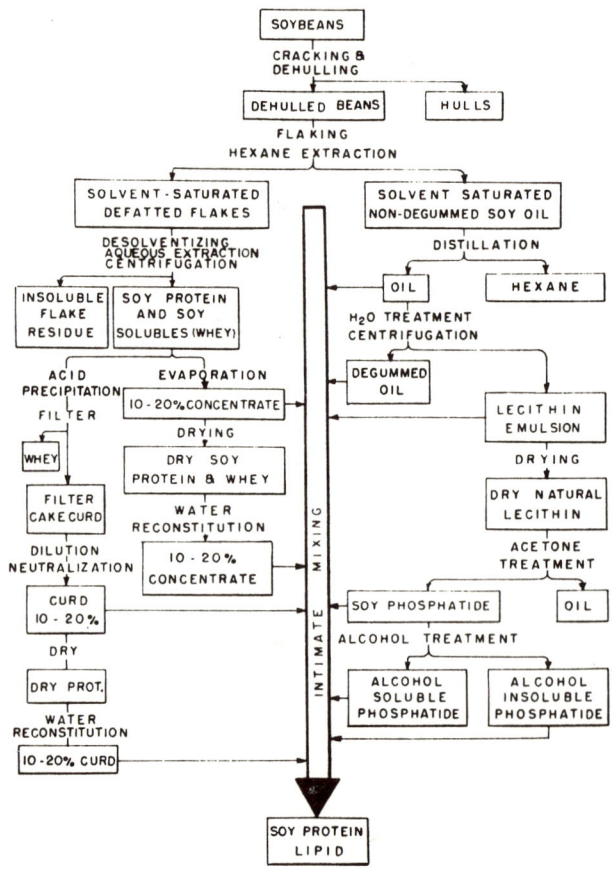

Source: S.J. Circle and E.W. Meyer; U.S. Patent 3,268,335; August 23, 1966

Example 1: 15 kg of soy protein curd at 30% solids were mixed with 30 kg of water to give a 10% solids dispersion, they neutralized with sodium hydroxide to a pH of about 7, as determined by using a Beckman glass electrode pH meter. 9 kg of a 50% emulsion of natural lecithin were diluted to 20% solids by the addition of 13.5 kg of water, and the diluted protein dispersion and diluted lecithin emulsion were mixed with a paddle stirrer running at about 300 rpm for 30 minutes and thereafter the mixture was spray dried in a

Turbulaire laboratory spray dryer. The dryer was type N, manufactured by the Western Precipitation Corporation of Los Angeles, California. The mixture entered the dryer at a temperature of 350°F, the exit temperature was 180°F, and the feed rate through the dryer was 10 kg per hour of diluted mix. The pneumatic atomizer nozzle of the dryer had a ⅛ inch feed opening and operated at 30 psi of atomizing air.

The resultant product was a free-flowing, slightly yellow powder. The particles appear to have a somewhat irregularly spherical shape, i.e., a hollow core, that shows up as a doughnut in cross section. The incorporated lipid material appears to be quite evenly dispersed throughout the particle, and in certain cases there is a slight concentration along the outside of the particle. The particles are about 50 microns in diameter.

Example 2: The procedure of Example 1 was followed, but in place of the 9 kg of the 50% lecithin emulsion, 4½ kg of dry commercial lecithin were employed, diluted to 20% concentration through the addition of 18 kg of water prior to mixing with the soy protein. Again, the product produced hereunder was substantially the same as that produced in Example 1.

Example 3: The procedure of Example 2 was followed, but in place of starting with 4½ kg of dry natural lecithin, 4½ kg of acetone-washed, oil-free phosphatides in dried forms were employed. These were diluted to a 20% emulsion and thereafter combined with the soy protein to provide a product similar to those of Example 1 and 2.

COMPANY INDEX

The company names listed below are given exactly as they appear in the patents, despite name changes, mergers and acquisitions which have, at times, resulted in the revision of a company name.

Agway Inc. - 60
Ajinomoto KK - 29
American Cyanamid Co. - 94, 239
Archer-Daniels-Midland Co. - 79, 139, 228, 229, 248, 250, 252
Ste d'Assistance Technique pour Produits Nestle SA - 135
Beatrice Food Co. - 290
Beloit Corp. - 3, 5
Bio-Technical Resources, Inc. - 37
C.C.D. Processes N.Y. Ltd. - 14, 16
CPC International Inc. - 170, 180
Cargill Inc. - 44, 63
Carnation Co. - 111
Central Soya Co., Inc. - 67, 87, 89, 299
Ciba Corp. - 210, 213
Coca-Cola Co. - 24
Corn Products Co. - 33, 182
Cornell Research Foundation, Inc. - 66
Dannen Mills, Inc. - 70
Etablissement Public: Institut National de la Recherche Agronomique - 27
Far-Mar-Co., Inc. - 172
Food Techniques, Inc. - 19, 154
Fuji Oil Co., Ltd. - 125
General Electric Co. - 12
General Foods Corp. - 77, 122, 224, 227, 236, 244, 262
General Mills, Inc. - 124, 131, 217, 218, 220, 226, 232, 255, 269
Glidden Co. - 107
Grain Processing Corp. - 91, 110
Griffith Laboratories, Inc. - 95, 99, 103, 129, 137, 296
Hercules Inc. - 35
Holtz & Willemsen - 58
Hong Kong Soya Bean Products Co., Ltd. - 62, 275, 276
Hunt-Wesson Foods, Inc. - 113
ICI America Inc. - 22
Ste Industrielle des Oleagineux - 56
Interchemical Corp. - 183
International Minerals & Chemical Corp. - 292
Kal Kan Foods, Inc. - 242
Kansas State University Research Foundation - 165
Keever Starch Co. - 174
Kikkoman Shoyu Co., Ltd. - 24, 284, 286
John Kraft Sesame Corp. - 193
Kraftco Corp. - 73
Lever Brothers Co. - 127, 128, 234, 259, 261, 277
Libby, McNeil & Libby - 280
Loma Linda, Foods - 264
Morinaga Milk Industry Co., Ltd. - 237
Morton-Norwich Products, Inc. - 141
Nisshin Oil Mills, Ltd. - 117, 233
Nutrilite Products, Inc. - 184
Ogilvie Flour Mills Co., Ltd. - 140, 177
Oil Seed Products, Inc. - 202
Pacific Vegetable Oil Corp. - 190
Pennwalt Corp. - 31
Pillsbury Co. - 161, 163
Procter & Gamble Co. - 75, 84, 105, 131, 186, 188, 298
Protein-Compagnie GmbH - 8
Proto, International Hygienic Food Co. - 159, 161
Quaker Oats Co. - 57
Ralston Purina Co. - 46, 48, 101, 118, 120, 222, 240, 242, 253, 257
Ranchers Cotton Oil - 150
Refining, Unincorporated - 152
Rohm and Haas Co. - 133, 136
J.R. Short Milling Co. - 108, 131

Company Index

Showa Sangyo KK - 65, 272
Soy Food Products, Inc. - 45
A.E. Staley Mfg. Co. - 92
Swift & Co. - 25, 86, 97, 98, 149, 153, 216, 220, 256, 278, 294
Takasago Perfumery Co., Ltd. - 267
Takeda Chemical Industries, Ltd. - 231, 293
Texas A & M Research Foundation - 146
Top-Scorer Products, Inc. - 176
Arthur C. Trask & Sons - 21

Harry Truax & Sons Co., Inc. - 49
Ueno Seiyaku Oyo Kenkyujo, KK - 281
U.S. Secretary of Agriculture - 6, 26, 42, 54, 74, 81, 82, 121, 144, 147, 151, 152, 155, 157, 166, 167, 171, 191, 198, 206, 207, 272, 283, 286, 288
University of Chile - 205
Wander Co. - 72
Worthington Foods, Inc. - 175
Zaidan Hojin Noda Sangyo KK - 291

INVENTOR INDEX

Akioka, A. - 117
Albrecht, W.J. - 268
Allen, M. - 12
Altschul, A.M. - 152
Anderson, P.T. - 19, 33, 154
Anker, C.A. - 226
Anson, M.L. - 127, 128, 259, 261
Aonuma, T. - 284, 286
Arai, A. - 284, 286
Arai, S. - 74
Arima, T. - 236
Ariyama, H. - 274
Asogawa, T. - 293
Atkinson, W.T. - 228, 229, 248, 250
Attebery, J.M. - 10
Baile, R.P. - 53
Barros, F.M. - 205
Beaber, N.J. - 108
Bedenk, W.T. - 105, 131, 298
Berardi, L.C. - 6
Bertullo, V.H. - 30
Betz, N.L. - 295
Bierly, G.M. - 174
Blanchon, E.M.J. - 20
Bock, H. - 8
Bonagura, A.G. - 122
Bookwalter, G.N. - 268
Boyer, R.A. - 216, 222, 253
Bradof, R.W. - 72
Brown, A.V. - 253
Bremer, J.W., Jr. - 149
Burchill, P.I. - 226
Calvert, F.E. - 101, 250
Carter, R. - 184
Cavanagh, G.C. - 150
Cavanagh, J.C. - 2
Chayen, I. - 14, 16
Chien, H.-C. - 73

Christianson, D.D. - 171
Circle, S.J. - 89, 107, 299
Cooke, R.R. - 84
Dannert, R.D. - 218
D'Aquin, E.L. - 147
Darzins, E. - 196
DeLapp, D.F. - 94
de Paolis, P.U. - 113, 115
Dienst, C.S. - 10
Dollbaum, W. - 58
Dudman, R.K. - 220
Dunning, H.N. - 255
Duren, J.T. - 46, 48
Eagle, E. - 149
Eaves, P.H. - 147
Ebisawa, M. - 233
Edwards, A.W. - 208
Edwards, G.W. - 208
Eldridge, A.C. - 121
Elmquist, L.F. - 232
Engel, E. - 224, 227
Faith, W.T., Jr. - 133
Fellers, D.A. - 166, 167
Finucane, T.P. - 262
Frampton, V.L. - 144, 152
Frazeur, D.R. - 110
Frederiksen, C.W. - 118, 120, 242
Freeman, J.E. - 169
Fujimaki, M. - 74
Fujita, K. - 267
Fukushima, D. - 284, 291
Funakoshi, Y. - 293
Galle, E.L. - 163
Gallo, T. - 177
Gastrock, E.A. - 147
Gaver, K.M. - 140
Geister, R.S. - 153
Gerrish, O.B., Sr. - 172

Inventor Index

Gibbs, B.M. - 277
Glicksman, M. - 254
Goering, K. - 202
Goldblatt, L.A. - 26
Goodban, A.E. - 191
Gould, M.R. - 57
Gracza, R. - 161
Greene, F.C. - 198
Griffin, E.L., Jr. - 54, 81, 82
Groschke, A.C. - 60
Guidarelli, E.J. - 44, 45
Habermann, W.F. - 3, 5
Hack, A.W. - 33
Hamdy, M.M. - 252
Hampton, R.J. - 177
Hara, H. - 117
Harada, Y. - 236
Harrel, C.G. - 161
Hartman, W.E. - 175
Hawley, R.L. - 46, 48, 118, 280
Hayes, L.P. - 92
Heegaard, E.V. - 292
Hemmings, B.D. - 277
Henderson, R.M. - 3, 5
Herzberg, C. - 58
Hesseltine, C.W. - 283, 286, 288
Heusdens, W. - 242
Hirsch, A. - 277
Hobbs, M.F. - 292
Hoer, R.A. - 101, 118, 240
Hoersch, T.M. - 25
Höhl, J. - 179
Hoover, W.J. - 165
Howland, D.W. - 182
Hunter, J.E. - 84
Huston, R.B. - 110
Iacobucci, G.A. - 23
Ikawa, S. - 29
Ishler, N.H. - 227
Jantzen, I. - 21
Jenkins, S.L. - 222
Johnson, D.C. - 86
Johnson, G.W. - 265
Johnson, R.A. - 19, 33, 154
Johnston, P.H. - 166
Julian, P. - 107
Kanno, T. - 272
Kato, H. - 74
Kaufmann, H.H. - 63
Kawamura, Y. - 29
Keen, J.L. - 220
Kenmotsu, Y. - 237
King, W.H. - 144, 151, 152, 155, 157
Kirk, L.D. - 206, 207
Kirkeby, R.D. - 254
Kjelson, N.A. - 226
Klose, R.E. - 254
Kneeland, J.A. - 190
Knight, J.W. - 180

Kobayashi, H. - 272
Kohler, G.O. - 191
Konishi, S. - 29
Kopas, G.A. - 190
Koshi, W.E. - 269
Kovásznay, E. - 40
Kovásznay, I. - 40
Kraft, J.H. - 193
Kruse, N.F. - 67
Kuramoto, S. - 131, 217, 220
Landfried, B.W. - 176
Lawrence, J.F. - 44, 63
Layton, L.L. - 198
Liepa, A.L. - 75
Liggett, J.J. - 194
Lippold, G.L. - 70
Lo, F.Y.-Y. - 275, 276
Lo, K.S. - 62
Loew, F.C. - 183
Long, C.L. - 244
Luce, S.B. - 294
Luksas, A.J. - 290
Lundstedt, E. - 275, 276
MacAllister, R.V. - 227, 262
Magnino, P.J., Jr. - 120
Manning, W.T. - 161
Manwaring, M.E. - 218
Martinelli, A., Jr. - 286
Martinez, W.H. - 6
Masri, M.S. - 26
Matsumoto, T. - 29
McAnelly, J.K. - 256
McCabe, E.M. - 135
Meadors, D.A. - 17
Meinke, W.W. - 145
Melcer, I. - 95, 103
Melnychyn, P. - 111, 199
Meyer, E.W. - 299
Miles, C.P. - 264
Miley, W.M. - 174
Miller, D.M. - 98
Mimoto, H. - 233
Mitchell, J.H., Jr. - 201
Mitchell, R.W. - 84
Mogi, K. - 284, 286
Moneymaker, J.R. - 176
Moritaka, S. - 230
Moroe, T. - 267
Morton, B.J. - 37
Moshy, R.J. - 77
Mullen, J.D. - 124
Muller, S.A. - 91
Mustakas, G.C. - 54, 81, 82, 206, 207, 268
Myers, D.V.B. - 23
Nagasawa, T. - 237
Nagata, A. - 29
Nakao, Y. - 231
Nakayama, S. - 125

Nash, A.M. - 121
Newsom, B.G. - 234
Noe, F.F. - 136
Norris, F.A. - 86, 153
Noznick, P.P. - 290
Obayashi, T. - 237
Obey, J.H. - 108
Oborsh, E.V. - 253
O'Connor, D.E. - 131, 186, 188
Ogrins, A. - 124
Ohan, W.J. - 244
Okubo, K. - 23
Okumura, G.K. - 246, 266
Olson, R.M. - 169
O'Neal, L. - 193
Onishi, S. - 29
Pader, M. - 127, 128, 259, 261
Palmer, H.C. - 242
Paulsen, T.M. - 79
Pavuk, J.S. - 140
Peng, A.C.-Y. - 97, 271
Pereira, C.R. - 30
Pichel, M.J. - 278
Pour-El, A. - 139
Purves, E.R. - 298
Rambaud, M. - 56
Rao, G.V. - 172
Raymond, C.A. - 39
Reck, D.R. - 184
Reiser, R. - 145
Robbins, F.M. - 122
Roberson, W.D. - 35
Rockland, L.B. - 42
Rolland, J.R. - 177
Rozsa, T.A. - 161
Rusch, D.T. - 22
Rusoff, I.I. - 244
Saewert, H.E. - 216
Sair, L. - 95, 99, 103, 129, 137, 296
Sakai, H. - 117
Sakita, T. - 233
Salbego, E.J. - 83
Satani, E. - 293
Sato, E. - 267
Sawada, K. - 230
Schatzman, E.A. - 222
Schmitt, E.E. - 239
Schulz, A.A. - 222, 253
Schweiger, R.G. - 91
Sfat, M.R. - 37
Shank, J.L. - 25
Sherba, S.E. - 133
Shibasaki, K. - 283
Simms, R.P. - 92
Smith, A.K. - 283
Smith, D.E. - 124, 269

Smith, M.L. - 288
Smythe, C.V. - 133
Staron, T.J. - 27
Steigerwalt, R.B. - 133
Steinkraus, K.H. - 66
Stepaniuk, N.J. - 295
Stringfellow, A.C. - 171
Strommer, P.K. - 255
Swartz, D.L. - 57
Swenson, T.C. - 139
Szczesniak, A.S. - 224, 227
Tamuki, T. - 29
Tamura, Y. - 237
Taniguchi, H. - 125
Tateishi, T. - 125
Terashima, S. - 29
Thomas, L.M. - 174
Thurman, B.H. - 152
Tombs, M.P. - 234
Tomita, M. - 237
Touba, A.R. - 269
Truax, H. - 49
Tsen, C.C. - 165
Tsukada, N. - 284
Ueno, R. - 281
Unger, L.G. - 182
Van Hulle, G.J. - 255
Vester, F. - 210, 213
Vidal, F.D. - 31
Vix, H.L.E. - 26
Wada, S. - 293
Waggle, D.H. - 257
Wall, J.S. - 171
Ward, A.B. - 161
Watanabe, H. - 65, 284
Weiss, T.J. - 278
Westeen, R.W. - 217, 220
Whitney, R.W. - 89, 107
Wilding, M.D. - 97, 98
Wilkinson, J.E. - 246, 266
Williams, M.A. - 87
Wintz, M. - 141
Witwicka, I.E. - 140
Wolcott, J.M. - 111
Yamamoto, H. - 29
Yamashita, M. - 74
Yamato, Y. - 125
Yare, R.S. - 122
Yasuda, A. - 284, 286
Yasumatsu, K. - 231
Yokotsuka, T. - 284, 286
Yoshimura, K. - 29
Young, H.H. - 294
Yuasa, T. - 286
Zacharia, T. - 159, 161

U.S. PATENT NUMBER INDEX

2,682,466 - 216
2,730,447 - 216
2,730,448 - 216
2,776,894 - 67
2,785,069 - 220
2,795,502 - 39
2,797,212 - 174
2,797,997 - 149
2,802,738 - 127
2,813,024 - 128
2,813,025 - 259
2,830,902 - 259
2,833,651 - 261
2,873,190 - 151
2,881,159 - 107
2,895,831 - 159
2,919,192 - 277
2,920,963 - 196
2,928,821 - 14
2,930,700 - 72
2,934,431 - 150
2,934,432 - 152
2,950,198 - 144
2,952,543 - 224
2,958,600 - 152
2,960,408 - 153
2,967,108 - 283
2,987,399 - 202
2,990,285 - 193
2,999,753 - 140
3,001,875 - 129
3,010,953 - 182
3,023,107 - 81
3,043,826 - 108
3,047,395 - 244
3,052,556 - 53
3,058,829 - 40
3,077,408 - 161

3,084,046 - 154
3,090,779 - 16
3,091,538 - 161
3,093,483 - 227
3,096,177 - 274
3,099,649 - 29
3,100,709 - 79
3,102,031 - 262
3,118,959 - 217
3,124,461 - 145
3,126,285 - 70
3,126,286 - 77
3,127,388 - 19
3,141,776 - 63
3,141,777 - 44
3,142,571 - 256
3,155,524 - 86
3,168,406 - 77
3,170,802 - 291
3,173,792 - 206
3,175,909 - 232
3,177,079 - 220
3,197,310 - 226
3,218,307 - 121
3,220,851 - 56
3,228,773 - 286
3,243,301 - 288
3,245,804 - 292
3,252,807 - 131
3,253,930 - 57
3,258,407 - 20
3,261,822 - 122
3,268,335 - 299
3,268,503 - 82
3,271,160 - 190
3,288,614 - 264
3,290,152 - 175
3,290,155 - 54

3,294,776 - 198
3,295,985 - 2
3,303,182 - 117
3,343,961 - 49
3,346,390 - 278
3,361,574 - 79
3,361,575 - 79
3,362,829 - 176
3,364,034 - 25
3,365,440 - 89
3,370,054 - 183
3,386,833 - 265
3,389,997 - 12
3,390,999 - 21
3,391,000 - 207
3,391,001 - 137
3,392,026 - 206
3,394,119 - 294
3,394,120 - 210
3,397,067 - 163
3,397,991 - 33
3,399,997 - 266
3,402,165 - 8
3,407,073 - 45
3,409,440 - 179
3,429,709 - 26
3,434,845 - 60
3,440,054 - 296
3,442,656 - 17
3,450,688 - 199
3,454,404 - 65
3,455,697 - 228
3,459,555 - 155
3,460,950 - 267
3,463,770 - 167
3,468,669 - 222
3,469,991 - 280
3,469,994 - 87

3,475,402 - 213	3,627,536 - 236	3,714,210 - 91
3,476,739 - 194	3,630,753 - 111	3,716,372 - 115
3,481,743 - 155	3,635,726 - 99	3,718,479 - 272
3,488,770 - 248	3,635,728 - 42	3,721,569 - 66
3,489,570 - 290	3,639,129 - 269	3,723,407 - 98
3,490,914 - 246	3,640,725 - 133	3,728,327 - 110
3,493,384 - 167	3,642,490 - 118	3,733,207 - 135
3,493,385 - 33	3,645,745 - 120	3,734,901 - 92
3,496,858 - 222	3,645,746 - 229	3,736,147 - 24
3,498,794 - 250	3,645,747 - 242	3,741,771 - 139
3,498,965 - 167	3,647,484 - 284	3,743,515 - 276
3,501,451 - 167	3,649,293 - 101	3,743,516 - 275
3,516,349 - 30	3,653,912 - 269	3,749,581 - 233
3,520,868 - 3	3,655,403 - 31	3,753,728 - 131
3,535,305 - 184	3,656,963 - 35	3,761,353 - 136
3,538,069 - 5	3,661,593 - 171	3,762,929 - 94
3,542,559 - 191	3,662,671 - 242	3,764,708 - 286
3,542,562 - 84	3,662,672 - 240	3,775,542 - 83
3,542,754 - 167	3,662,673 - 253	3,780,183 - 208
3,547,900 - 10	3,669,677 - 103	3,780,188 - 165
3,563,762 - 62	3,669,678 - 193	3,782,964 - 180
3,574,180 - 166	3,674,500 - 237	3,782,968 - 58
3,579,496 - 6	3,674,501 - 295	3,790,553 - 172
3,583,872 - 97	3,682,646 - 113	3,792,175 - 239
3,585,047 - 74	3,682,647 - 298	3,793,464 - 22
3,586,662 - 186	3,684,521 - 254	3,794,731 - 218
3,594,184 - 46	3,684,522 - 226	3,794,735 - 234
3,594,185 - 48	3,687,686 - 105	3,798,339 - 271
3,594,192 - 124	3,687,687 - 75	3,800,056 - 201
3,604,123 - 157	3,689,277 - 37	3,803,328 - 27
3,607,860 - 125	3,698,912 - 141	3,806,611 - 231
3,615,648 - 205	3,704,131 - 177	3,809,767 - 95
3,615,655 - 169	3,707,380 - 255	3,809,771 - 272
3,615,657 - 147	3,711,301 - 293	3,810,764 - 257
3,622,556 - 188	3,712,823 - 281	3,810,997 - 73
3,623,885 - 252		

NOTICE

Nothing contained in this Review shall be construed to constitute a permission or recommendation to practice any invention covered by any patent without a license from the patent owners. Further, neither the author nor the publisher assumes any liability with respect to the use of, or for damages resulting from the use of, any information, apparatus, method or process described in this Review.